Triboelectric Nanogenerators

Triboelectric Nanogenerators

Technology, Applications, and Commercialization

Yunlong Zi, Hengyu Guo, Jie Wang, Chi Zhang, Xiangyu Chen, and Qing Zhao

WILEY-VCH

Authors

Yunlong Zi
Hong Kong University of Science and Technology (Guangzhou)
Guangzhou, Guangdong
China

Hengyu Guo
Chongqing University
Chongqing
China

Jie Wang
Beijing Institute of Nanoenergy
Beijing
China

Chi Zhang
Beijing Institute of Nanoenergy
Beijing
China

Xiangyu Chen
Beijing Institute of Nanoenergy
Beijing
China

Qing Zhao
Chongqing University
Chongqing
China

Cover Image: © Carlos Barros/Getty Images

Library of Congress Card No.: applied for

British Library Cataloguing-in-Publication Data
A catalogue record for this book is available from the British Library.

Bibliographic information published by the Deutsche Nationalbibliothek
The Deutsche Nationalbibliothek lists this publication in the Deutsche Nationalbibliografie; detailed bibliographic data are available on the Internet at <http://dnb.d-nb.de>.

The manufacturer's authorized representative according to the EU General Product Safety Regulation is WILEY-VCH GmbH, Boschstr. 12, 69469 Weinheim, Germany, e-mail: Product_Safety@wiley.com.

Print ISBN: 978-3-527-35042-1
ePDF ISBN: 978-3-527-83788-5
ePub ISBN: 978-3-527-83789-2
oBook ISBN: 978-3-527-83790-8

Typesetting Straive, Chennai, India
Druck und Bindung: CPI books GmbH Leck, Germany

Contents

Preface

Since its invention, triboelectric nanogenerator (TENG) has experienced development for over 10 years, like a big tree grown from a tiny seed. 10 years ago, we were still working on the basic modes, fundamental theories, evaluation standards, etc. Five years ago, we studied mechanisms and designs that can enhance the output performance of TENGs. Nowadays, TENG has become a well-recognized energy harvesting technology with output energy density of 10^5 J/m^3 level while the peak power density can achieve 15 MW/m^2, and scientists are working on TENG's applications in various scenarios. As recommended by authors, we believe the topics selected are important research topics in TENG, which may not only contribute to the development of science in TENG, but also promote the applications. The significance and the related questions of these chapters are stated below:

1) Chapters 1 and 2 state how the triboelectric charge was generated and discharged, which highlights the charge transfer flow in TENG. In the meanwhile, an issue worthy to discuss is how the energy conversion in TENG was completed at the beginning, and how a large portion of the electrostatic energy was wasted if they were not used in applications.
2) Chapter 3 regarding the standardization of TENG is touching a key question of TENG, that is how much energy can TENG provide? The provided standards are all related the output energy density of TENG, which is now around 10^5 J/m^3. As comparison, the lithium ion battery can achieve around 10^9 J/m^3, while the limit of the breakdown discharge of materials is around 10^8 J/m^3, which means we may still have some gaps to catch.
3) Chapters 4–8 discuss about different methods to promote and/or config the output performance of TENGs. The gas environment can impact the breakdown discharge limits as well as equivalent resistance of among triboelectric layers and electrodes. Liquid–solid interface may provide a way for triboelectric charge generation with little energy waste in friction. Electromechanical structures may provide means to excite the charge in circles until breakdown limit. DC-TENG provides a new design that can directly supply direct current output. The power-management and energy storage systems provide solutions to manage and store the generated power, which is also necessary in the system.

4) Chapters 9–12 outlook four new related research areas, which are all related to potential applications. Tribotronics utilizes TENG to serve as a gate voltage, which can be triggered by simply touching. Tribophotonics couples the triboelectricity as the power source and photonics as the wireless transmission means, targeting at self-powered wireless systems. Combined with artificial intelligence (AI), sensing systems based on TENG mechanisms will become more accurate, with advantages of high signal-to-noise ratio. High-voltage output from TENG, with thousands or even tens of thousands volts in voltage, may be a new application area that is promising to trigger breakdown discharge, electrospray, field emission, etc.

In light of the four parts as stated above, we can see that TENG is experiencing a rapid development in the past five years. Compared with the book *Triboelectric Nanogenerators* published in 2016 by Springer, the contents of this book cover more in-depth mechanism studies, discuss more methods to promote/regulate the output performance, and expands more applications with new research directions proposed. Hopefully with this new book, readers can understand the progress in this exciting field, and facilitate mutual communications among scientists, engineers, students, and public, which may end up with better utilization of TENG technology to serve the people's life.

Introduction of Triboelectric Nanogenerator

I.1 What is a Triboelectric Nanogenerator (TENG)?

To meet the rapidly increased energy demands of the Internet of Things (IoT) and modern smart cities, a new type of energy harvesting technology, nanogenerator, has been invented to provide sustainable power source by collecting energy from the ambient environment. In them, the triboelectric nanogenerator (TENG), which is based on the coupling of triboelectrification and electrostatic induction effects, is focused in recent years [1–6]. This emerging technology was predicted to play a critical role in harvesting low-frequency energy such as body motion energy and ocean-wave energy [7–9]. Due to its advantages of lightweight, low cost, and high efficiency, plenty of research has demonstrated the great potential of TENGs on numerous applications [10–13].

The term "nanogenerator" is defined as an emerging type of technology that can convert small-scale mechanical and thermal energy into electricity. Different from traditional generators, nanogenerator usually utilizes Maxwell's displacement current initiated by the static charges, which were generated by triboelectric, piezoelectric, and pyroelectric effects, to drive effective energy conversion [14, 15]. TENG is the major type of nanogenerator which utilizes the charge generated in triboelectric effect, which is also the most powerful energy harvester in nanogenerators.

I.2 First-Principle Theoretical Model

Traditionally, it is believed that TENG is operated based coupling effects of tribo-electrification and electrostatic induction. Triboelectrification (triboelectric effect) describes the origin of the static charges, while electrostatic induction explains the power generation. However, further theories from fundamental physics are still required to understand the operation of TENG.

Triboelectric Nanogenerators: Technology, Applications, and Commercialization, First Edition.
Yunlong Zi, Hengyu Guo, Jie Wang, Chi Zhang, Xiangyu Chen, and Qing Zhao.

Wang demonstrated the first-principle theoretical models from Maxwell's displacement current [14, 15]. Maxwell's equations are shown below:

$$\begin{aligned}
&\nabla \cdot \boldsymbol{D} = \rho \\
&\nabla \cdot \boldsymbol{B} = 0 \\
&\nabla \times \boldsymbol{E} = -\frac{\partial \boldsymbol{B}}{\partial t} \\
&\nabla \times \boldsymbol{H} = \boldsymbol{J} + \frac{\partial \boldsymbol{D}}{\partial t}
\end{aligned} \tag{I.1}$$

Here, $\boldsymbol{D} = \varepsilon \boldsymbol{E}$ is the electric displacement field, $\boldsymbol{B}$ is the magnetic field vector, $\boldsymbol{E}$ is the electric field, $\boldsymbol{H}$ is the magnetic field strength, $\boldsymbol{J}$ is the current density, and ρ is the volume charge density. The term $\partial \boldsymbol{D}/\partial t$ can be also written as $\boldsymbol{J}_\mathrm{D}$, as named Maxwell's displacement current. Wang proposed an additional polarization term $\boldsymbol{P}_\mathrm{S}$ in the electric displacement field due to the static charge generation in nanogenerator, and thus $\boldsymbol{J}_\mathrm{D}$ can be written as:

$$J_\mathrm{D} = \frac{\partial \boldsymbol{D}}{\partial t} = \varepsilon \frac{\partial \boldsymbol{E}}{\partial t} + \frac{\partial \boldsymbol{P}_\mathrm{S}}{\partial t} \tag{I.2}$$

Therefore, the volume charge density ρ and current density $\boldsymbol{J}$ can be redefined as ρ' and $\boldsymbol{J}'$, respectively:

$$\rho' = \rho - \nabla \cdot \boldsymbol{P}_\mathrm{S} \tag{I.3}$$

$$\boldsymbol{J}' = \boldsymbol{J} + \frac{\partial \boldsymbol{P}_\mathrm{S}}{\partial t} \tag{I.4}$$

And they still satisfy the charge conversion and continuation equation:

$$\nabla \cdot \boldsymbol{J}' + \frac{\partial \rho'}{\partial t} = 0 \tag{I.5}$$

Through this displacement current as the driving force in nanogenerators, the conduction current can be driven on the external load, forming a complete loop. And then, the output characteristics of TENGs can be further discussed in theoretical models such as Displacement Current Theory model [16, 17], quasi-electrostatic model [18–20], and Distance-Dependent-Electric-Field (DDEF) mode [21–23], which can be used to calculate electric potential and power generation accurately.

I.3 Equivalent Circuit Models and Basic Operation Modes

I.3.1 Equivalent Circuit Models

As stated above, the TENG is operated based on the additional displacement current term, originating from the static charge generated from triboelectric effect. The maintained opposite static charge in triboelectric surfaces determines the inherent capacitive behavior of the TENG. At the open-circuit condition, the potential difference V_OC accumulated between electrodes originates from the additional polarization term $\boldsymbol{P}_\mathrm{S}$, as a function of the displacement x of the moving part in TENG. In the meanwhile, if we assume no charge generation in TENG, the device can be

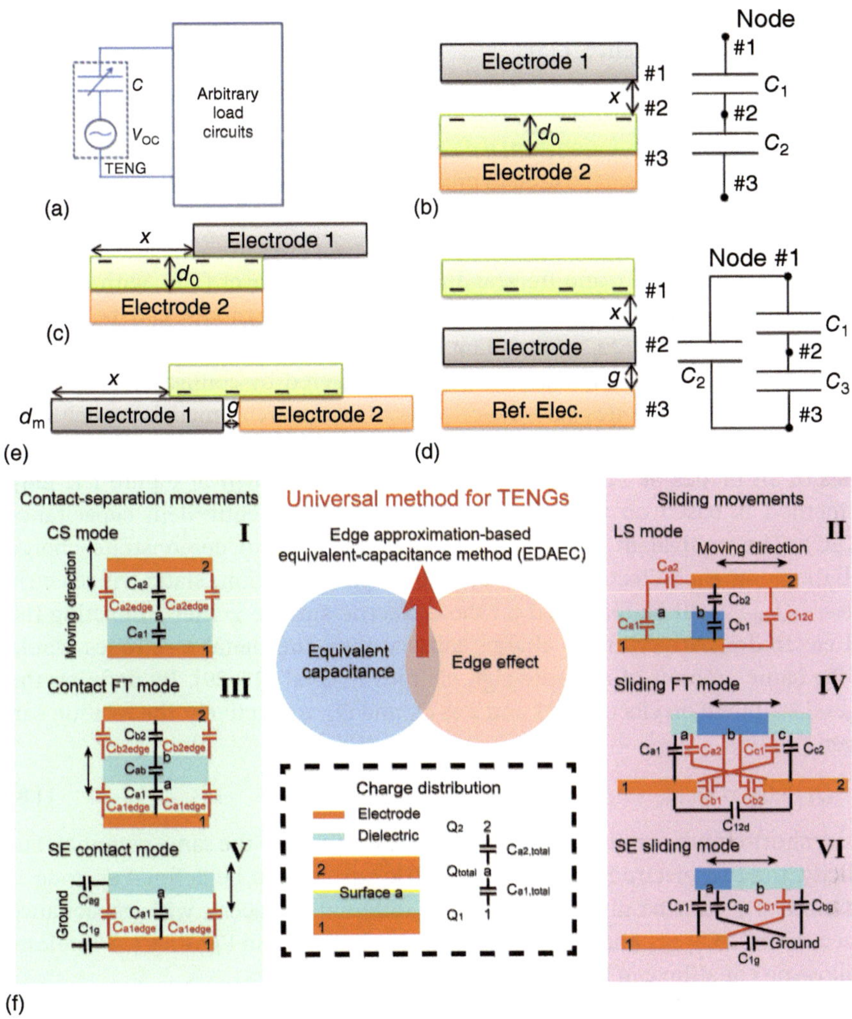

Figure I.1 Equivalent circuit models and basic operation modes of TENGs. (a) Basic equivalent circuit of TENG. Source: Reproduced with permission of [2], 2015 © Elsevier. (b–e) The theoretical models of contact-separation (CS), lateral-sliding (LS), single-electrode (SE), freestanding triboelectric-layer (FT) modes TENGs. (f) The EDAEC method for quantitatively analysis of all modes of TENG. Source: Reproduced with permission of [26], 2019 © Royal Society of Chemistry.

treated as a pure capacitor with capacitance C determined by x as well. So in the generalized case, the output voltage V and the charge transfer Q between electrodes are determined by the equivalent circuit of a voltage source $V_{OC}(x)$ in parallel with a capacitor $C(x)$ as shown in Figure I.1a, with the governing equation described by: [2, 24]

$$V = -\frac{1}{C(x)}Q + V_{OC}(x) \tag{I.6}$$

In short-circuit condition, full charge transfer (Q_{SC}) can be obtained as driven by the voltage source, and hence: [25]

$$Q_{SC}(x) = C(x)V_{OC}(x) \tag{I.7}$$

From such the V-Q-x relationship, we can derive the lumped parameter equivalent circuit mode, which can be used to predict characteristics of TENGs in different modes.

TENG has four basic operation modes: contact-separation (CS), lateral sliding (LS), single electrode (SE), and freestanding triboelectric layer (FT), with detailed structures shown in Figure I.1b–e. In them, SE and FT modes can be further divided into SE contact (SEC), SE sliding (SES), sliding FT (SFT), and contact FT (CFT) modes, depending on whether they are triggered by contact separation or sliding motions. These modes have their own equivalent circuit models available for simulations and theoretical calculations. Here, a universal method for quantitative analysis of all modes of TENG is used for analysis, as shown in Figure I.1f [26]. This method is based on the edge approximation-based equivalent capacitance (EDAEC). The equivalent capacitance models are used to demonstrate charge distributions on each electrode. Due to contact electrification, static triboelectric charges – $Q_{\text{tribo},\chi}$ will be dispersed on the dielectric surface χ after contacting the metal electrodes. According to charge conservation, the metal electrodes would have the same amount of opposite-sign charges in total [27–29]. By defining the charges distributed on electrodes 1 and 2 as Q_1 and Q_2, respectively, the relation can be given below:

$$Q_1 + Q_2 = Q_{\text{tribo}} \tag{I.8}$$

Under short-circuit conditions, two electrodes would have the same potential. For simplicity, the two electrodes and the dielectric surface can be defined as node 1, 2 and surface χ (can be different letters for different surfaces), with capacitance between them as $C_{\chi1,\text{total}}$ and $C_{\chi2,\text{total}}$, respectively, as shown in Figure I.1. Therefore, the following equation can be obtained.

$$V_1 = \frac{Q_1}{C_{\chi1,\text{total}}} = V_2 = \frac{Q_2}{C_{\chi2,\text{total}}} \tag{I.9}$$

Thus, the short-circuit equilibrium charges Q_1 and Q_2 on electrodes 1 and 2, respectively, are given as: [30]

$$\begin{cases} Q_1 = \sum \frac{1}{1+\frac{C_{\chi2,\text{total}}}{C_{\chi1,\text{total}}}} Q_{\text{tribo},\chi} \\ Q_2 = \sum \frac{1}{1+\frac{C_{\chi1,\text{total}}}{C_{\chi2,\text{total}}}} Q_{\text{tribo},\chi} \end{cases} \tag{I.10}$$

Sum symbol $\sum$ is used here to indicate the charge contributions from different surfaces. And then:

$$Q_{SC}(x) = |Q_1(x) - Q_1(0)| \tag{I.11}$$

From the equations above, the working mechanism of TENGs can be easily illustrated. When the distance between surface a and electrode 2 is zero the capacitance across them would be much larger than that across electrode 1 ($C_{\chi 2,\text{total}} \gg C_{\chi 1,\text{total}}$). Most of the positive tribo-charges would be attracted to electrode 2. Q_2 is close to Q_{tribo} and Q_1 is approximately zero. On the other hand, when the distance is quite large, the capacitance across surfaces a and electrode 2 would be much smaller than that across electrode 1 instead ($C_{\chi 1,\text{total}} \gg C_{\chi 2,\text{total}}$). So Q_1 is close to Q_{tribo} and Q_2 is approximately zero. Thus, $Q_{\text{SC}}(x)$ can be calculated by the difference between $Q_1(x)$ and $Q_1(0)$ or between $Q_2(x)$ and $Q_2(0)$, and the total capacitance $C(x) = C_{\chi 1,\text{total}}C_{\chi 2,\text{total}}/(C_{\chi 1,\text{total}} + C_{\chi 2,\text{total}})$. According to this EDAEC method the charge distributed on each electrode could then be quantitatively calculated.

I.3.2 CS Mode TENG

CS mode TENG was the original type TENG demonstrated in early-stage studies [1, 3]. Based on triboelectric materials used, the CS-mode TENG can be divided into two types: dielectric/dielectric contact and dielectric/metal contact. However, since there is no fundamental difference in operation mechanism between these two types, we just take the dielectric/metal contact type as an example, with the model shown in Figure I.1b. The dielectric layer with a thickness of d and relative dielectric constant ε_r is directly facing metal surface as one electrode, and another electrode is attached on the backside of the dielectric layer. Under external excitation, the displacement x is defined as vertical distance between two surfaces. After triboelectric effect, the surface charge density $-\sigma$ was generated in the dielectric surface a with area $S = w \times L$, making $Q_{\text{tribo}} = S\sigma$. By choosing the minimum achievable charge reference state (MACRS) [30], with the charge transfer of Q, the top electrode (electrode 2) is with total charge of $Q_2 = S\sigma - Q$, and the bottom electrode (electrode 1) is with total charge of $Q_2 = Q$. As shown in Figure I.1f-I, the two electrodes and the dielectric layer surface form two capacitances: $C_{\text{a2,total}}(x)$ is the capacitance between electrode 2 and the dielectric surface a, depending on x; and $C_{\text{a1,total}}$ is the capacitance between the dielectric surface a and electrode 1, which is a fixed value which can be considered as parallel-board capacitance. If $S \gg x^2$, the capacitance $C_{\text{a2,total}}(x)$ can be also considered as parallel-board capacitance $C_{\text{a2}}(x)$ without considering the edge effect, and we can derive:

$$Q_{\text{SC}}(x) = \frac{S\sigma x}{d/\varepsilon_r + x} \tag{I.12}$$

$$C(x) = \frac{\varepsilon_0 S}{d/\varepsilon_r + x} \tag{I.13}$$

$$V_{\text{OC}}(x) = \frac{\sigma x}{\varepsilon_0} \tag{I.14}$$

$$V = -\frac{1}{C(x)}Q + V_{\text{OC}}(x) = -\frac{d/\varepsilon_r + x}{\varepsilon_0 S}Q + \frac{\sigma x}{\varepsilon_0} \tag{I.15}$$

While if the edge effect is considered, the total capacitance of $C_{\text{a2,total}}(x)$ should include capacitance contributions from edge effects $C_{\text{a2edge}}(x)$ in parallel, as shown

Figure I.1f-I. Based on previous studies, this capacitance can be estimated as: [31]

$$C_{a2edge} = \varepsilon_r\varepsilon_0 \left\{ \frac{L}{\pi} \left[1 + \ln\left(1 + 2\pi\frac{w}{d} + \ln\left(1 + 2\pi\frac{w}{d}\right)\right)\right] + \frac{w}{\pi}\left[1 + \ln\left(1 + 2\pi\frac{L}{d} + \ln\left(1 + 2\pi\frac{L}{d}\right)\right)\right]\right\} \tag{I.16}$$

With that, we can calculate numerous values for $Q_{SC}(x)$, $C(x)$, and $V_{OC}(x)$, obtaining the governing equation.

I.3.3 LS Mode TENG

Traditionally, by assuming the parallel-board capacitance model, the total capacitance $C(x)$ fully depends on the capacitance of the covered part between electrodes 1 and 2, as described in Figures I.1c and I.2a [24, 32, 33]. The length L and the width w of the electrodes and dielectric layers are all defined as 100 mm. The dielectric layer with the thickness of d_0 is stacked on electrode 1 and the gap between the dielectric layer and the sliding electrode 2 is defined as d. The total capacitance $C(x)$, which is based on a traditional unlimitedly large plane, can be described as the equations below:

$$C(x) = \frac{\varepsilon_0(L-x)w}{d_0/\varepsilon_r + d} \tag{I.17}$$

And the equivalent capacitance of $C(x)$ can also be described as the series connection of $C_{b1}(x)$ and $C_{b2}(x)$, in which:

$$\begin{cases} C_{b1}(x) = \frac{\varepsilon_r\varepsilon_0(L-x)w}{d_0} \\ C_{b2}(x) = \frac{\varepsilon_0(L-x)w}{d} \end{cases} \tag{I.18}$$

Here, a new analytical EDAEC model is built, taking the capacitance of edge effect into consideration, to quantitatively describe the TENG's outputs. The real equivalent capacitance of $C(x)$ is shown in Figure I.2b, which includes three parts. Part 1 shows the edge capacitance $C_{a2}(x)$ between the surface a (dielectric layer) and electrode 2, and the capacitance $C_{a1}(x)$ between surface a and electrode 1. $C_{a2}(x)$, which is almost a constant, is quite small when comparing to $C_{a1}(x)$. Therefore $C_{a2(x)}$ can be approximated as a small constant through:

$$\begin{cases} C_{a1}(x) = \frac{\varepsilon_r\varepsilon_0 xw}{d_0} \\ C_{a2}(x) = C_a \end{cases} \tag{I.19}$$

Part 2 in Figure I.2b shows the edge capacitance C_{12d} between electrode 1 and electrode 2, which is also considered as a small constant for easy calculation. And Part 3 shows the capacitance C_{1b2} based on traditional parallel-board capacitance model between the covered part of electrode 1 and electrode 2.

Therefore, the total capacitance $C(x)$ can be calculated as:

$$C(x) = C_{1a2} + C_{12d} + C_{1b2} = \frac{\varepsilon_r\varepsilon_0 xw \cdot C_a}{\varepsilon_r\varepsilon_0 xw + C_a d_0} + C_{12d} + \frac{\varepsilon_r\varepsilon_0(L-x)w}{\varepsilon_r d + d_0} \tag{I.20}$$

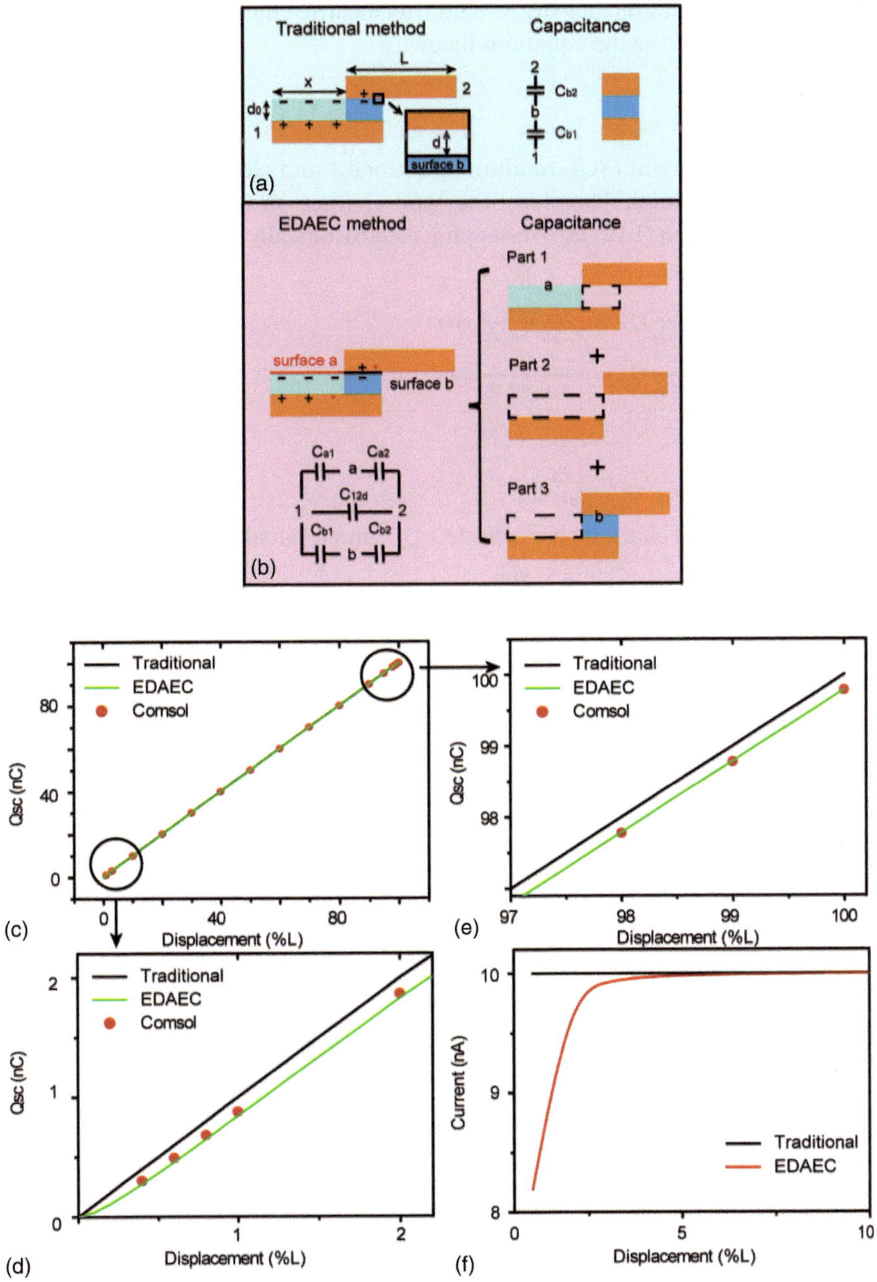

Figure I.2 The traditional mode (a) and the mode of EDAEC method (b) of the LS mode TENG. The comparison of transferred charge by traditional and EDAEC methods (c–e). The comparison of short-circuit current by traditional and EDAEC methods. Source: Reproduced with permission of [26], 2019 © Royal Society of Chemistry.

With triboelectric charge density σ, total triboelectric charge on surface sections a and b can be defined as the equations below:

$$\begin{cases} Q_a = \sigma w(x) \\ Q_b = \sigma w(L-x) \end{cases} \tag{I.21}$$

Under the short-circuit (SC) condition, electrode 1 and electrode 2 will have the same electrical potential. Therefore, the total charges on each electrode can be derived from Equation (I.10) by overlapping electrostatically induced charges from Q_a and Q_b:

$$\begin{cases} q_{a1}(x) = \frac{C_{a1}}{C_{a1}+C_{a2}} Q_a = \frac{\varepsilon_r \varepsilon_0 xw}{\varepsilon_r \varepsilon_0 xw + d_0 C_a} \sigma wx \\ q_{a2}(x) = \frac{C_{a2}}{C_{a1}+C_{a2}} Q_a = \frac{d_0 C_a}{\varepsilon_r \varepsilon_0 xw + d_0 C_a} \sigma wx \\ q_{b1}(x) = \frac{C_{b1}}{C_{b1}+C_{b2}} Q_b = \frac{\varepsilon_r d}{\varepsilon_r d + d_0} \sigma w(L-x) \\ q_{b2}(x) = \frac{C_{b1}}{C_{b1}+C_{b2}} Q_b = \frac{d_0}{\varepsilon_r d + d_0} \sigma w(L-x) \end{cases} \tag{I.22}$$

Therefore, the total charges on electrode 1 Q_1 can be calculated as:

$$\begin{aligned} Q_1(x) = q_{a1} + q_{b1} &= \frac{\varepsilon_r \varepsilon_0 xw}{\varepsilon_r \varepsilon_0 xw + d_0 C_a} \sigma wx + \frac{\varepsilon_r d}{\varepsilon_r d + d_0} \sigma w(L-x) \\ &= \left(\sigma w \frac{d_0}{\varepsilon_r d + d_0} \right) x + \frac{\left(\frac{d_0 C_a}{\varepsilon_r \varepsilon_0 w} \right)^2 \sigma w}{x + \frac{d_0 C_a}{\varepsilon_r \varepsilon_0 w}} + \sigma w L \frac{\varepsilon_r d}{\varepsilon_r d + d_0} - \sigma \frac{d_0 C_a}{\varepsilon_r \varepsilon_0} \end{aligned} \tag{I.23}$$

When $x = 0$ is assigned to be the initial state, the initial Q_1 is defined as:

$$Q_1(0) = \sigma w L \frac{\varepsilon_r d}{\varepsilon_r d + d_0} \tag{I.24}$$

The transferred charges Q_{SC} is equal to the change of the charge on electrode 1, therefore the transferred charge Q_{SC} can be calculated as the difference between the induced charges $Q_1(x)$ and the initial state charges $Q_1(0)$ on electrode 1. The short-circuit charge Q_{SC} can be calculated by the equation below.

$$Q_{SC}(x) = |Q_1(x) - Q_1(0)| = \left(\sigma w \frac{d_0}{\varepsilon_r d + d_0} \right) x + \frac{{Z_a}^2 \sigma w}{x + Z_a} - \sigma w Z_a \tag{I.25}$$

In the equation above, Z_a is a constant when the structure is fixed:

$$Z_a = \frac{C_a d_0}{\varepsilon_r \varepsilon_0 w} \tag{I.26}$$

It should be noticed that the first item of Q_{SC} has a linear relationship with displacement x, representing the part of σwx from traditional method. The second item of Q_{SC} is a non-linear item, and the third item is a constant.

Traditionally, the graph of traditional Q_{SC} has a strictly linear relationship with x. However, in the new EDAEC method, as we consider the edge effect, the Q_{SC} does not have a strictly linear relationship with the displacement. As shown in Figure I.2c, the graph of Q_{SC} has a flatter start at the beginning (Figure I.2d), and then the slope of Q_{SC} will increase until it has almost the same slope with the traditional graph when

$d = 0$ (Figure I.2e). In order to validate the above theoretical derivation, an LS mode TENG model is simulated by finite element method (by COMSOL Multiphysics software package), and the corresponding results are plotted in Figure I.2c–e. It should be emphasized that the theoretically calculated Q_{SC} by COMSOL simulation is closer to that from our EDAEC model than traditional Q_{SC}. According to the results of Q_{SC}, the graph of current and displacement is also plotted with a constant sliding speed of 0.01 m/s. Unlike the traditional graph which gives constant current, the EDAEC method shows that the current would be much lower when x is small and then sharply increase to the traditional value (in Figure I.2f). These results demonstrate that our EDAEC model which considers the edge effect is more suitable when describing the LS mode TENG.

With the total capacitance $C(x)$ and the short-circuit charge Q_{SC} known, the open-circuit voltage can be obtained. The capacitance calculated by the EDAEC method is demonstrated to be more accurate than traditional values (Figure I.3a) as compared to the COMSOL results. The open-circuit voltages are also plotted in Figure I.3b, which gradually increase with displacement x, and increase sharply when x is quite close to L. The total capacitance of the LS TENG will be zero according to the traditional method at $x = L$. At this position, the open-circuit voltage will be infinitely large, which is far away from the real situation. The EDAEC method takes the edge capacitance into consideration, and thus the open-circuit voltage can be accurately calculated even when $x = L$. Compared to traditional calculation, the EDAEC method shows that the voltage will give a result which is more consistent with that by COMSOL, as shown in Figure I.3b.

According to Equation (I.24), the transferred charge Q_{SC} is also affected by the TENG's structural parameters, for example, the distance d between the dielectric material and electrode 2, and the thickness d_0 of the dielectric material. Therefore, the short-circuit transferred charges and open-circuit voltages with different gap d is also demonstrated in Figure I.3c,d. It is obvious that the slopes of Q_{SC} and V_{OC} against the change of x become much smaller when the gap d increases. This result can explain experimental phenomenon of why the outputs (Q_{SC} and V_{OC}) of the LS TENG become quite low when the gap between the dielectric material and electrode 2 is larger than a threshold value. The thickness d_0 of the dielectric material will also affect the LS TENG's outputs, as shown in Figure I.3e,f. The Q_{SC} against x will become a bit smaller when the dielectric thickness d_0 increases, while the V_{OC} will become much larger due to the decrease in the total capacitance. These results are also compared with the simulation results by COMSOL, which shows a high degree of agreement. Based on the above analysis, the EDAEC method, by considering the edge effect, shows high accuracy in calculating the capacitance $C(x)$, short-circuit charge Q_{SC} and open-circuit voltage V_{OC} of LS mode TENGs under different conditions. It should be emphasized that this EDAED method is quite important for the grating structured TENGs. Because the electrode sizes of the grating structured TENGs are much smaller than other sliding TENGs, the electrodes could not be regarded as the traditional unlimitedly large plane anymore. Therefore, the influence of the side-effect capacitance will become much more

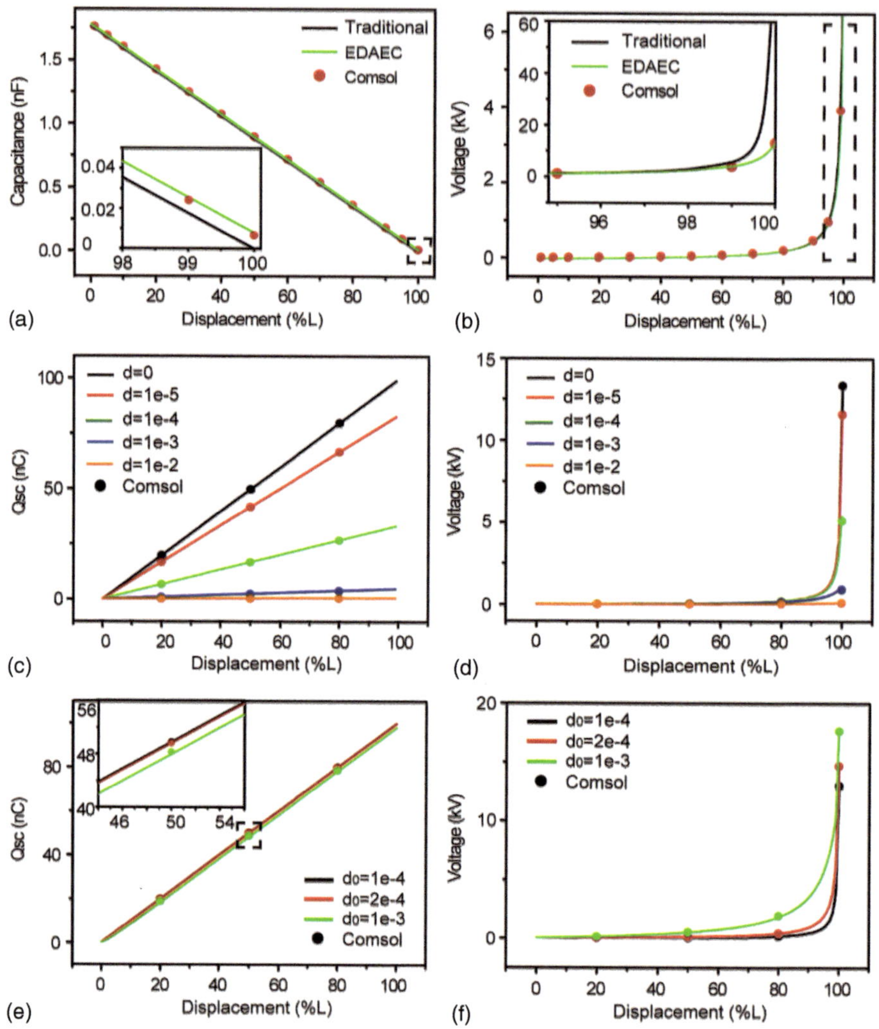

Figure I.3 Comparison of traditional and the EDAEC method on capacitance (a) and voltage (b). The insets are the magnified graph of the dotted part. (c) The relationship between transferred charges at short-circuit and the separation distance *d*. (d) The relationship between open-circuit voltage and the separation distance *d*. (e) The relationship between transferred charges at short-circuit and the thickness of dielectric layer d_0. (e) The relationship between open-circuit voltage and the thickness of dielectric layer d_0. Source: Reproduced with permission of [26], 2019 © Royal Society of Chemistry.

significant. The short-circuit charge of the grating structured TENGs is no longer a straight line [34].

I.3.4 SE Mode TENG

SE mode TENG is a design that utilizes only one electrode to collect output and connect to the external device, getting rid of the moving electrode which may bring limitations in applications. However, in practical applications, the other connection node of the external devices is always connected to the ground, so the ground should be considered as a reference electrode for SE-TENGs. Here we will introduce models for SEC and SES modes TENGs.

(1) SEC mode TENG

We take the conduct-to-dielectric SEC TENG as an example, with the structure shown in Figure I.1d. The output of dielectric-to-dielectric SEC TENG will be slightly different, but they follow similar characteristics. The top dielectric serves as the moving part, with the bottom surface a taking charge density of -σ, and the surface area of $S = w \times L$, making $Q_{tribo} = S\sigma$. The electrodes in TENG and the ground are marked as nodes "1" and "2", respectively. The gap between the electrode and the ground is g, while the displacement of the dielectric is x which is defined as the gap between the dielectric bottom surface a and the electrode. By choosing the minimum achievable charge reference state (MACRS), with the charge transfer of Q, the electrode is with total charge of $Q_1 = S\sigma - Q$, and the ground is with total charge of $Q_2 = Q$. As shown in Figure I.1f-V, the electrode, dielectric surface a, and reference electrode forms 2 capacitances: C_{a1} is the capacitance between the electrode and the dielectric surface a, depending on x, which is infinite at $x = 0$; C_{a2} is the capacitance between the ground and the dielectric surface a, depending on x. In this mode, the direct capacitance C_3 between the electrode and ground should also be considered, except the short-circuit condition [25]. The actual capacitance between the electrode and ground can be obtained by considering the connections among the three capacitances:

$$C(x) = C_3 + \frac{C_{a1}C_{a2}}{C_{a1} + C_{a2}} \tag{I.27}$$

Therefore, according to Eqs. I.7, I.10, and I.11, the short-circuit charge can obtained as:

$$Q_{SC}(x) = |Q_1(x) - Q_1(0)| = \left| \frac{\sigma S}{1 + \frac{C_{a2}}{C_{a1}}} - \frac{\sigma S}{1 + \frac{C_{a2}(0)}{C_{a1}(0)}} \right| = \sigma S \frac{C_{a2}}{C_{a1} + C_{a2}} \tag{I.28}$$

$$V_{OC}(x) = \frac{Q_{SC}(x)}{C(x)} = \frac{\sigma S C_{a2}}{C_{a1}C_{a2} + C_{a1}C_3 + C_{a2}C_3} \tag{I.29}$$

Please note that in SEC TENG, if we use the parallel-board capacitance for these capacitances, there will be no output, since the edge effects play an important role in SEC TENG. We can use Equation (I.16) to calculate these capacitances, obtaining numerous values for $Q_{SC}(x)$, $C(x)$, and $V_{OC}(x)$, and the governing equation.

(2) SES mode TENG

Owing to only one electrode, it is difficult for researchers to quantitatively describe output performances of the SES TENG [25]. The EDAEC method mentioned above can solve this problem. The mode in Figure I.4a is designed to demonstrate the working principle of the SES TENG, as the ground could be regarded as the second

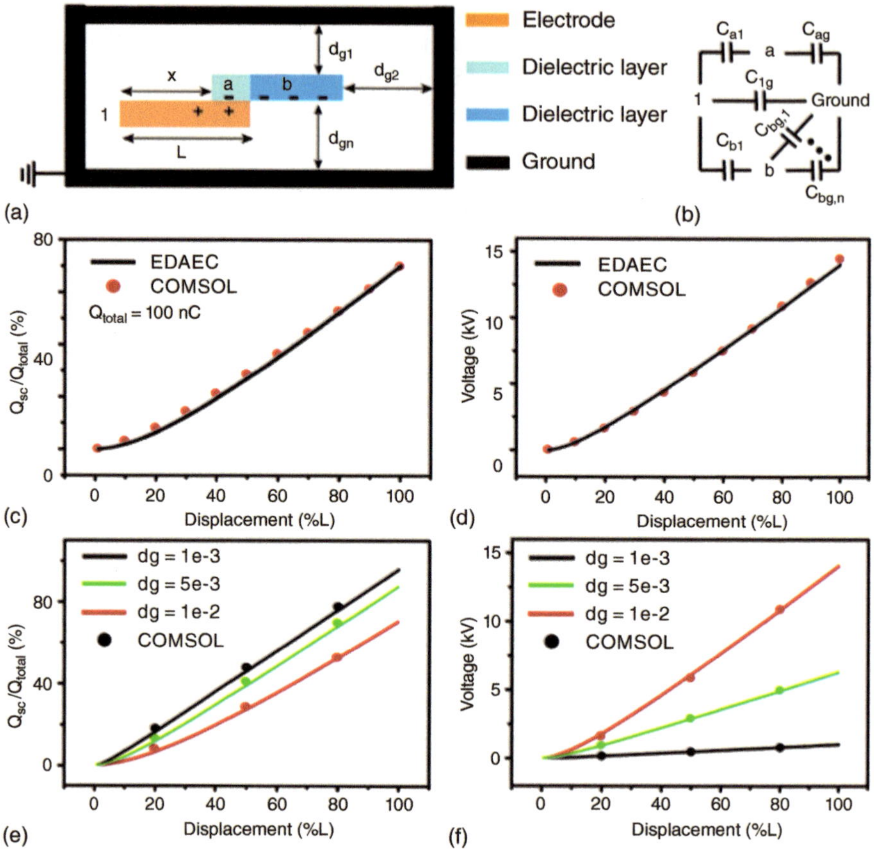

Figure I.4 The schematic (a) and equivalent capacitance (b) of SES TENG. (c) The relationship of transferred charges with the displacement. (d) The relationship of open-circuit voltage with the displacement. (e,f) The relationship between transferred charges at short-circuit (e) and open-circuit voltage (f) with the distance between the dielectric layer and the ground d_g. Source: Reproduced with permission of [26], 2019 © Royal Society of Chemistry.

electrode of the single-electrode TENG. Similar to the method described before, the metal electrode 1 and the dielectric layer are defined with length $L = 100$ mm and width $w = 100$ mm. The distance between dielectric layer and electrode 1 is defined as d, and the distance between the dielectric layer and the ground is defined as d_g. The surrounding environment connected to the ground will have an effect to the d_g, which could be regarded as a big conductive box containing the SES TENG.

The equivalent capacitance of C_{total} is shown in Figure I.4b, which includes three parts, including the capacitances between electrode 1 and the ground electrode *via* surfaces a (C_{a1} and C_{ag}), and b (C_{b1} and C_{bg}). Because the ground electrode is surrounding the dielectric layer, therefore there are several in-parallel capacitances $C_{bg,n}$ between the dielectric layer and each surrounding surface, and C_{bg} is the total capacitance. And C_{1gd} is the direct capacitance between electrode 1 and the ground electrode. The capacitance can be defined by the following equations.

$$\begin{cases} C_{a1(x)} = \frac{\varepsilon_0 w(L-x)}{\mathrm{d}} \\ C_{ag(x)} = \frac{\varepsilon_0 w(L-x)}{\mathrm{dg}} \\ C_{bg(x)} = \frac{\varepsilon_0 w(x)}{\mathrm{dg}} \\ \frac{1}{d_g} = \sum_{i=1}^{n} \frac{1}{d_{g_i}} \end{cases} \tag{I.30}$$

According to the EDAEC we described before, the edge capacitance C_{b1} is quite small as compared with other capacitances. Therefore, C_{b1} can be approximated as a small constant. The total capacitance could be defined as:

$$C_{\text{total}} = C_{1ag} + C_{1gd} + C_{1bg} = \frac{C_{a1} * C_{ag}}{C_{a1} + C_{ag}} + C_{1gd} + \frac{C_{b1} * C_{bg}}{C_{b1} + C_{bg}} \tag{I.31}$$

After contact electrification, the upper surface of the dielectric layer would be filled with triboelectric charges. Under short-circuit conditions, the total charges on electrode 1 can be calculated by the following equations.

$$\begin{aligned} Q_1(x) = q_{a1} + q_{b1} &= \frac{1}{1 + \frac{C_{ag}}{C_{a1}}} Q_a + \frac{1}{1 + \frac{C_{bg}}{C_{b1}}} Q_b \\ &= \left(\frac{d_g}{d_g + d} \right) \sigma w(L - x) + Z\sigma w \left(1 - \frac{z}{Z + x} \right) \end{aligned} \tag{I.32}$$

In the equation above, to facilitate the calculation, the constant is defined as:

$$Z = \frac{C_{b1} d_g}{w \varepsilon_0} \tag{I.33}$$

Thus, the short-circuit charge Q_{SC} can be calculated by the equation below.

$$Q_{SC}(x) = |q_1(0) - q_1(x)| = \left(\frac{d}{d_g + d} \right) \sigma w x + z^2 \sigma w \frac{1}{Z + x} - Z\sigma w \tag{I.34}$$

Due to the characteristics of the single electrode TENGs, the short-circuit charge Q_{SC} is noticeably lower than the total charges on the electrode 1 Q_{total}, which is equal to the total triboelectric charge of the dielectric layer. From Figure I.4c, at $x = L$, the short-circuit charge Q_{SC} only reaches around 70% of the total movable charge Q_{total}. This phenomenon could explain why the output power or the efficiency of SE TENG is quite lower than LS or SFT TENG. To validate the above equations, a SES TENG model is simulated by COMSOL. The transferred charge from the equation and the numerical results simulated from COMSOL are plotted in Figure I.4c which are quite consistent with our model.

With the total capacitance C_{total} and the short-circuit charge Q_{SC} known, the open-circuit voltage can be obtained by Equation (I.7). The open-circuit voltage from the equation and the results simulated from COMSOL are plotted on the same graph to compare, in Figure I.4d. The theoretical calculation coincides with numerical simulation, which demonstrates the accuracy of the proposed theoretical model.

According to Equation (I.34), the distance d_g between the dielectric material and the ground electrode will significantly affect the transferred charge Q_{SC}. Therefore, the transferred charge and open-circuit voltage with different distance d_g is also demonstrated in Figure I.4e,f. It is interesting to notice that the slope of Q_{SC} will be much smaller when the distance d_g increases, while V_{OC} will be much larger.

I.3.5 FT Mode TENG

FT TENG is another two-electrode TENG design that can make both electrodes keep static, while the high output performance can be guaranteed [30]. It can also be distinguished from CFT and SFT modes.

(1) CFT mode TENG

The CFT can also be divided by the different material selections as dielectrics or conductors, while their output characteristics are quite similar. Here, we just introduce a typical one with the dielectric freestanding layer and metal electrodes, as shown in Figure I.1f-III. The dielectric freestanding layer serves as the moving part, with the top and bottom surfaces a and b taking charge density of $-\sigma$ for each, and the surface area of $S = w \times L$, making $Q_{\text{tribo,a}} = Q_{\text{tribo,b}} = S\sigma$. The top and bottom metal electrodes are marked as nodes "1" and "2", respectively. The gap between the electrodes is g, while the displacement of the dielectric layer is x, which is defined as the distance between the bottom dielectric surface b and the bottom electrode 2. The dielectric thickness is d with relative dielectric constant ε_r. As shown in Figure I.1f-III, the electrodes 1, 2, and dielectric surfaces a, b can form 3 capacitances: C_{a1} is the capacitance between the top electrode 1 and the dielectric surface a, depending on g–x, which is infinite at $x = g$; C_{ab} is the capacitance of the dielectric layer, which is a constant estimated by the parallel-board capacitance; C_{b2} is the capacitance between the electrode 2 and the dielectric surface b, depending on x, which is infinite at $x = 0$.

Q_1 and Q_2 of this CFT TENG can be derived as the sum of charge contributed from two charged surfaces a and b, and then:

$$\begin{cases} Q_1 = \dfrac{1}{1+\frac{C_{a2,\text{total}}}{C_{a1,\text{total}}}} Q_{\text{tribo},a} + \dfrac{1}{1+\frac{C_{b2,\text{total}}}{C_{b1,\text{total}}}} Q_{\text{tribo},b} = \sigma S \dfrac{\frac{1}{C_{ab}}+\frac{2}{C_{b2}}}{\frac{1}{C_{a1}}+\frac{1}{C_{ab}}+\frac{1}{C_{b2}}} \\ Q_2 = \dfrac{1}{1+\frac{C_{a1,\text{total}}}{C_{a2,\text{total}}}} Q_{\text{tribo},a} + \dfrac{1}{1+\frac{C_{b1,\text{total}}}{C_{b2,\text{total}}}} Q_{\text{tribo},b} = \sigma S \dfrac{\frac{1}{C_{ab}}+\frac{2}{C_{a1}}}{\frac{1}{C_{a1}}+\frac{1}{C_{ab}}+\frac{1}{C_{b2}}} \end{cases} \tag{I.35}$$

And then:

$$Q_{SC}(x) = |Q_1(x) - Q_1(0)| = \sigma S \left(\frac{\frac{1}{C_{ab}} + \frac{2}{C_{b2}}}{\frac{1}{C_{a1}} + \frac{1}{C_{ab}} + \frac{1}{C_{b2}}} - \frac{C_{a1}(0)}{C_{a1}(0) + C_{ab}} \right) \tag{I.36}$$

$$C(x) = C_{a1} \| C_{ab} \| C_{b2} \tag{I.37}$$

So $V_{OC}(x)$ can be also calculated correspondingly. If $S \gg g^2$, the capacitance $C_{a1}(x)$ and $C_{b2}(x)$ can be also considered as parallel-board capacitances without considering the edge effect, and we can derive:

$$Q_{SC}(x) = \frac{2S\sigma x}{d/\varepsilon_r + g} \tag{I.38}$$

$$C(x) = \frac{\varepsilon_0 S}{d/\varepsilon_r + g} \tag{I.39}$$

$$V_{OC}(x) = \frac{2\sigma x}{\varepsilon_0} \tag{I.40}$$

$$V = -\frac{1}{C(x)}Q + V_{OC}(x) = -\frac{d/\varepsilon_r + g}{\varepsilon_0 S}Q + \frac{2\sigma x}{\varepsilon_0} \tag{I.41}$$

If the edge effect is considered, the total capacitance of these two capacitors should include capacitance contributions from edge effects in parallel, as shown in Figure I.1f-III. Based on previous studies, this capacitance can be estimated by Equation (I.16). With that, we can calculate numerous values for $Q_{SC}(x)$, $C(x)$, and $V_{OC}(x)$, obtaining the governing equation.

(2) SFT mode TENG

The capacitance caused by edge effect in sliding FT (SFT) TENG is significant and brings a much more obvious influence than that for LS TENG. The simplest structure of the SFT TENG, shown in Figure I.1e, consists of two adjacent metal electrodes (length: $L = 100$ mm, width: $w = 100$ mm) with a gap g and a movable dielectric layer with identical size to each electrode. The dielectric layer is placed above the electrodes at a distance of *d*. Due to impact of the edge capacitance, the sliding displacement of the TENG can be divided into three sections according to x, $0 < x < g$ (Part 1, Figure I.5a,b), $g \le x \le L$ (Part 2, Figure I.5c,d), and $L < x < L + g$ (Part 3), which is symmetric to Part 1. Therefore, both the short-circuit transferred charge Q_{SC} (Figure I.5e,f) and open-circuit voltage (Figure I.5g,h) can be divided into 3 parts correspondingly: the first flat part corresponding to $0 < x < g$, the second linear-like part corresponding to $g \le x \le L$ and the third flat part corresponding to $L < x < L + g$.

When $g \le x \le L$, the total capacitance (Figure I.5c) consists of the capacitances between electrodes *via* surfaces a (C_{a1} and C_{a2}), b (C_{b1} and C_{b2}), and c (C_{c1} and C_{c2}), and the direct capacitance between electrodes (C_{12d}). Traditionally, only C_{a1} and C_{c2} are considered, which can be given by the following equations.

$$\begin{cases} C_{a1(x)} = \frac{\varepsilon_0 w(L-x)}{\text{d}} \\ C_{c2(x)} = \frac{\varepsilon_0 w(x-g)}{\text{d}} \end{cases} \tag{I.42}$$

Similar to the Equation (I.19) of LS TENG, C_{a2} and C_{c1} are quite small, which can be approximated as small constants. Due to the structural symmetry, C_{b1} is equal to C_{b2}, thus $Y_b = C_{b2}/C_{b1} = 1$. Therefore, according to the charge distribution Equation (I.10), the charge distributed on the electrode 1 can be defined as:

$$Q_1(x) = q_{a1} + q_{b1} + q_{c1} = \frac{1}{1 + \frac{C_{a2}}{C_{a1}}}Q_a + \frac{1}{1 + \frac{C_{b2}}{C_{b1}}}Q_b + \frac{1}{1 + \frac{C_{c2}}{C_{c1}}}Q_c \cdot (g \le x \le L) \tag{I.43}$$

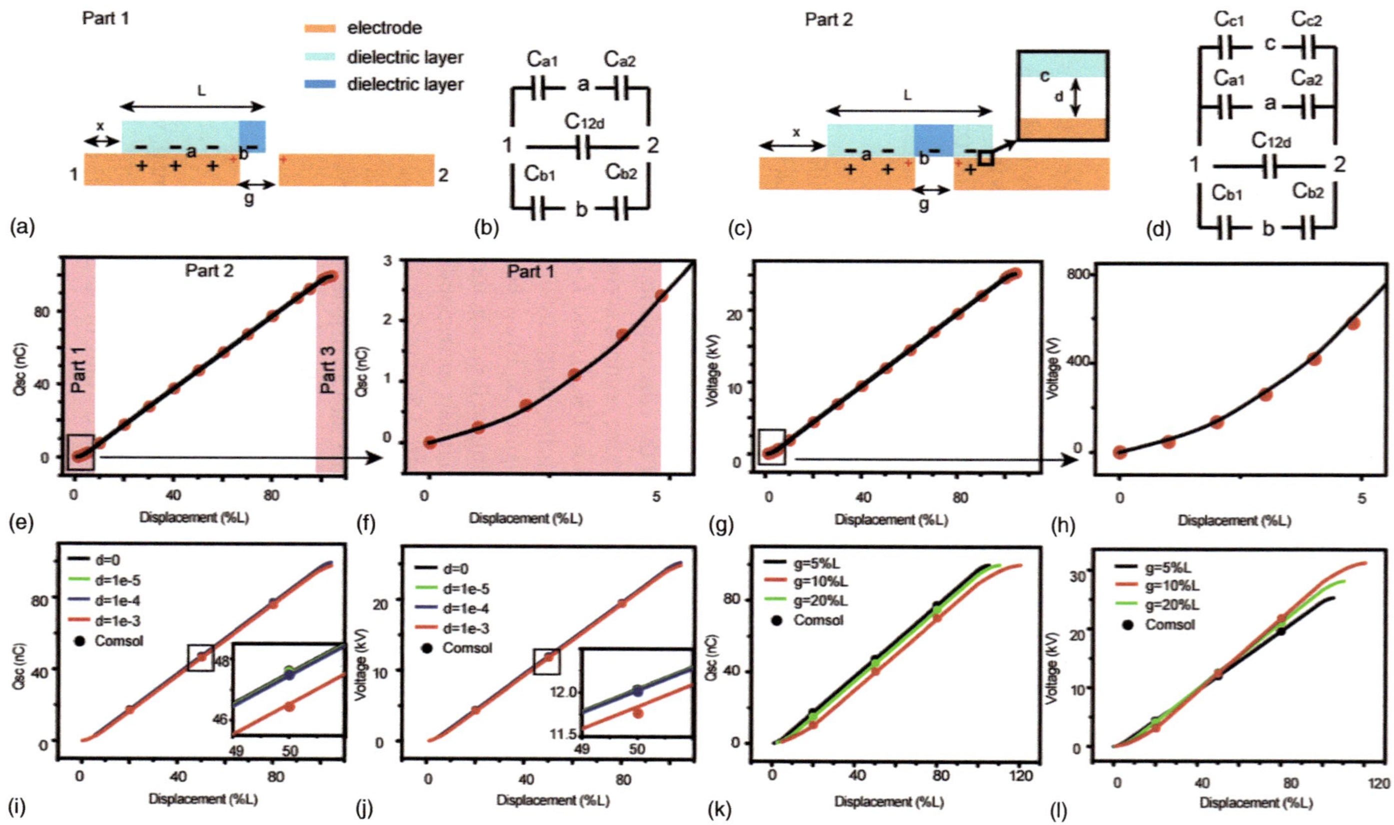

Figure I.5 The schematic (a) and equivalent capacitance (b) of SFT TENG part 1. The schematic (c) and equivalent capacitance (d) of SFT TENG part 2. (e) The total transferred charges of three parts with the displacement. (f)The magnified graph of the dotted part 1 of (e). (g) The total open-circuit voltage of three parts with the displacement. (h) The magnified graph of the dotted part 1 of (g). (i,j) The relationship between transferred charges at short-circuit (i) and open-circuit voltage (j) with the separation distance *d*. (k,l) The relationship between transferred charges at short-circuit (k) and open-circuit voltage (l) with the gap *g*. Source: Reproduced with permission of [26], 2019 © Royal Society of Chemistry.

When the distance d is fixed, Y_b, Z_a, and Z_c are considered as constants, for convenience of the calculation:

$$\begin{cases} Y_b = \frac{C_{b2}}{C_{b1}} \\ Z_a = \frac{C_{a2}d}{w\varepsilon_0} \\ Z_c = \frac{C_{c1}d}{w\varepsilon_0} \end{cases} \tag{I.44}$$

Therefore, the charge distributed on the electrode 1 can be defined as:

$$\begin{aligned} Q_1(x) &= \sigma w\left(L - Z_a + \frac{g}{1+Yb} + Z_c\right) \\ &-\sigma w\left(x - \frac{Z_a{}^2}{(L-x)+Z_a} + \frac{Z_c{}^2}{(x-g)+Z_c}\right) \cdot (g \le x \le L) \end{aligned} \tag{I.45}$$

With $x = 0$ assigned to be the reference state, the short-circuit charge Q_{SC} can be calculated by the equation below.

$$\begin{aligned} Q_{SC}(x) &= |q_{1(0)} - q_{1(x)}| \\ &= \sigma w\left(\frac{Z_a{}^2}{L+Z_a} - \frac{g}{1+Yb} - Z_c\right) \\ &+\sigma w\left(x - \frac{Z_a{}^2}{(L-x)+Z_a} + \frac{Z_c{}^2}{(x-g)+Z_c}\right) \end{aligned} \tag{I.46}$$

It should be emphasized that the second part of Q_{SC} does not have a strict linear relationship with the displacement. Because the constants Z_a and Z_c are quite small, thus the items $Z_a{}^2/(L-x+Z_a)$ and $Z_c{}^2/(x-g+Z_c)$ do not have a significant influence on the Q_{SC}, resulting that the Q_{SC} seems to have a linear relationship with the displacement when $(g \le x \le L)$.

When the distance $d = 0$, the Z_a and Z_c are zero and Q_{SC} can be simplified as.

$$Q_{SC}(x) = \sigma w(x - g/2) \tag{I.47}$$

This equation demonstrates that when the distance d is zero, part 2 of Q_{SC} has a linear relationship with the displacement x.

When $0 < x < g$, the total capacitance consists of the capacitances between electrodes *via* surfaces a (C_{a1} and C_{a2}) and b (C_{b1} and C_{b2}) and the direct capacitance between electrodes (C_{12d}). Figure I.5f shows that the Q_{SC} of part 1, as the framed part in Figure I.5e, increases slowly at the beginning and then much faster. The EDAEC method mentioned above can be used to quantitatively describe the non-linear relationship of the Q_{SC}. The equivalent capacitance of the SFT TENG when $0 < x < g$ is demonstrated in Figure I.5b. Due to the edge capacitance C_{b2} or C_{a2} between the surface b or a of the dielectric material and electrode 2, quite a bit of charges will be induced to distribute on electrode 2, causing the slow non-linear increase. According to the charge distribution Equation (I.3), as the x increases, the ratio of C_{b2}/C_{b1} will be bigger, causing more and more charges transferred to electrode 2. This non-linear increasing phenomenon could not be quantitatively explained without considering the edge capacitance. The charge distributed on the electrode 1 could be given as:

$$\begin{aligned} Q_1(x) &= q_{a1} + q_{b1} \\ &= \frac{1}{1+\frac{C_{a2}}{C_{a1}}} Q_a + \frac{1}{1+\frac{C_{b2}}{C_{b1}}} Q_b \\ &= \sigma w(L - Z_a) - \sigma wx\left(1 - \frac{1}{1+Y_b}\right) + \frac{\sigma w {Z_a}^2}{(L-x)+Z_a} \ (0 < x < g) \end{aligned} \tag{I.48}$$

The Q_{SC} for $L < x \leq L + g$ (Part 3) can be directly given due to the symmetry of it to that for $0 \leq x < g$ (Part 1), and then the charge distributed on the electrode 1 could be given as:

$$Q_1(x) = \begin{cases} \sigma w(L - Z_a) - \sigma wx\left(1 - \frac{1}{1+Y_b}\right) + \frac{\sigma w {Z_a}^2}{(L-x)+Z_a} \ 0 \leq x < g \\ \sigma w\left(L - Z_a + \frac{g}{2} + Z_c\right) - \sigma w\left(x - \frac{{Z_a}^2}{(L-x)+Z_a} + \frac{{Z_c}^2}{(x-g)+Z_c}\right) \ g \leq x \leq L \\ \sigma w\left(\frac{L+g}{1+Y_b} + Z_c\right) - \sigma wx\left(\frac{1}{1+Y_b}\right) - \frac{\sigma w {Z_c}^2}{(x-g)+Z_c} \ L < x \leq L+g \end{cases} \tag{I.49}$$

Thus, the short-circuit charge Q_{SC} can be calculated by the equation below.

$$Q_{sc}(x) = \begin{cases} \frac{\sigma w {Z_a}^2}{L+Z_a} + \sigma wx\left(1 - \frac{1}{1+Y_b}\right) - \frac{\sigma w {Z_a}^2}{(L-x)+Z_a} \ 0 \leq x < g \\ \sigma w\left(\frac{{Z_a}^2}{L+Z_a} - \frac{g}{2} - Z_c\right) + \sigma w\left(x - \frac{{Z_a}^2}{(L-x)+Z_a} + \frac{{Z_c}^2}{(x-g)+Z_c}\right) \ g \leq x \leq L \\ \sigma w\left(L + \frac{{Z_a}^2}{L+Z_a} - \frac{L+g}{1+Y_b} - Z_a - Z_c\right) + \sigma wx\left(\frac{1}{1+Y_b}\right) + \frac{\sigma w {Z_c}^2}{(x-g)+Z_c} \ L < x \leq L+g \end{cases} \tag{I.50}$$

In order to validate the above theoretical calculation, an SFT TENG model is simulated by COMSOL, which also takes the edge effect into consideration, and the corresponding results are plotted in Figure I.5e,f. It should be emphasized that the Q_{SC} from Equation (I.23) is consistent with the simulated Q_{SC} by COMSOL.

When the distance $d = 0$, the Z_a and Z_c are zero and Q_{SC} can be simplified as.

$$Q_{sc}(x) = \begin{cases} \sigma wx\left(1 - \frac{1}{1+Y_b}\right) \ 0 \leq x < g \\ \sigma wx - \sigma wg/2 \ g \leq x \leq L \\ \sigma w\left(L - \frac{L+g}{1+Y_b}\right) + \sigma wx\left(\frac{1}{1+Y_b}\right) \ L < x \leq L+g \end{cases} \tag{I.51}$$

The total capacitance C_{total} and the short-circuit transferred charge Q_{SC} have already been quantitatively given, and thus the open-circuit voltage can be obtained. The voltage is also divided into three parts according to x, as shown in Figure I.5g. Due to the edge capacitance, there is a slow-increase part of Part 1 ($0 < x < g$) as shown in Figure I.5h. The voltage of Part 2 ($g \leq x \leq L$) has a linear-like relationship with the x. Due to the increasing edge capacitance effect, the increase of the voltage of Part 3 ($L < x < L + g$) will turn flat again.

According to Equation (I.50), the short-circuit transferred charge Q_{SC} is also affected by the TENG's structural parameters, for example, the distance d between the dielectric material and the electrodes, and the gap g between the two electrodes. Therefore, the transferred charge and open-circuit voltage with different distance d

and gap g are also demonstrated in Figure I.5i–l. The slope of Q_{sc} and V_{oc} become a little smaller when the distance *d* increases, as shown in Figure I.5i,j.

When the gap g becomes larger, the ratio of part 1 and part 3 become larger. And thus the influence of edge capacitance becomes more significant. As a result, the non-linear part becomes more obvious, and the flat parts of Q_{SC} at both ends are much more noticeable. Besides, the total capacitance will decrease due to the increase of the gap g. The total transferred charge remains the same during the whole process. Therefore, the open-circuit voltage will increase, as shown in Figure I.5l. Meanwhile, the portions of parts 1 and 3 also become larger. The results simulated by COMSOL are also presented, which are well-consistent. Based on the above analysis, the new method shows high accuracy in describing the short-circuit charge Q_{SC} and open-circuit voltage V_{OC} of SFT TENGs.

I.4 Energy Conversion and Electromechanical Coupling Models

As an energy system, TENG can either collect tiny energy in the environment to provide power for some electronic devices or act as a self-driven sensor [35–38]. It is crucial to study how to improve the efficiency and performance of TENG. In 2019, Xu et al. proposed a concept to study the efficiency of TENG from the perspective of energy flow [5]. As shown in Figure I.6a, the energy of TENG starts from harvesting external mechanical energy as input. Then it is converted into internal electrostatic energy in TENG and finally transformed into electrical power. It should be noted that the force (*F*) integral to the displacement (*x*) represents the mechanical energy input, and the *V*-*Q* plot demonstrates the energy output. Therefore, the total efficiency of TENG can be given by:

$$\eta = \frac{\oint VdQ}{\oint Fdx \mid Fdx \geq 0} \tag{I.52}$$

Moreover, the TENG energy harvesting system mainly involves three parts: an excitation source such as human motion or vehicle vibration, an energy harvesting device, and an interface circuit. As a dynamic system, a TENG system experiences oscillating energy flow, which means that studying the efficiency of TENG should fully consider the energy input and output simultaneously. In Figure I.6b, we choose the most widely used CS mode TENG as an example [1], where the whole system consists of a CS-TENG fixed at the excitation source. Here, spring is necessary because it can ensure that the upper electrodes of the TENG can be separated smoothly after contact. This electromechanical coupling system is modeled as an equivalent lumped-parameter one degree of freedom (DOF) dynamic system (Figure I.6c), whose mechanical governing equation is shown below:

$$m\ddot{x} + c\dot{x} + kx + F_e = -m\ddot{Z} \tag{I.53}$$

Where m, c, k, and F_e are the equivalent mass, damping, stiffness, and electrostatic force. And the method explaining how to calculate the electrostatic force has been

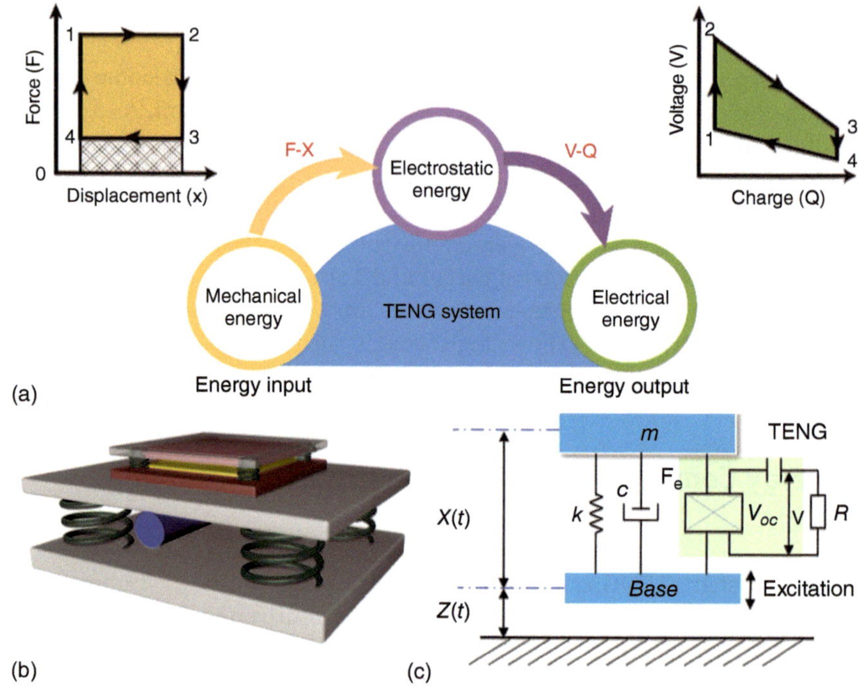

Figure I.6 Energy flow in the TENG system. (a) The TENG system's energy flow, force-displacement (*F-x*), and voltage-charge (*V-Q*) plots. Source: Reproduced with permission of [5], 2019 © Elsevier. (b) 3D model of an oscillator triggers a contact separation mode TENG. (c) Schematic of a TENG system involving mechanical and electrical parameters. Source: Reproduced with permission of [54], 2022 © John Wiley & Sons.

introduced in our work [5]. Solving the above Eqs. (I.53) and (I.6) and can guide us to calculate the TENG efficiency and optimize its output. Based on such electromechanical coupling models, multiple efforts have been devoted to studying dynamic behaviors of the TENG systems, as well as output promotion [6, 39–43].

However, the actual situation is much more complicated. For example, (1) The electrostatic force is relatively weak compared to other parties, which is mainly because the surface charge density of TENG is not high enough [13, 44]. (2) The friction force needs to be carefully considered. On the one hand, TENG needs contact electrification to maintain the surface charge; on the other hand, friction as a dissipative force results in low output efficiency of the TENG [45, 46]. Therefore, a trade-off in contact intimacy is needed. (3) The structure of TENG is not limited to the basic four fundamental modes but also involves hybrid modes whose dynamic behavior is more complex [47–49]. (4) Collision issues are also important, as the strong nonlinear collision force wastes energy and makes our model invalid [40, 50]. (5) The vibration source in the natural environment is usually irregular, while the theoretical simulation is based on simple harmonic vibration, which may also cause inaccurate results [51–53].

I.5 Summary

This introduction discusses the basic theoretical models of TENG, which are considered simple but effective based on equivalent capacitances. Compared to models based on electric fields such as displacement current model, quasi-electrostatic model, and DDEF model, the EDAEC model may lack some details in a local electric field or a little variation on the calculated voltage, while the calculated accuracy from EDAEC model has been high enough to be close to the simulation results demonstrated by COMSOL Multiphysics [26]. In the meanwhile, EDAEC model shows a much simpler in calculation process compared to electric field-based models, even though the equivalent capacitances from parasitic effects need to be carefully determined.

On the other hand, the energy conversion process of TENG is a key mechanism to understand the physical principles of TENG, and how to promote the output performances. However, the introduction here is still a very ideal situation without considering energy waste during conversion. In the following chapters, we will cover some key energy waste pathways including discharge and frictional energy waste, and discuss how to prevent this energy waste, ultimately making a breakthrough on the output limit of TENGs.

References

1 Niu, S.M., Wang, S.H., Lin, L. et al. (2013). Theoretical study of contact-mode triboelectric nanogenerators as an effective power source. *Energy Environ. Sci.* 6: 3576–3583.

2 Niu, S. and Wang, Z.L. (2015). Theoretical systems of triboelectric nanogenerators. *Nano Energy* 14: 161–192.

3 Fan, F.-R., Tian, Z.-Q., and Wang, Z.L. (2012). Flexible triboelectric generator. *Nano Energy* 1: 328–334.

4 Wang, S., Xie, Y., Niu, S. et al. (2014). Freestanding triboelectric-layer-based nanogenerators for harvesting energy from a moving object or human motion in contact and non-contact modes. *Adv. Mater.* 26: 2818–2824.

5 Xu, G., Li, X., Xia, X. et al. (2019). On the force and energy conversion in triboelectric nanogenerators. *Nano Energy* 59: 154–161.

6 Xu, G., Fu, J., Li, C. et al. (2022). A nonlinear triboelectric nanogenerator with a broadened bandwidth for effective harvesting of vibration energy. *iEnergy* 1 (2): 236–242.

7 Zi, Y., Guo, H., Wen, Z. et al. (2016). Harvesting low-frequency (<5 Hz) irregular mechanical energy: a possible killer application of triboelectric nanogenerator. *ACS Nano* 10: 4797–4805.

8 Chen, J., Yang, J., Li, Z. et al. (2015). Networks of triboelectric nanogenerators for harvesting water wave energy: a potential approach toward blue energy. *ACS Nano* 9 (3): 3324–3331.

9 Wen, Z., Yeh, M.-H., Guo, H. et al. (2016). Self-powered textile for wearable electronics by hybridizing fiber-shaped nanogenerators, solar cells, and supercapacitors. *Sci. Adv.* 2 (10): e1600097.

10 Chen, J., Zhu, G., Yang, W. et al. (2013). Harmonic-resonator-based triboelectric nanogenerator as a sustainable power source and a self-powered active vibration sensor. *Adv. Mater.* 25 (42): 6094–6099.

11 Pu, X., Liu, M., Chen, X. et al. (2017). Ultrastretchable, transparent triboelectric nanogenerator as electronic skin for biomechanical energy harvesting and tactile sensing. *Sci. Adv.* 3 (5): e1700015.

12 Chen, X., Gao, L., Chen, J. et al. (2020). A chaotic pendulum triboelectric-electromagnetic hybridized nanogenerator for wave energy scavenging and self-powered wireless sensing system. *Nano Energy* 69: 104440.

13 Fu, J., Xu, G., Li, C. et al. (2020). Achieving ultrahigh output energy density of triboelectric nanogenerators in high-pressure gas environment. *Adv. Sci.* 7 (24): 2001757.

14 Wang, Z.L. (2017). On Maxwell's displacement current for energy and sensors: the origin of nanogenerators. *Mater. Today* 20 (2): 74–82.

15 Wang, Z.L. (2020). On the first principle theory of nanogenerators from Maxwell's equations. *Nano Energy* 68: 104272.

16 Shao, J. and Wang, Z.L. (2023). Displacement current theory of triboelectric nanogenerators. In: *Handbook of Triboelectric Nanogenerators* (ed. Z.L. Wang et al.), 1–64. Cham: Springer International Publishing.

17 Shao, J., Willatzen, M., and Wang, Z.L. (2020). Theoretical modeling of triboelectric nanogenerators (TENGs). *J. Appl. Phys.* 128 (11): 111101.

18 Shao, J., Willatzen, M., Shi, Y. et al. (2019). 3D mathematical model of contact-separation and single-electrode mode triboelectric nanogenerators. *Nano Energy* 60: 630–640.

19 Shao, J., Liu, D., Willatzen, M. et al. (2020). Three-dimensional modeling of alternating current triboelectric nanogenerator in the linear sliding mode. *Appl. Phys. Rev.* 7 (1): 011405.

20 Lu, M., Yin, W., Peyton, A. et al. (2019). A model for the triboelectric nanogenerator with inductive load and its energy boost potential. *Nano Energy* 63: 103883.

21 Dharmasena, R.D.I.G., KDGI, J., Mills, C.A. et al. (2017). Triboelectric nanogenerators: providing a fundamental framework. *Energy Environ. Sci.* 10 (8): 1801–1811.

22 Dharmasena, R.D.I.G., KDGI, J., Mills, C.A. et al. (2018). A unified theoretical model for triboelectric nanogenerators. *Nano Energy* 48: 391–400.

23 Deane, J.H.B., Dharmasena, R.D.I.G., and Gentile, G. (2018). Power computation for the triboelectric nanogenerator. *Nano Energy* 54: 39–49.

24 Niu, S., Liu, Y., Wang, S. et al. (2013). Theory of sliding-mode triboelectric nanogenerators. *Adv. Mater.* 25 (43): 6184–6193.

25 Niu, S., Liu, Y., Wang, S. et al. (2014). Theoretical investigation and structural optimization of single-electrode triboelectric nanogenerators. *Adv. Funct. Mater.* 24 (22): 3332–3340.

26 Li, X., Lau, T.H., Guan, D. et al. (2019). A universal method for quantitative analysis of triboelectric nanogenerators. *J. Mater. Chem. A* 7 (33): 19485–19494.

27 Wang, S.H., Zi, Y.L., Zhou, Y.S. et al. (2016). Molecular surface functionalization to enhance the power output of triboelectric nanogenerators. *J. Mater. Chem. A* 4 (10): 3728–3734.

28 Guo, W.X., Li, X.Y., Chen, M.X. et al. (2014). Electrochemical cathodic protection powered by triboelectric nanogenerator. *Adv. Funct. Mater.* 24 (42): 6691–6699.

29 Chen, M.X., Li, X.Y., Lin, L. et al. (2014). Triboelectric nanogenerators as a self-powered motion tracking system. *Adv. Funct. Mater.* 24 (32): 5059–5066.

30 Niu, S.M., Liu, Y., Chen, X.Y. et al. (2015). Theory of freestanding triboelectric-layer-based nanogenerators. *Nano Energy* 12: 760–774.

31 Li, Y., Li, Y.H., Li, Q.X. et al. (2003). Computation of electrostatic forces with edge effects for non-parallel comb-actuators. *J. Tsinghua Univ. (Sci & Tech)* 43 (8): 1024–1026. 1030.

32 Shao, J.J., Jiang, T., Tang, W. et al. (2018). Studying about applied force and the output performance of sliding-mode triboelectric nanogenerators. *Nano Energy* 48: 292–300.

33 Shao, J.J., Jiang, T., Tang, W. et al. (2018). Structural figure-of-merits of triboelectric nanogenerators at powering loads. *Nano Energy* 51: 688–697.

34 Niu, S.M., Wang, S.H., Liu, Y. et al. (2014). A theoretical study of grating structured triboelectric nanogenerators. *Energy Environ. Sci.* 7: 2339.

35 Feng, X., Zhang, Y., Kang, L. et al. (2021). Integrated energy storage system based on triboelectric nanogenerator in electronic devices. *Front. Chem. Sci. Eng.* 15 (2): 238–250.

36 Li, Q., Dai, K., Zhang, W. et al. (2021). Triboelectric nanogenerator-based wearable electronic devices and systems: Toward informatization and intelligence. *Digit. Signal Process.* 113: 103038.

37 Liao, W., Liu, X., Li, Y. et al. (2022). Transparent, stretchable, temperature-stable and self-healing ionogel-based triboelectric nanogenerator for biomechanical energy collection. *Nano Res.* 15 (3): 2060–2068.

38 Hou, W., Tang, X., Fang, L. et al. (2022). Self-driven real-time angle vector sensor as security dialer based on bi-directional backstop triboelectric nanogenerator. *Nano Energy* 99: 107430.

39 Yang, J., Chen, J., Yang, Y. et al. (2014). Broadband vibrational energy harvesting based on a triboelectric nanogenerator. *Adv. Energy Mater.* 4 (6): 1301322.

40 Hu, Y., Yang, J., Jing, Q. et al. (2013). Triboelectric nanogenerator built on suspended 3D spiral structure as vibration and positioning sensor and wave energy harvester. *ACS Nano* 7 (11): 10424–10432.

41 Bhatia, D., Hwang, H.J., Huynh, N.D. et al. (2019). Continuous scavenging of broadband vibrations via omnipotent tandem triboelectric nanogenerators with cascade impact structure. *Sci. Rep.* 9 (1): 1–9.

42 Tang, G., Cheng, F., Hu, X. et al. (2019). A two-degree-of-freedom cantilever-based vibration triboelectric nanogenerator for low-frequency and broadband operation. *Electronics* 8 (12): 1526.

43 Guan, D., Xu, G., Xia, X. et al. (2021). Boosting the output performance of the triboelectric nanogenerator through the nonlinear oscillator. *ACS Appl. Mater. Interfaces* 13 (5): 6331–6338.
44 Wu, H., Fu, S., He, W. et al. (2022). Improving and quantifying surface charge density via charge injection enabled by air breakdown. *Adv. Funct. Mater.* 32: 2203884.
45 Wang, P., Pan, L., Wang, J. et al. (2018). An ultra-low-friction triboelectric–electromagnetic hybrid nanogenerator for rotation energy harvesting and self-powered wind speed sensor. *ACS Nano* 12 (9): 9433–9440.
46 Wen, J., He, H., Niu, C. et al. (2022). An improved equivalent capacitance model of the triboelectric nanogenerator incorporating its surface roughness. *Nano Energy* 96: 107070.
47 Zou, Y., Sun, M., Yan, F. et al. (2022). A high-performance flag-type triboelectric nanogenerator for scavenging wind energy toward self-powered IoTs. *Materials* 15 (10): 3696.
48 Qu, Z., Huang, M., Chen, C. et al. (2022). Spherical triboelectric nanogenerator based on eccentric structure for omnidirectional low frequency water wave energy harvesting. *Adv. Funct. Mater.* 32: 2202048.
49 Yang, W., Chen, J., Jing, Q. et al. (2014). 3D stack integrated triboelectric nanogenerator for harvesting vibration energy. *Adv. Funct. Mater.* 24 (26): 4090–4096.
50 Sun, W., Jiang, Z., Xu, X. et al. (2021). Electromechanical coupling modeling and analysis of contact-separation mode triboelectric nanogenerators. *Int. J. Non-Linear Mech.* 136: 103773.
51 Rodrigues, C., Nunes, D., Clemente, D. et al. (2020). Emerging triboelectric nanogenerators for ocean wave energy harvesting: state of the art and future perspectives. *Energy Environ. Sci.* 13 (9): 2657–2683.
52 Wei, X., Zhao, Z., Zhang, C. et al. (2021). All-weather droplet-based triboelectric nanogenerator for wave energy harvesting. *ACS Nano* 15 (8): 13200–13208.
53 Ren, Z., Wu, L., Pang, Y. et al. (2022). Strategies for effectively harvesting wind energy based on triboelectric nanogenerators. *Nano Energy* 100: 107522.
54 Xu, G., Li, C., Chen, C. et al. (2022). Dynamics of triboelectric nanogenerators: a review. *Int. J. Mech. Syst. Dyn.* 2 (4): 311–324.

1

Models of Triboelectric Effect

1.1 Introduction

Contact electrification (CE) exists universally between any materials. It can be widely applied for energy harvesting [1, 2], electrostatic printing [3], etc., but may also cause electric breakdown in electronics and even fires [4, 5]. Understanding CE mechanisms will be of great significance to accurately control the surface charge level upon requirements of applications. Although CE has been discovered for more than 2600 years, its mechanism continues to elude the scientific community due to lack of proper systematic studies. Long-term debates exist on the charge carriers of CE: some scientists demonstrated the electron transfer dominant mechanism, while others favored material or ion transfer mechanism. An early study in 1980 by Lowell and Rose-Innes concluded CE mechanism between metal and dielectric as the electron transfer, with the surface states model proposed [6]. Through redox chemical reactions induced by triboelectric static charges, Bard et al. further confirmed electrons as charge carriers in CE [7–10]. The thermionic emission behaviors observed by Xu et al. indicated electron transfer dominant mechanism, with the electron-cloud-potential-well model developed [11–14]. Through atomic force microscopy (AFM) studies conducted by Lin et al., Zhou et al., and Li et al., further evidences were provided for CE on microscale to support electron transfer mechanism [15–17]. In the meanwhile, the transfer mechanisms by materials or ions were proposed by other scientists. Prof. Whitesides' group conducted a series of works on CE in ionic electrets from 2003 [18–21], concluding the ion transfer dominant mechanism with a model proposed. Prof. Grzybowski's group discovered the mosaic of surface charge generated by CE as attributed to material transfer [22–25]. Giuseppe Pezzotti and Šutka et al. attributed electrification as the output of bond cleavage and chemical changes at the molecular scale, implying ion/material transfer mechanism in CE [26, 27]. Therefore, to provide an answer to this millennium puzzle, further systematic studies are still required on CE in various material systems with well-designed methodology.

Triboelectric Nanogenerators: Technology, Applications, and Commercialization, First Edition.
Yunlong Zi, Hengyu Guo, Jie Wang, Chi Zhang, Xiangyu Chen, and Qing Zhao.

1.2 Thermionic Emission Method

A method to quantitatively investigate real-time charge transfer in CE via triboelectric nanogenerator (TENG) as a function of temperature was reported based on thermionic emission in 2018 [11]. The conventional metal–polymer or polymer–polymer structure of TENGs was not adopted in this study, mainly because polymers could not withstand at high temperatures. Here, two kinds of high-temperature-resistant contact-separation (CS) mode TENGs, a Ti-SiO_2 TENG and a Ti-Al_2O_3 TENG, were designed and fabricated, which were able to withstand a maximum temperature of 673 K and operate stably for a long period of time. Each TENG was placed in a heating cabinet, which could provide the desired temperature with an accuracy of ±5 K. The structure of the Ti-SiO_2 TENG is shown in Figure 1.1a. The charges increasingly accumulate with the CS operation of the TENG and then tend to balance at 293 K. It is worth noting that both the short-circuit charge transfer Q_{SC} (0.45 nC) and the open-circuit voltage V_{OC} (1.3 V) are rather low, which means that the CS Ti-SiO_2 TENG in the experiment can only generate limited charges during the CS operation. Thus, to investigate the influence of temperature on the tribocharges on the surfaces, SiO_2 was first rubbed

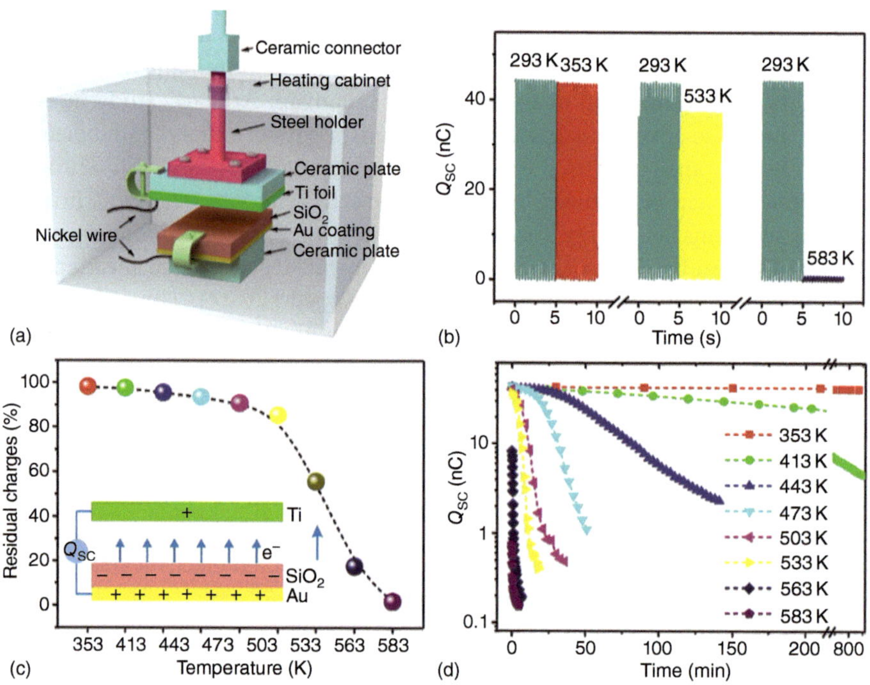

Figure 1.1 Performances of the Ti-SiO_2 TENG at different temperatures: (a) setup of the measurement platform, (b) the change of Q_{SC} at various temperatures, (c) the residual charges of the TENG at different temperatures (Inset is the diagram of the working model of the TENG) and (d) Q_{SC} evolution with time under high temperatures. Source: Reproduced with permission of [11], 2018 © John Wiley & Sons.

by polyurethane (PU) foam to introduce initial surface charges on SiO_2, as reflected by a Q_{SC} of around 45 nC. The cabinet was heated up to the desired temperature and then held for five minutes to measure the variation of Q_{SC}. Figure 1.1b shows the change of Q_{SC} at temperatures of 353, 533, and 583 K, demonstrating that the charge density decreased more rapidly at higher temperatures. When the temperature rose to 583 K, the charges quickly disappeared and the total charges Q_{SC} were less than 1 nC. Figure 1.1c shows the residual charges on the TENG after five minutes of measurement at different temperatures and the inset is the diagram of the working model of the CS Ti-SiO_2 TENG. The residual charges decreased more rapidly with the increase in the temperature, and it is interesting to note that they started to decrease more rapidly once the temperature was higher than 533 K.

It is worthwhile to note that the aforementioned experiments are the results after holding the TENG at different temperatures for five minutes. In order to further systematically explore the effect of high temperature on Q_{SC}, long-term measurements were conducted on the TENG under various temperatures. Figure 1.1d shows long-term charge decay under high temperatures, which indicates that the increase in temperature facilitates charge decay. It also reveals that the Q_{SC} response is analogous to exponential decay under high temperatures. In addition, taking the temperature of 503 K as an example, all of Q_{SC}, V_{OC}, and short-circuit current I_{SC} share the same decay characteristics, which feature a slow-fast-slow trend in decay speed.

Ruckdeschel and Hunter studied the CE of Al_2O_3-glass in the range of 293–393 K and observed the charge decay at high temperatures [28]. However, they only attributed this phenomenon to the thermal desorption of water on the Al_2O_3 surface. Here, since the Q_{SC} decay is analogous to exponential decay at higher temperatures, it is assumed that they may be consistent with the electron thermionic emission model, which is always applied for charge transport through Schottky diodes [29, 30]. Thus, the measured Q_{SC} values are fitted according to the thermionic emission equation [31, 32]:

$$J = \lambda A_0 T^2 e^{\frac{W}{kT}} \left[e^{\frac{\Delta W}{kT}} - 1 \right] \tag{1.1}$$

where J is current density, λ is the material-specific correction factor, A_0 is Richardson constant of a free electron, T is temperature, W is height of the potential barrier, k is Boltzmann constant, and ΔW is the potential barrier height variation due to the surface electric field E. Such potential barrier height W might be related to the work function of materials [12]. Since $E \propto \frac{\sigma}{\varepsilon_0} \propto Q_{SC}$, we may assume that $\Delta W = \lambda_1 Q_{SC}/\lambda$. ($\lambda_1$ is a constant). When $\Delta W \ll kT$, $e^{\frac{\Delta W}{kT}} - 1 \approx \frac{\Delta W}{kT}$, then

$$J = \frac{\lambda_1 A_0}{k} T e^{\frac{eV}{kT}} Q_{SC} \tag{1.2}$$

or equivalently:

$$\ln\left(\frac{J}{A_0 T}\right) = -\frac{eV}{kT} + \ln\left(\frac{\lambda_1}{k} Q_{SC}\right) \tag{1.3}$$

where e is the electronic charge. By assuming $J = \frac{1}{A}\frac{dQ_{SC}}{dt} = SQ_{SC}$, where A is the surface area:

$$Q_{SC} = e^{-SAt}Q_{SC0} \tag{1.4}$$

where Q_{SC0} is the initial value of Q_{SC}. According to Equation (1.4), Q_{SC} follows an exponential decay during thermionic emission. For tested metal/Kapton pairs, such exponential decay relations are shown in Figure 1.2a,d,g. Evolution of the measured Q_{SC} in the Al/Kapton pair at different temperatures was summarized in Figure 1.2b, similar to that of Ti/Kapton and SLS/Kapton pairs as shown in Figure 1.2e,h. Residual ratios of metal/Kapton pairs at different temperatures are summarized in Figure S2, where around 60% Q_{SC} remained on the surface with T of 420 K, but rapidly

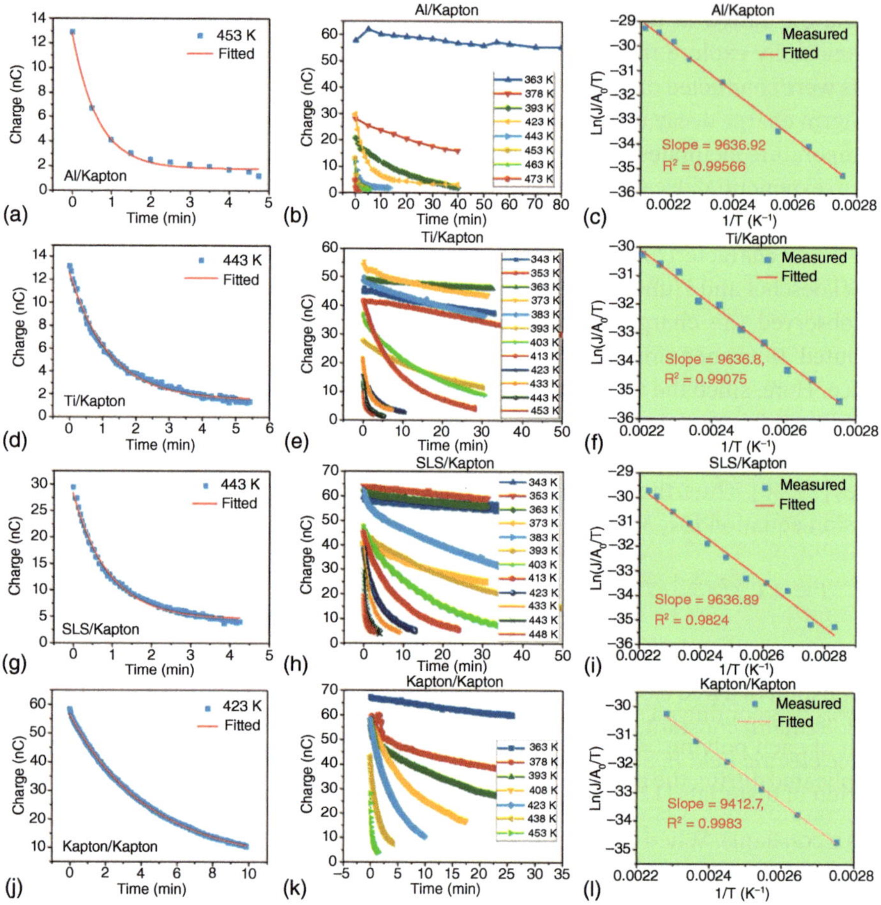

Figure 1.2 Results of metal/Kapton and Kapton/Kapton tribo-pairs. The measured (dots) and fitted (line) Q_{SC} as a function of the time at various temperatures, evolution of Q_{SC} with time at different temperatures, plots of $\ln(J/A_0/T)$ against $1/T$ with measured results in dots and linear fitted results in line, for (a–c) Al/Kapton; (d–f) Ti/Kapton; (g–i) SLS/Kapton; (j–l) Kapton/Kapton, respectively. Fitted slope of the plots of $\ln(J/A_0/T)$ against $1/T$ and the R^2 are marked next to the line. Source: Reproduced with permission of [33], 2020 © Elsevier.

decreased to 20% when T reached 450 K. Plots of $\ln(J/A_0/T)$ against $1/T$ with measured results in dots for Al/Kapton, Ti/Kapton, and SLS/Kapton pairs were shown in Figure 1.2c,f,i, respectively. Slopes of the fitted lines of $\ln(J/A_0/T)$ against $1/T$ and the coefficients of determination R^2 were marked next to the line. It should be noted that all slopes for metal/Kapton pairs with different metals are around 9600, with very little variation. This reflects that the material for thermionic emission is identical for these pairs, which should be Kapton. To confirm this conclusion, the tribo-pair of Kapton/Kapton was employed in experiments to rule out the potential influence of metals and served as a reference experiment. The exponential decay with fitted curves at 423 K, evolution of Q_{SC} with time at different temperatures, and plots of $\ln(J/A_0/T)$ against $1/T$ for the Kapton/Kapton pair were shown in Figure 1.2j–l, respectively. The slope in Figure 1.2l is similar to that of metal/Kapton pairs, verifying Kapton as the only possible thermionic emission material in metal/Kapton pairs. The extracted W of 0.82745 ± 0.0173 eV from these tribo-pairs should be that of the Kapton surface.

1.3 Material-Dependent Charge Transfer Mechanism and Model

According to the previous results of metal/Kapton in Figure 1.2, thermionic emission electrons carried by Kapton as material B were originally transferred from metals as material A, confirming the electron transfer dominant mechanism. To further study the dominant charge transfer mechanism during CE, experiments of nine other metal/polymer pairs were conducted, where Al, Ti, and SLS were applied as the positive surfaces to contact with three polymers of fluorinated ethylene propylene (FEP), polycarbonate (PC), and polyether ether ketone (PEEK). These polymers were employed due to their relatively high working temperature limits, which are all around or over 473 K. Residual ratio of metal/FEP pairs gradually decreased with temperature increase, similar to that of metal/Kapton pairs. Such trends were quite different in residual ratio of metal/PC and metal/PEEK pairs due to the effects of glass transition in PC and PEEK, indicating possibly different W at the temperature below (part I) and above (part II) the glass transition, as marked W_{I} and W_{II}. Thus, analysis was conducted separately for temperatures below and above the glass transition of each polymer. Considering the surface properties of polymers are usually complicated during the glass transition, the data collected during the glass transition temperature range was not analyzed. We may conclude from the results in Table 1.1 that W extracted from metal/polymer pairs is usually consistent with that of negative polymers, as measured from identical polymer/polymer pairs.

Besides metal/polymer pairs, studies about four polymer/polymer pairs were conducted, with results shown in Table 1.1. For polymer/polymer pairs, the extracted W is more complicated, which is all quite different from that of material B, and besides only a few of them are similar to that of material A. It should be noticed that W extracted from the Kapton/PC pair is between that of Kapton and PC for both temperature ranges (Parts I & II), indicating the co-existence of both electron transfer

Table 1.1 Summary and comparison of different tribo-pairs.

Tribo-pairs (Positive/Negative)		Part I ($T < T_g$)		Part II ($T > T_g$)	
		Slope	W_I (eV)	Slope	W_{II} (eV)
	Al/Kapton	9803.2	0.844778	N/A	N/A
Metal/Kapton	Ti/Kapton	9649.8	0.831559		
	SLS/Kapton	9542.9	0.822347		
	Al/PC	3763	0.324271	20 097	1.7318
Metal/PC	Ti/PC	3562.5	0.306994	19 498	1.6802
	SLS/PC	3210.3	0.276639	18 773.6	1.6178
	Al/PEEK	6563.2	0.565575	24 600.1	2.1199
Metal/PEEK	Ti/PEEK	5288.5	0.455729	26 695	2.3004
	SLS/PEEK	6819	0.587618	24 258	2.0904
	Al/FEP			6737.2	0.5806
Metal/FEP	Ti/FEP	N/A	N/A	6477.8	0.5582
	SLS/FEP			6368.2	0.5488
Polymer/polymer	Kapton/FEP	N/A	N/A	3772.3	0.3251
	Kapton/PC	6393.7	0.550969	15 642.3	1.3480
	Kapton/Kapton	9412.7	0.811127	N/A	N/A
	PEEK/PC	6886.65	0.593448	24 967.65	2.1516
	PC/FEP	3520	0.303331281	11519	0.9926
	FEP/FEP	N/A	N/A	6234.5	0.5372

and material/ion transfer mechanisms according to Figure 1.1d. Similar trends can be observed in PC/FEP pair, where W_{II} of PC/FEP pair is between PC and FEP and W_I extracted is similar to that of PC. Additionally, W_{II} extracted from Kapton/FEP was 0.3251 eV, which is lower than that of both Kapton and FEP, indicating possibly varied surface chemical status. W_I and W_{II} from PEEK/PC were always far away from that of PC as the negative material but similar to W_I and W_{II} of PEEK, respectively, indicating a possibly material transfer mechanism. Thus, it can be concluded that the W extracted from polymer/polymer pairs is usually different from that of negative polymers, indicating that material/ion transfer, electron transfer, and even chemical reactions may co-exist in CE of polymer/polymer pairs. And besides, the dominant charge transfer mechanism in CE is highly dependent on the type of material pairs.

The model to illustrate the material-dependent charge transfer mechanisms in CE was established in this work. As illustrated in Figure 1.3a, CE in metal/dielectric pairs usually follows electron transfer dominant mechanism, as demonstrated in this article as well as numerous previous studies. Figure 1.3b shows the potential CE mechanisms of dielectric/dielectric pairs. When the pairs are contacted together, electron transfer along with possibly ion/material transfer happens simultaneously,

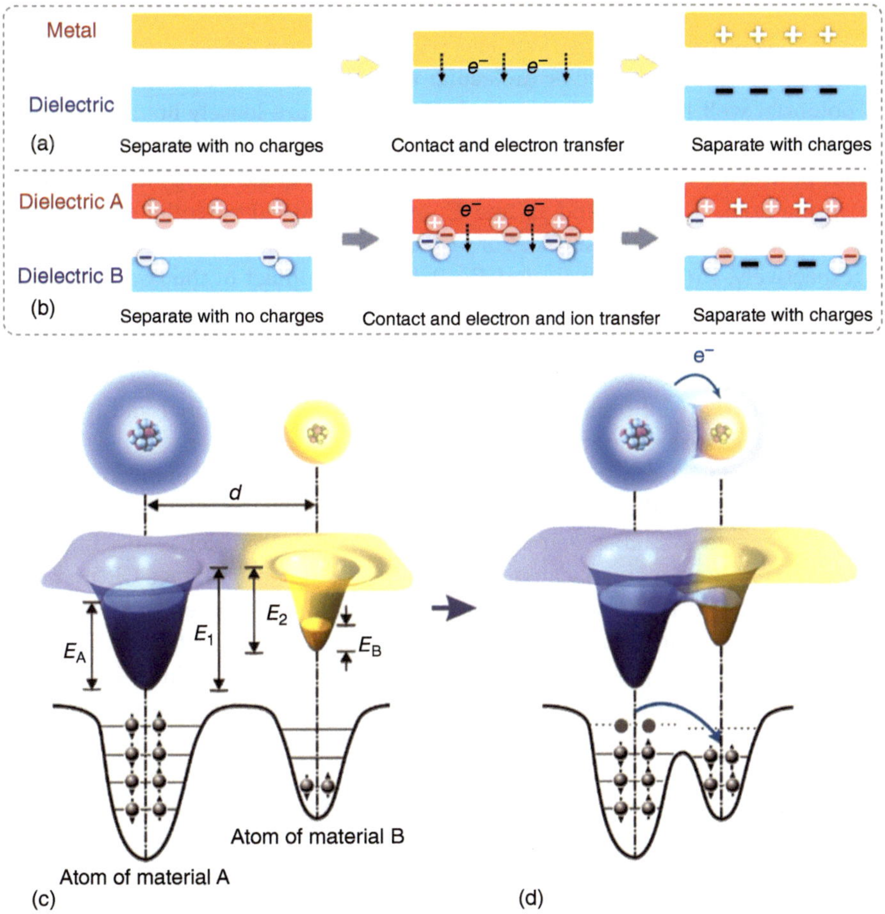

Figure 1.3 Mechanisms and model for contact electrification. (a) Illustration for electron transfer dominant mechanism in CE of metal/dielectric pairs. (b) Illustration for charge transfer mechanisms in CE of dielectric/dielectric pairs. Source: Reproduced with permission of [33], 2020 © Elsevier. Schematic of the electron cloud and potential energy profile (3D and 2D) of two atoms belonging to two materials A and B, respectively, when they are: (c) before contact and (d) in contact. Source: Reproduced with permission of [11], 2018 © John Wiley & Sons.

while the ratio of transferred charge contributed by each part varies a lot in different pairs, which is still waiting for further studies. After separation, both negative and positive charges may co-exist on the surface, consistent with the mosaic patterns observed before, while the net charge of a certain material may be either positive or negative. This model is also consistent with the covalent bond cleavage mechanism proposed previously [26, 34, 35], where the cleavage between interfacial bonds may induce electron transfer while that within a certain material may contribute to material/ion transfer.

To explain the mechanism of electron transfer, an electron cloud/potential model based on fundamental electron cloud interaction is proposed to explain all types of

CE phenomena for general materials (Figure 1.3c). Here, the electron clouds are formed by electrons that are spatially localized within specific atoms or molecules, and occupying specific atomic or molecular orbitals. An atom can be represented by a potential well in which the outer shell electrons are loosely bounded, forming an electron cloud of the atom or molecule. As shown in Figure 1.3c, d is the distance between electron clouds, E_A and E_B are the occupied energy levels of electrons in materials A and B, and E_1 and E_2 are the required potential energies for electrons to escape from the surfaces of materials A and B, respectively. E_A and E_B are, respectively, smaller than E_1 and E_2. Before the contact of the two materials, the electrons cannot transfer due to the local trapping effect of the potential wells. When material A contacts with material B, the electron clouds overlap due to the "screening" between the two materials introduced in physical contact, and the initial single potential well becomes asymmetric double-well potential and then the electron could hop from material A to material B (Figure 1.3d). Later on, Li et al. discovered the interface interatomic electron transition as induced by photon emission in CE, which confirms the existence of electron transfer in CE [36].

Although previous studies demonstrated that electron transfer is dominant in metal/polymer CE, the complicated mechanism of polymer/polymer CE remains unclear. To address this issue, a quantified model was proposed to investigate the CE, with the contribution ratios from different charge transfer mechanisms identified for different polymers. Since the potential barrier height Φ is affected by the electric field, different charge transfer mechanisms can be distinguished by the polarity of the charge transfer and the corresponding emission materials [37]. Considering the random breakage of chemical bonds of both surfaces, the transferred materials were found to usually include both positive and negative charges. Additionally, considering the different electron affinity of local atoms in molecular chain, the electron transfer was also bidirectional, and thus the polymer/polymer CE model with six potential mechanisms was established. A quantified model for calculating the contribution ratios of different charge transfer mechanisms was thus derived, and the contribution ratios were calculated. These contribution ratios were usually larger than 1, indicating that the total generated charge in CE is actually much larger than the net surface charge, which is also consistent with the previous studies. The relationship between the positive/negative charge generation ratio and the relative wear rate ratio, as well as the relationship between the relative potential barrier height and the relative wear rate ratio, were investigated, further confirming the material-dependent CE mechanism in polymer/polymer CE.

1.4 Liquid–Solid Contact Electrification Mechanism

CE may actually happen in all possible interfaces, including not only all types of solid–solid interfaces, but also liquid–solid [38–40], liquid–liquid [41–43], and even potentially solid–gas or liquid–gas interfaces [44]. Specifically, as compared to CE in solid–solid interfaces, CE in liquid–solid interfaces (LSI) show advantages of low friction (for liquid with low viscosity), minimum wearing, and high fluidity, with high potential applications in irregular energy harvesting [40]. Considering the potential of chemical reactions and physical processes, other potential applications of LSI include electrochemistry [45, 46], catalysis [47, 48], electrowetting [49, 50], etc., which attract broad attention from scientists and industry. However, the mechanism of CE in LSI still requires further systematic studies.

Usually, the charge transfer in LSI is believed to be related to the formation of the electric double-layer (EDL), which indicates that the ions from liquid attracted to solid surface dominate the charge distribution in LSI [51]. Such ions were distributed in a Stern layer in which ions were adsorbed next to the solid surface and a diffuse layer where the ion concentration decays with the distance to the solid surface [52]. However, the formation of EDL could be just a result instead of reason for the charge transfer. Even though the EDL was formed by ions, there is still not enough evidence to demonstrate that these ions originate from ion transfer. In fact, it was theoretically calculated that in deionized (DI) water, the ion density may not be able to provide enough ions to obtain observed charge density in LSI [53].

Lin et al. demonstrate a new method to reveal the origin of the charge in CE of LSI [54]. With a droplet of liquid drops on the surface of solid material, both ion and electron transfers may happen on the interface to transfer charge. If the high temperature was applied after the droplet was removed from the surface, thermionic emission of electrons will happen, which is similar to the process in solid–solid interface [11]. However, the thermionic emission of ions is usually quite difficult, considering the high potential barrier height for ions (8.5 eV for OH^- and 20 eV for H^+ on SiO_2 surface [55, 56]). As a comparison, the potential barrier height for electrons is usually <2.5 eV as shown in Table 1.1. Therefore, the electron will be emitted, leaving the residue charge due to ions, as shown in Figure 1.4a. Through Kelvin probe force microscopy (KPFM) equipped in atomic force microscope (AFM), the variation of the surface charge density can be monitored, as shown in Figure 1.4b. The droplet-charging and thermionic-emission processes are repeated many times as shown in Figure 1.4c, while the residue charge due to ions (red dots) increases gradually until saturation at around $-800\,\mu C/m^2$. It should be noted that most of such charge transfer was initiated by electrons, which can already achieve around $-800\,\mu C/m^2$ in the first cycle, and then more charge was transferred to more ions in following cycles. In other words, the charge transfer in CE of LSI was mostly initiated by electron transfer at the beginning, while ion accumulation on the surface becomes a stable status of charge in LSI. Similar results were also obtained in other experimental and theoretical studies [53, 57].

Therefore, as summarized by Prof. Zhong Lin Wang, the CE in LSI follows a "two-step" formation process with a hybrid EDL model [58]. As shown in Figure 1.5a, the first step of CE in LSI is similar to that in solid–solid interface, in which the electron transfer happens while the electron clouds of solid interface and liquid molecules overlap due to liquid pressures and thermal effects. The transferred electrons may remain in the charged solid surface due to the potential barrier height, as the very beginning electron transfer step to initiate the CE process. As shown in Figure 1.5b, in the second step, the ions in the liquid are attracted electrostatically to the LSI interface to form EDL. Such a migration of ions from

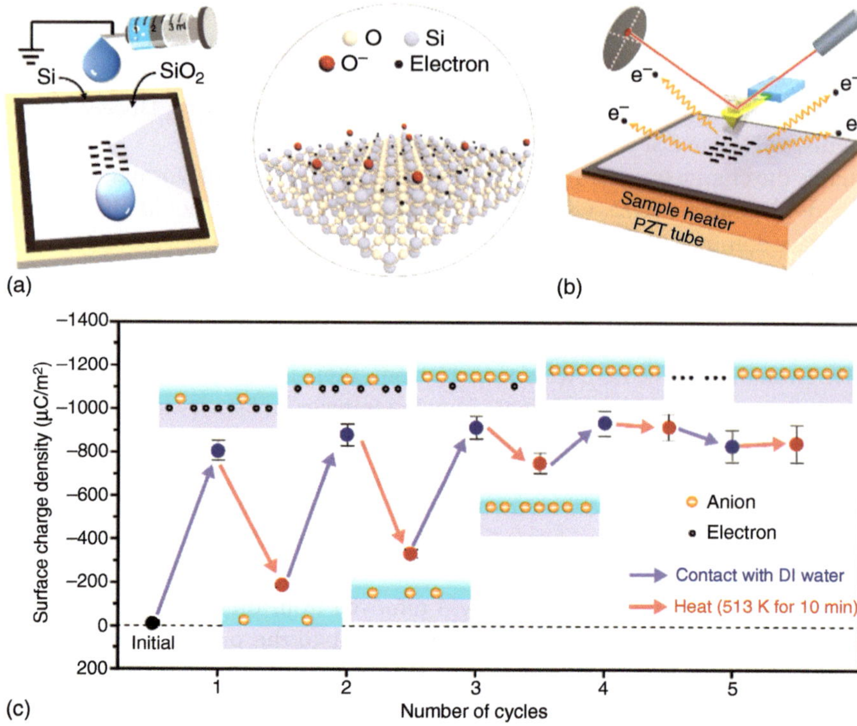

Figure 1.4 The experiment by Lin et al. to determine the charge transfer mechanism in LSI. (a) the experiment step and ion/electron formation in LSI. (b) the test method by KPFM after thermionic emission. (c) The tested charge density after repeated DI water contact and thermionic emission. Source: Reproduced with permission of [54], 2020 © Springer Nature.

liquid body to the interface may also facilitate the ionization reaction, which already exists in the liquid such as water, which further facilitates the formation of EDL. In this hybrid EDL model, the charge transfer process was initiated in the first step and dominated by electron transfer. The ion accumulation in EDL formation in the second step is just like a result of electron-dominated charge transfer. Different from solid–solid interface CE, we may assume the material transfer based on wearing is minimal in LSI interface.

1.5 Environmental and Material Effects on Charge Transfer

As a long-term mysterious phenomenon with debate over centuries, the CE mechanism still requires further explorations, which is due to the complexity brought by many impact factors. Typically, environmental factors such as temperature, humidity, stress/strain status, and atmosphere as well as material factors such as surface functional groups and strain statues, highly affect the CE process. Here, we summarized several recent studies in these aspects.

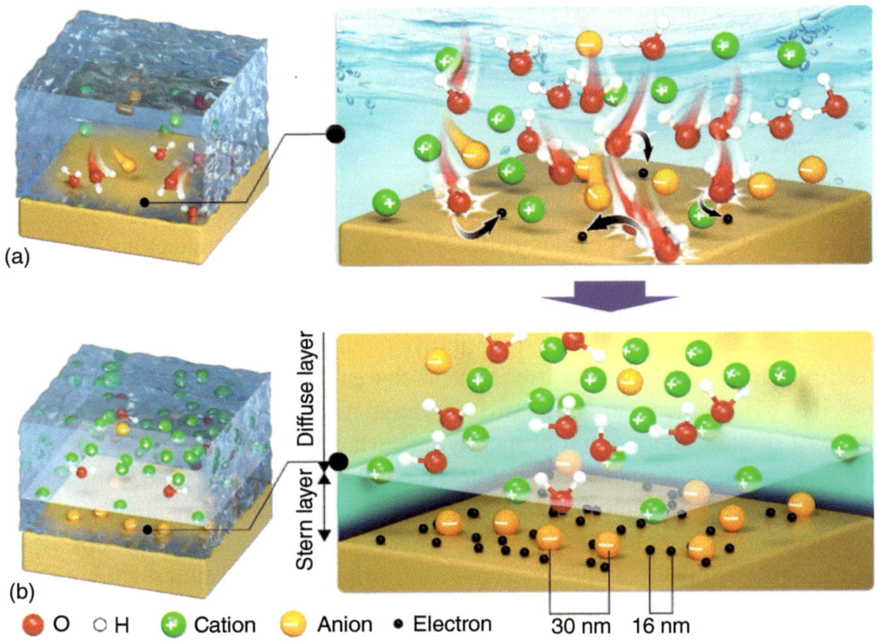

Figure 1.5 The hybrid EDL model with the "two-step" formation process for CE in LSI. (a) The electron transfer in the first step. (b) the EDL formation due to ion migration and redistribution. Source: Reproduced with permission of [44], 2022 © American Chemical Society.

1.5.1 Temperature Effect on the CE

As stated above, high temperature may facilitate the thermionic emission of electrons, as an experimental tool to reveal the CE mechanism, and in the meanwhile, it is also a pathway to release triboelectric charge, which may limit the application scenario of triboelectricity. The TENG used in previous studies is mainly based on CS motions, in which the two triboelectric surfaces are separated mostly. It has been discovered that the charge decay during contact status is much slower than that in separation status, which indicates that more charges can be kept at contact status [11]. Therefore, Wang et al. demonstrated sliding-based CE to generate more charge in high temperatures [59]. It has been discovered that sliding-based CE can facilitate CE charge transfer even in the highest temperature tested. Such a CE mechanism may comprehensively involve surface charge transfer, thermionic emission, and surface contact area changes. Based on sliding CE, Xu et al. developed the preannealing strategy to further raise the working temperature of CE, to be up to 673 K [60]. The mechanism of preannealing strategy to promote charge transfer is attributed to the atomic thermal vibration, which increases with the temperature significantly and partially balances the thermionic emission. Such annealing strategy may even change the polarity of the triboelectric charges in some materials, as carefully examined by KPFM [61]. The annealing strategy may also be unevenly applied on both sides or only on one side, which may facilitate the electron transfer from the heated side to the unheated side, due to thermionic emission of electrons [62]. Such an effect may be used to generate and modify triboelectric charge on purpose, even in identical material pairs, considering the unevenly heated areas in contact surfaces due to frictional heat.

1.5.2 Impact of Material Surface

As an interface effect, the characteristics of CE or triboelectric effect are strongly related to the material surface. The triboelectric series, as the empirical sequence of triboelectric polarity, was usually used to select and compare triboelectric materials qualitatively. Zou et al. developed a quantitative method to understand the triboelectric series by characterizing the triboelectric charge density (TECD) as a material "gene" [63, 64]. In the meanwhile, surface of materials also plays key roles in CE. In the past decade, the method of modifying surface to promote TECD has been intensively investigated, including chemical modification method [65–67] and introduction of surface micro-/nanostructures [68–70]. However, more systematic studies are still required to quantify the promotion effect of surface functional groups.

Li et al. demonstrated that the electron-withdrawing (EW) ability of the functional groups is the key to determining their polarity and resulting TECD [71]. In various functional groups, the −F group shows the best EW ability, and increase of the −F groups can greatly enhance the negative TECD, which explains why polymers that are fully fluorinated (such as FEP and polytetrafluoroethylene, PTFE) demonstrate the best TECD compared to other common polymers. The EW ability for other common functional groups is ranked as $-CH_3 < -H < -OH < -Cl < -F$. Besides, it was discovered that unsaturated groups can enhance the EW of fluorinated polymers, such as $-CF_2{=}CF_2-$ and $-CF{\equiv}CF-$, which may be introduced to the original fluorinated polymers through thin firm sputtering, UV-ozone treatment, etc. In the meanwhile, the density of surface states (DOSS) may also contribute to the resulting TECD [72]. A model based on DOSS was proposed to explain the abnormal TECD generation between ternary materials, with an experimental method established to characterize DOSS of materials. Through this DOSS model, it can be predicted that TECD of FEP may achieve $\sim 560\,\mu C/m^2$, which is close to the reported maximum values. These studies may inspire a more complete mechanism model for CE and TECD in the near future.

1.5.3 Stress/Strain States and Others

Besides the chemical composition of the material surface, the surface stress or strain states may also impact the TECD. However, even for identical material pairs in contact, CE may also happen. Xu et al. demonstrated that the charge transfer is also related to the curvature of surfaces, while positive curvature tends to take negative charge and negative curvature tends to take positive charge [73]. The mechanism of CE between the curved surfaces may be attributed to the shift of occupied level of surface states. In light of that, mutual CE may even happen between any identical material pairs in contact considering the surface roughness, which may result in uneven stress/strain distribution on the surface. This discovery may explain the mosaic distribution of triboelectric charge [22]. Furthermore, the applied stress/strain can also vary the electron transfer amount during CE, which may be contributed by the induced flexoelectricity [74].

For some stretchable polymer materials, the crystallization level may be induced by strain, which can greatly impact the band diagram. Liu et al. demonstrated that strain-induced crystallization can shift the position of polymers in triboelectric series, and even induce polarity reversal of TECD [75]. With the crystallization process, the electron cloud on the surface can be totally redistributed, which can greatly impact the CE process by assuming the electron-dominated electron-cloud-potential-well model. CE can also be promoted by UV light, as attributed to photoexcitation effect on electrons [15, 76], which demonstrated the existence of threshold photon energy to assist electrons to overcome the potential barrier heights. The TECD can even be modified by atmosphere, especially the oxygen atmosphere, which may both impact the DOSS and occupied energy levels [77].

1.6 Potential Applications

When two different materials or surfaces are in physical contact or sliding against one another, charges will be transferred from one surface to the other, making them positively or negatively charged, respectively. This is the triboelectric effect (also called triboelectrification or CE) well-known for thousands of years, which was first recorded by Thales of Miletus (c. 624–546 BC), an ancient Greek philosopher. Understanding and discovering methods to tailor the triboelectric effect by controlling the surface charge density will benefit a number of applications. As examples:

- *Prevention of electrostatic damage (ESD)*: The electrostatic charge produced by the triboelectric effect may cause sparks which can ignite flammable materials or vapors, interfering in the operation of electronics in the household, workplace, and societal infrastructures including public transportation, aircraft and spacecraft. It has been reported that related losses in the electronics industry are up to **US$5 billion**. (https://www.durablecorp.com/the-costs-of-esd-damage) Therefore, these industries would benefit from the ability to downregulate or decrease the triboelectric effect.
- *Power generation*: TENG is an emerging technology for low-frequency energy harvesting with demonstrated advantages of being lightweight, facile fabrication, high performance, and low cost [78]. It has been predicted that TENG and the resulting applications will be a **US$480 million market** by 2028 (https://www.idtechex.com/en/research-report/triboelectric-energy-harvesting-teng-2018-2028/577). As indicated by the TENG figure-of-merit (FOM) developed by the PI, TECD is the key parameter to maximize output performance [38]. Therefore, these applications would require that the triboelectric effect could be upregulated, maximized, or increased.

- *High-voltage applications*: For applications which require high-voltage input, such as the TENG-triggered electrospray for mass spectrometry (MS) demonstrated by the PI, quantitatively controlled charge generation is critical to achieving nanocoulomb level and ultrasensitive chemical analysis [79]. Similar research conducted by the PI includes field emission of electrons [78], and microplasma generation [80]. Here, the triboelectric effect should be tailored precisely according to the applications.
- *Self-powered wireless solution through tribophotonics*: Tribophotonics is coupling of triboelectricity as a self-powered solution and photonics as a wireless transmission solution, which was proposed in recent years [81]. The developed technologies under tribophotonics include tribo-induced EM-wave generation (TIEG), tribo-induced light propagation tuning (TILPT), tribo-induced electroluminescence (TIEL), and tribo-assisted spectrometry (TAS), which was detailed in Chapter 9. As induced by photon emission from CE, Li et al. also proposed CE-induced interface photon emission spectroscopy (CEIIPES), which can be considered a new type of TAS, and may be expanded as a new self-powered spectroscopy tool [36].

1.7 Summary

This chapter summarized the studies on CE mechanisms in recent years, as facilitated by the emerging field of TENG. Through experiments based on thermionic emission, intensive studies have been conducted on charge transfer mechanisms, with universal existence of electron transfer confirmed in CE through various studies. An electron-cloud-potential-well model was established, as key model for CE studies. In the meanwhile, the contributions from material transfer and ion transfer were studied in solid–solid interface and LSI CEs, respectively. The impacts of materials, environment, stress/strain, etc., were investigated, which greatly improved our understanding and provided charge modification tools for CE. It is expected that a more detailed CE mechanism will be revealed in the near future considering the intensive studies in recent years.

References

1. Wu, C., Wang, A.C., Ding, W. et al. (2019). Triboelectric nanogenerator: a foundation of the energy for the new era. *Adv. Energy Mater.* 9 (1): 1802906.
2. Hinchet, R., Yoon, H.-J., Ryu, H. et al. (2019). Transcutaneous ultrasound energy harvesting using capacitive triboelectric technology. *Science* 365 (6452): 491–494.
3. Zhu, G., Pan, C., Guo, W. et al. (2012). Triboelectric-generator-driven pulse electrodeposition for micropatterning. *Nano Lett.* 12 (9): 4960–4965.
4. Lebedev, M. and Akedo, J. (2002). What thickness of the piezoelectric layer with high breakdown voltage is required for the microactuator? *Jpn. J. Appl. Phys.* 41 (5B): 3344–3347.

5 Wang, J., Wu, C., Dai, Y. et al. (2017). Achieving ultrahigh triboelectric charge density for efficient energy harvesting. *Nat. Commun.* 8 (1): 88.
6 Lowell, J. and Rose-Innes, A.C. (2006). Contact electrification. *Adv. Phys.* 29 (6): 947–1023.
7 Liu, C. and Bard, A.J. (2008). Electrostatic electrochemistry at insulators. *Nat. Mater.* 7 (6): 505–509.
8 Liu, C.Y. and Bard, A.J. (2009). Chemical redox reactions induced by cryptoelectrons on a PMMA surface. *J. Am. Chem. Soc.* 131 (18): 6397–6401.
9 Liu, C.Y. and Bard, A.J. (2009). Electrons on dielectrics and contact electrification. *Chem. Phys. Lett.* 480 (4–6): 145–156.
10 Liu, C.-Y. and Bard, A.J. (2010). Electrostatic electrochemistry: nylon and polyethylene systems. *Chem. Phys. Lett.* 485 (1–3): 231–234.
11 Xu, C., Zi, Y., Wang, A.C. et al. (2018). On the electron-transfer mechanism in the contact-electrification effect. *Adv. Mater.* 30 (15): e1706790.
12 Xu, C., Zhang, B., Wang, A.C. et al. (2019). Effects of metal work function and contact potential difference on electron thermionic emission in contact electrification. *Adv. Funct. Mater.* 29 (29): 1903142.
13 Lin, S., Xu, L., Xu, C. et al. (2019). Electron transfer in nanoscale contact electrification: effect of temperature in the metal-dielectric case. *Adv. Mater.* 31 (17): e1808197.
14 Zhang, S.L., Xu, M., Zhang, C. et al. (2018). Rationally designed sea snake structure based triboelectric nanogenerators for effectively and efficiently harvesting ocean wave energy with minimized water screening effect. *Nano Energy* 48: 421–429.
15 Lin, S., Xu, L., Zhu, L. et al. (2019). Electron transfer in nanoscale contact electrification: photon excitation effect. *Adv. Mater.* 31 (27): e1901418.
16 Zhou, Y.S., Wang, S., Yang, Y. et al. (2014). Manipulating nanoscale contact electrification by an applied electric field. *Nano Lett.* 14 (3): 1567–1572.
17 Li, S., Zhou, Y., Zi, Y. et al. (2016). Excluding contact electrification in surface potential measurement using kelvin probe force microscopy. *ACS Nano* 10 (2): 2528–2535.
18 Wiles, J.A., Grzybowski, B.A., Winkleman, A., and Whitesides, G.M. (2003). A tool for studying contact electrification in systems comprising metals and insulating polymers. *Anal. Chem.* 75 (18): 4859–4867.
19 Thomas, S.W. 3rd, Vella, S.J., Kaufman, G.K., and Whitesides, G.M. (2008). Patterns of electrostatic charge and discharge in contact electrification. *Angew. Chem. Int. Ed.* 47 (35): 6654–6656.
20 McCarty, L.S. and Whitesides, G.M. (2008). Electrostatic charging due to separation of ions at interfaces: contact electrification of ionic electrets. *Angew. Chem. Int. Ed.* 47 (12): 2188–2207.
21 Thomas, S.W. 3rd, Vella, S.J., Dickey, M.D. et al. (2009). Controlling the kinetics of contact electrification with patterned surfaces. *J. Am. Chem. Soc.* 131 (25): 8746–8747.
22 Baytekin, H.T., Patashinski, A.Z., Branicki, M. et al. (2011). The mosaic of surface charge in contact electrification. *Science* 333 (6040): 308–312.

23 Baytekin, B., Baytekin, H.T., and Grzybowski, B.A. (2012). What really drives chemical reactions on contact charged surfaces? *J. Am. Chem. Soc.* 134 (17): 7223–7226.

24 Baytekin, H.T., Baytekin, B., Incorvati, J.T., and Grzybowski, B.A. (2012). Material transfer and polarity reversal in contact charging. *Angew. Chem. Int. Ed.* 51 (20): 4843–4847.

25 Siek, M., Adamkiewicz, W., Sobolev, Y.I., and Grzybowski, B.A. (2018). The influence of distant substrates on the outcome of contact electrification. *Angew. Chem. Int. Ed.* 57 (47): 15379–15383.

26 Pezzotti, G., Camara, C., Marin, E. et al. (2019). Physical chemistry insights into surface charge phenomena during frictional coupling in triboelectric X-ray sources. *J. Mater. Chem. C* 7 (25): 7708–7724.

27 Šutka, A., Mālnieks, K., Lapčinskis, L. et al. (2019). The role of intermolecular forces in contact electrification on polymer surfaces and triboelectric nanogenerators. *Energy Environ. Sci.* 12 (8): 2417–2421.

28 Ruckdeschel, F.R. and Hunter, L.P. (2008). Contact electrification between insulators: phenomenological aspects. *J. Appl. Phys.* 46 (10): 4416–4430.

29 Crowell, C.R. and Sze, S.M. (1966). Current transport in metal-semiconductor barriers. *Solid State Electron.* 9 (11): 1035–1048.

30 Ravinandan, M., Rao, P.K., and Rajagopal Reddy, V. (2009). Analysis of the current–voltage characteristics of the Pd/Au Schottky structure on n-type GaN in a wide temperature range. *Semicond. Sci. Technol.* 24 (3): 035004.

31 Crowell, C.R. (1965). The Richardson constant for thermionic emission in Schottky barrier diodes. *Solid State Electron.* 8 (4): 395–399.

32 Racko, J., Grmanová, A., and Breza, J. (1996). Extended thermionic emission-diffusion theory of charge transport through a Schottky diode. *Solid State Electron.* 39 (3): 391–397.

33 Xia, X., Wang, H., Guo, H. et al. (2020). On the material-dependent charge transfer mechanism of the contact electrification. *Nano Energy* 78: 105343.

34 Li, A., Liu, Y., and Szlufarska, I. (2014). Effects of interfacial bonding on friction and Wear at silica/silica interfaces. *Tribol. Lett.* 56 (3): 481–490.

35 Zhang, J., Rogers, F.J.M., Darwish, N. et al. (2019). Electrochemistry on tribocharged polymers is governed by the stability of surface charges rather than charging magnitude. *J. Am. Chem. Soc.* 141 (14): 5863–5870.

36 Li, D., Xu, C., Liao, Y. et al. (2021). Interface inter-atomic electron-transition induced photon emission in contact-electrification. *Sci. Adv.* 7 (39): eabj0349.

37 Xia, X., Wang, H., and Zi, Y. (2022). Field-assisted thermionic emission toward quantitative modeling of charge-transfer mechanisms in contact electrification. *SmartMat* 3 (4): 619–631.

38 Zi, Y.L., Niu, S.M., Wang, J. et al. (2015). Standards and figure-of-merits for quantifying the performance of triboelectric nanogenerators. *Nat. Commun.* 6: 8376.

39 Xu, W., Zheng, H., Liu, Y. et al. (2020). A droplet-based electricity generator with high instantaneous power density. *Nature* 578 (7795): 392–396.

40 Wu, H., Wang, Z., and Zi, Y. (2021). Multi-mode water-tube-based triboelectric nanogenerator designed for low-frequency energy harvesting with ultrahigh volumetric charge density. *Adv. Energy. Mater.* 11 (16): 2100038.

41 Jiang, P., Zhang, L., Guo, H. et al. (2019). Signal output of triboelectric nanogenerator at oil–water–solid multiphase interfaces and its application for dual-signal chemical sensing. *Adv. Mater.* 31 (39): 1902793.

42 Wang, P., Zhang, S., Zhang, L. et al. (2020). Non-contact and liquid–liquid interfacing triboelectric nanogenerator for self-powered water/liquid level sensing. *Nano Energy* 72: 104703.

43 Nie, J., Wang, Z., Ren, Z. et al. (2019). Power generation from the interaction of a liquid droplet and a liquid membrane. *Nat. Commun.* 10 (1): 2264.

44 Lin, S., Chen, X., and Wang, Z.L. (2022). Contact electrification at the liquid–solid interface. *Chem. Rev.* 122 (5): 5209–5232.

45 Magnussen, O.M. and Groß, A. (2019). Toward an atomic-scale understanding of electrochemical Interface structure and dynamics. *J. Am. Chem. Soc.* 141 (12): 4777–4790.

46 Zaera, F. (2012). Probing liquid/solid interfaces at the molecular level. *Chem. Rev.* 112 (5): 2920–2986.

47 Lee, S.H., Nedrygailov, I.I., Oh, S., and Park, J.Y. (2018). Hot electron flux at solid–liquid interfaces probed with Pt/Si catalytic nanodiodes: effects of pH during decomposition of hydrogen peroxide. *Catal. Today* 303: 282–288.

48 Tao, F. and Crozier, P.A. (2016). Atomic-scale observations of catalyst structures under reaction conditions and during catalysis. *Chem. Rev.* 116 (6): 3487–3539.

49 Wang, J., Wang, H., Li, X., and Zi, Y. (2019). Self-powered electrowetting optical switch driven by a triboelectric nanogenerator for wireless sensing. *Nano Energy* 66: 104140.

50 Li, X., Tian, H., Shao, J. et al. (2016). Decreasing the saturated contact angle in electrowetting-on-dielectrics by controlling the charge trapping at liquid–solid interfaces. *Adv. Funct. Mater.* 26 (18): 2994–3002.

51 Helmholtz, H. (1853). Ueber einige Gesetze der Vertheilung elektrischer Ströme in körperlichen Leitern mit Anwendung auf die thierisch-elektrischen Versuche. *Ann. Phys.* 165 (6): 211–233.

52 Stern, O. (1924). The theory of the electric double layer. *Electrochemistry* 30: 508.

53 Nie, J., Ren, Z., Xu, L. et al. (2020). Probing contact-electrification-induced electron and ion transfers at a liquid–solid interface. *Adv. Mater.* 32 (2): 1905696.

54 Lin, S., Xu, L., Chi Wang, A., and Wang, Z.L. (2020). Quantifying electron-transfer in liquid-solid contact electrification and the formation of electric double-layer. *Nat. Commun.* 11 (1): 399.

55 von Burg, K. and Delahay, P. (1981). Photoelectron emission spectroscopy of inorganic anions in aqueous solution. *Chem. Phys. Lett.* 78 (2): 287–290.

56 Takakuwa, Y., Niwano, M., Nogawa, M. et al. (1989). Photon-stimulated desorption of H+ ions from oxidized Si(111) surfaces. *Jpn. J. Appl. Phys.* 28 (12R): 2581.

57 Zhan, F., Wang, A.C., Xu, L. et al. (2020). Electron transfer as a liquid droplet contacting a polymer surface. *ACS Nano* 14 (12): 17565–17573.

58 Wang, Z.L. and Wang, A.C. (2019). On the origin of contact-electrification. *Mater. Today* 30: 34–51.

59 Wang, A.C., Zhang, B., Xu, C. et al. (2020). Unraveling temperature-dependent contact electrification between sliding-mode triboelectric pairs. *Adv. Funct. Mater.* 30 (12): 1909384.

60 Xu, C., Wang, A.C., Zou, H. et al. (2018). Raising the working temperature of a triboelectric nanogenerator by quenching down electron thermionic emission in contact-electrification. *Adv. Mater.* 30 (38): 1803968.

61 Zhu, Y., Lin, S., Gao, W. et al. (2021). Reversal of triboelectric charges on sol–gel oxide films annealed at different temperatures. *Appl. Phys. Lett.* 118 (24): 243902.

62 Lin, S., Xu, L., Xu, C. et al. (2019). Electron transfer in nanoscale contact electrification: effect of temperature in the metal–dielectric case. *Adv. Mater.* 31 (17): 1808197.

63 Zou, H., Guo, L., Xue, H. et al. (2020). Quantifying and understanding the triboelectric series of inorganic non-metallic materials. *Nat. Commun.* 11 (1): 2093.

64 Zou, H., Zhang, Y., Guo, L. et al. (2019). Quantifying the triboelectric series. *Nat. Commun.* 10 (1): 1427.

65 Wang, S., Zi, Y., Zhou, Y.S. et al. (2016). Molecular surface functionalization to enhance the power output of triboelectric nanogenerators. *J. Mater. Chem. A* 4 (10): 3728–3734.

66 Zhang, X.-S., Han, M.-D., Wang, R.-X. et al. (2014). High-performance triboelectric nanogenerator with enhanced energy density based on single-step fluorocarbon plasma treatment. *Nano Energy* 4: 123–131.

67 Wang, J., Li, S., Yi, F. et al. (2016). Sustainably powering wearable electronics solely by biomechanical energy. *Nat. Commun.* 7 (1): 12744.

68 Zhong, J., Zhong, Q., Fan, F. et al. (2013). Finger typing driven triboelectric nanogenerator and its use for instantaneously lighting up LEDs. *Nano Energy* 2 (4): 491–497.

69 Wang, J., Wen, Z., Zi, Y. et al. (2016). All-plastic-materials based self-charging power system composed of triboelectric nanogenerators and supercapacitors. *Adv. Funct. Mater.* 26 (7): 1070–1076.

70 Cui, N., Gu, L., Lei, Y. et al. (2016). Dynamic behavior of the triboelectric charges and structural optimization of the friction layer for a triboelectric nanogenerator. *ACS Nano* 10 (6): 6131–6138.

71 Li, S., Nie, J., Shi, Y. et al. (2020). Contributions of different functional groups to contact electrification of polymers. *Adv. Mater.* 32 (25): e2001307.

72 Xu, G., Guan, D., Fu, J. et al. (2022). Density of surface states: another key contributing factor in triboelectric charge generation. *ACS Appl. Mater. Interfaces* 14 (4): 5355–5362.

73 Xu, C., Zhang, B., Wang, A.C. et al. (2019). Contact-electrification between two identical materials: curvature effect. *ACS Nano* 13 (2): 2034–2041.

74 Lin, S., Zheng, M., Xu, L. et al. (2022). Electron transfer driven by tip-induced flexoelectricity in contact electrification. *J. Phys. D. Appl. Phys.* 55 (31): 315502.

75 Liu, Z., Li, S., Lin, S. et al. (2022). Crystallization-induced shift in a triboelectric series and even polarity reversal for elastic triboelectric materials. *Nano Lett.* 22 (10): 4074–4082.

76 Tao, X., Nie, J., Li, S. et al. (2021). Effect of photo-excitation on contact electrification at liquid–solid interface. *ACS Nano* 15 (6): 10609–10617.

77 Lin, S., Xu, L., Tang, W. et al. (2019). Electron transfer in nano-scale contact electrification: atmosphere effect on the surface states of dielectrics. *Nano Energy* 65: 103956.

78 Wang, Z.L., Lin, L., Chen, J. et al. (2016). *Triboelectric Nanogenerators*. Springer.

79 Li, A., Zi, Y., Guo, H. et al. (2017). Triboelectric nanogenerators for sensitive nano-coulomb molecular mass spectrometry. *Nat. Nanotechnol.* 12 (5): 481.

80 Cheng, J., Ding, W., Zi, Y. et al. (2018). Triboelectric microplasma powered by mechanical stimuli. *Nat. Commun.* 9 (1): 3733.

81 Wang, H., Xia, X., Fu, J. et al. (2023). A general self-powered wireless sensing solution based on triboelectric-discharge effect. *Nano Energy* 105: 107982.

2

Discharge Effects in TENG

2.1 Introduction

2.1.1 Fundamental Knowledge About the Discharge

The discharge effect is generally about the release of the electric charge through a certain medium. From the definition, it may include multiple forms of charge release such as breakdown discharge, field emission, and charge diffusion/drift. Here, due to the high-voltage output from triboelectric nanogenerator (TENG), the most frequent discharge in TENG is the breakdown discharge. In the meanwhile, compared with other effects which all take a relatively long time to decrease the charge density, the breakdown-discharge effect usually results in immediate discharging of the surface charge, and hence we consider it the key limitation of the output performance of TENG. So, here we will mainly talk about the breakdown-discharge effects or the discharge effects that were generated due to the large potential difference between surfaces.

In TENGs, the discharge usually occurs in the medium of air gaps due to the lower threshold breakdown electric field than that of solids/liquids. Therefore, we primarily focus on the discharge behaviors in the air, which follows Townsend's electron avalanche model [1–3] as shown in Figure 2.1a. When the voltage surpasses the threshold breakdown voltage V_b, or the electric field exceeds the threshold breakdown electric field, electrons emitted from the cathode will be accelerated to high kinetic energy, which will constantly collide with the gas molecules and cause collision ionization. And then the secondary electrons will be formed during the collision to initiate the chain reaction called electron avalanche.

The voltage-current (V-I) curve of the electric discharge is usually highly nonlinear, with multiple distinguished regimes. A typical V-I curve of the electrical discharge in Neon at 1 torr is shown in Figure 2.1b [7] as an example. These regimes are stated as follows:

(1) Regime below onset (A–C): At the voltage below point A, since the voltage is weak, only very little current was generated by the background radiation from cosmic rays or other radioactive sources. While the voltage is beyond point A, the available electrons and ions are swept away by the electric field, and the current

Triboelectric Nanogenerators: Technology, Applications, and Commercialization, First Edition.
Yunlong Zi, Hengyu Guo, Jie Wang, Chi Zhang, Xiangyu Chen, and Qing Zhao.

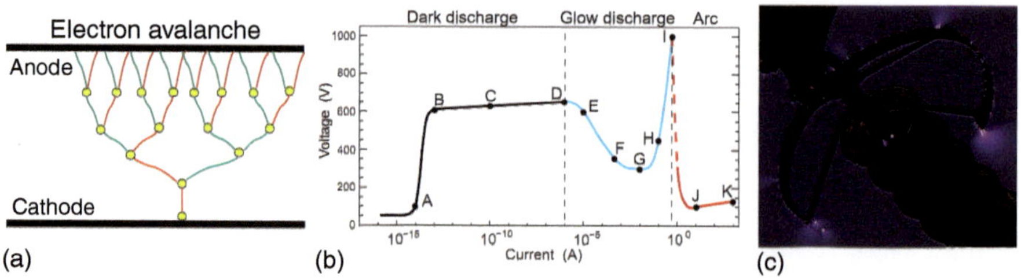

Figure 2.1 Fundamental mechanisms, behaviors, and studies about discharge effects. (a) The electron avalanche model for the discharge effect. Source: Reproduced with permission of [4], 2020 © John Wiley & Sons. (b) The regions show the different types of discharge according to I-V relationship. Source: [5]/Unknown Author/wikimedia commons /CC BY-SA 3.0. (c). The corona discharge captured on an insulator string of a 500 kV overhead power line. Nitromethane / Wkimedia Commans / CC BY SA 3.0.

increases until point B as the saturation point. After that, the current remains almost constant.

(2) Dark discharge regime above onset (C,D): At point C, when the electric field and the mean free path of electrons allow free electrons to acquire enough energy to induce impact ionization of molecules, the Townsend discharge/avalanche will occur. However, at this stage, the discharge still cannot be sustained until point D. Corona (Figure 2.1c) and/or streamer (i.e. filamentary sparks) discharges may be observed, with an onset voltage (V_{on}) around point C. At this stage, electromagnetic (EM) waves are emitted, with the spectrum determined by the gas type [5].

(3) Glow and arc/spark discharge regime (D–K): At point D, since ionization of molecules can generate enough electrons to induce discharge of new molecules, the discharge becomes self-sustained. The voltage at point D is usually called the breakdown voltage (V_{br}). Usually, in a low-pressure atmosphere, the electron energy will be high enough to emit visible light, generating luminous plasma, and thus the glow discharge is achieved. With an even higher current, an ionized, conductive channel can be generated in the gas medium, forming an arc/spark, which usually occurs at atmosphere or higher pressures.

In these key parameters, the breakdown voltage V_{br} is a key one which indicates the point to start the rapid discharge when the voltage surpasses V_{br}. Usually, this V_{br} highly relies on the atmosphere (gas composition, pressure, humidity, temperature, etc.), the shape of the surfaces which can impact the local electric field a lot, and even the materials. However, for the case of parallel plates, the V_{br} can be determined by the equation of Paschen's law [8]:

$$V_b = \frac{Bpd}{\ln(Apd) - \ln\left[\ln\left(1 + \frac{1}{\gamma_{se}}\right)\right]} \tag{2.1}$$

Here V_b is the threshold breakdown voltage, p is the pressure, d is the gap distance between the boards, constant γ_{se} is the secondary electron emission coefficient, constant A is the saturation ionization in the gas, and constant B is related to the

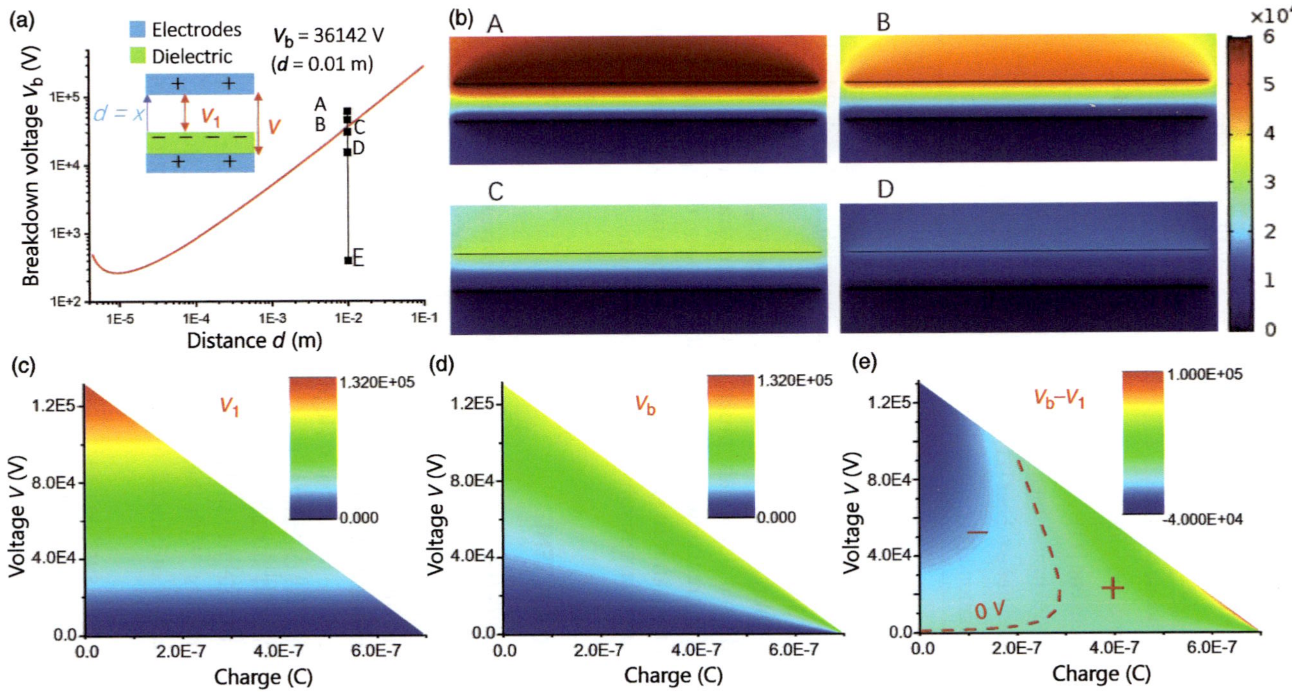

Figure 2.2 Simulations and calculations of the breakdown area. (a) Paschen's law to determine the breakdown discharge voltage. Source: Reproduced with permission of [9], 2017 © John Wiley & Sons. (b) The simulated potential distribution in TENG with different charge transfer. (c) The voltage V_1 between two triboelectric surfaces under different V-Q conditions. (d) The corresponding breakdown voltage V_b under different V-Q conditions. (e) The subtraction $V_b - V_1$ under different V-Q conditions, plotting the breakdown (–) and nonbreakdown (+) areas. Source: Reproduced with permission of [9], 2017 © John Wiley & Sons.

Table 2.1 The dielectric strength for breakdown in selected substances.

Substance type	Name of substance	Dielectric strength (10^6 V/m)
Gas	Air [10]	3
Gas	Helium [11]	0.15
Gas	SF_6 (sulfur hexafluoride) [11]	8.5–9.8
Liquid	Silicone oil [11]	10–15
Liquid	Distilled water [11]	65–70
Liquid	Benzene [11]	163
Solid	PTFE insulating film [11]	60–173
Solid	Neoprene rubber [11]	15.7–26.7
Solid	Diamond [11]	2000

excitation and ionization energies. The threshold breakdown voltage V_b calculated by Paschen's law for 1 atm air is plotted in Figure 2.2a, in which the threshold voltage for distance d of 0.01 m is 36 142 V.

In solids and liquids, there is also a breakdown voltage V_{br} existing. However, usually, the V_{br} in solids and liquids are much higher than that in gases since the threshold electric field for breakdown (also called dielectric strength in dielectrics) is usually much higher than that in gases. Table 2.1 lists the threshold electric field (or dielectric strength) in different substances for breakdown.

2.1.2 Discharge Effects in TENG

The discharge inside TENG (i.e. in-situ discharge) exists universally in all types of TENGs due to the high voltage (HV) generation [9]. The impacts of the discharge effects on the output performance of TENGs were first reported by Wang et al. in 2014 [12], right after the invention of the TENG technology. In that, the authors utilized the ion injection method to introduce the additional surface charge, until the maximized surface charge density of ~240 μC/m² was achieved for CS-TENG, in the short-circuit condition. As calculated from Paschen's law, the theoretical studies revealed that this maximized surface charge density is limited by the breakdown discharge between the triboelectric surfaces, while the threshold value is consistent with the experimental results. With the decrease of the thickness, the maximized surface charge density can be enhanced, further improving the output performance of TENG.

However, this maximized surface charge density cannot reflect the real limitation of the output energy because there is no energy output in the short-circuit condition. During practical operations, the air breakdown may happen with charge density below 240 μC/m² since, under the external load with a higher impedance, it is possible to produce a much higher potential difference between triboelectric surfaces than that in short-circuit conditions [13]. The maximized effective energy output as limited by air breakdown, which is the most crucial limitation for output performance of TENGs, is still required to be revealed by studies of air breakdown in TENGs under various circuit conditions through output voltage–charge transfer (V-Q) plots [14]. To illustrate characteristics of other TENGs, such as contact TENGs including CFT and SEC modes, and sliding TENGs including LS and SFT modes, the breakdown discharge effects in them are also to be studied.

2.2 Theoretical Studies of Breakdown Discharge in Contact-Separation-Based TENGs

Systematic theoretical studies of the breakdown discharge effect in contact-separation-based TENGs were conducted by Zi et al. in 2017 [9]. To prove the existence of air breakdown in CS mode TENG with low surface charge density, we consider one typical device as shown in inset of Figure 2.2a, with both length l and width w of 0.1 m, displacement x (equivalent to d) of 0.01 m, dielectric thickness d_r of 0.1 mm, relative dielectric constant ε_r of 0.5, and charge density σ of 70 μC/m^2. The voltage V_1 between the triboelectric surfaces at $d = 0.01$ m in short-circuit condition is calculated to be only about ~390 V, as reported previously [15] (all the calculations are in 1 atm air atmosphere unless otherwise specified). This voltage is far lower than V_b, as denoted by dot E in Figure 2.2a. However, if TENG is operated under high-impedance-load condition, i.e. there is a nonzero voltage V between two electrodes due to arbitrary charge transfer Q (here we use the minimum achievable charge reference state, MACRS [16]), V_1 might be higher (as denoted by dots A–D, corresponding to different Q). The potential distributions of TENG in statuses A–D were simulated by finite element method (FEM) using COMSOL Multiphysics software (Figure 2.2b), which indicates V_1(A) and V_1(B) are both higher than V_b. Therefore, to reveal all the circumstances when air breakdown happens in the TENG, the V_1 and V_b under various (Q, V) were simulated and plotted in Figure 2.2c,d. The difference between the two voltages $V_b - V_1$ is also calculated, as plotted in Figure 2.2e. Thus, the negative value means $V_b < V_1$ and indicates a breakdown area. We can notice that in this V-Q plot, the area with low charge transfer Q and HV V between electrodes is with $V_b < V_1$, which is easily achieved in the high-impedance-load condition. The contour with $V_b - V_1 = 0$ V is also plotted inside, outlining the boundary of this breakdown area. These results indicate the existence of the air breakdown effect under charge density of 70 μC/m^2.

2.3 Experimental Verification and Quantitative Measurements

2.3.1 Experimental Verification of the Threshold Air Breakdown Charge Density σ_t

To verify the existence of the threshold air breakdown charge density σ_t, Zi et al. designed an experiment to monitor the charge transfer in short-circuit and

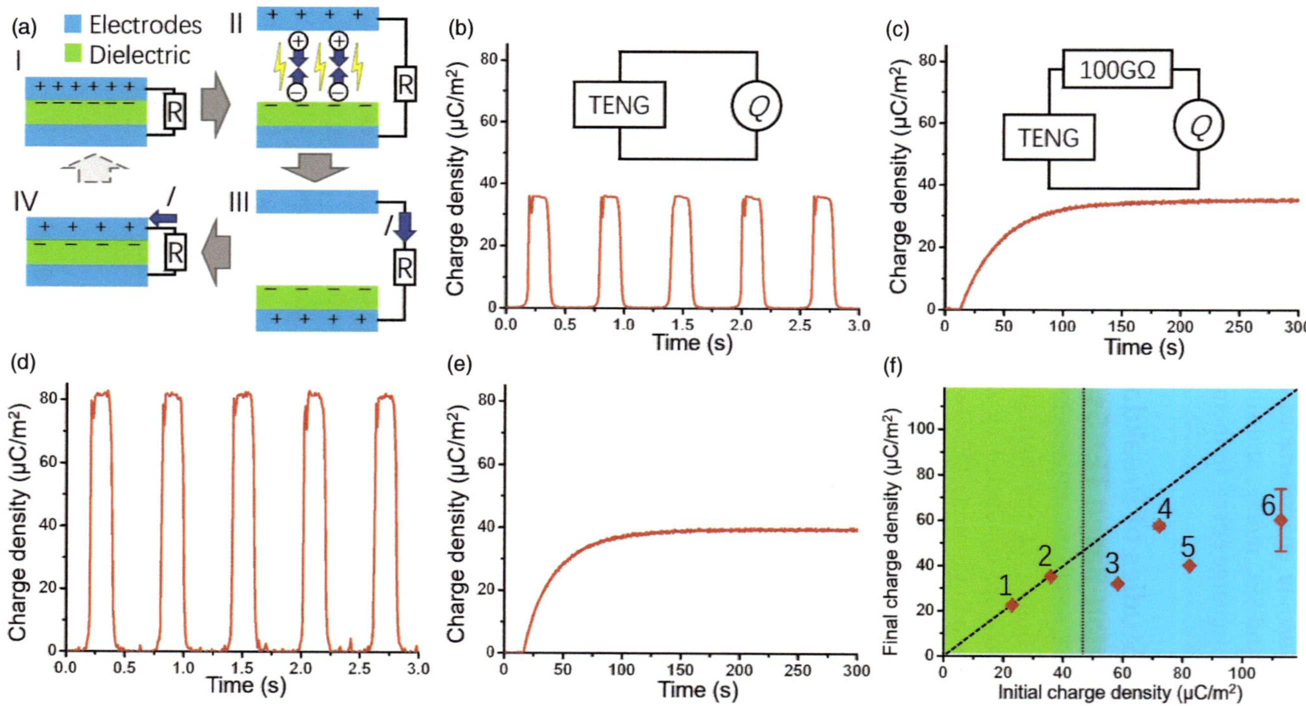

Figure 2.3 The experiment to demonstrate the existence of air breakdown with a low threshold surface charge density. (a) The process of the air breakdown during CS mode TENG operation and the mechanism of the final charge density measurement. (b,c) The initial and final charge density measurements with initial charge density of ≈35.8 μC/m^2. The insets show the corresponding circuits. (d,e) The initial and final charge density measurements with initial charge density of ≈82.3 μC/m^2. (f) The test results of final charge densities for six TENGs. The dashed inclined line indicates equal initial and final charge densities, and the dotted vertical line indicates σ_t which separates σ_0 without breakdown (green) and with breakdown (blue). Source: Reproduced with permission of [9], 2017 © John Wiley & Sons.

high-impedance-load conditions [9]. The mechanism of the designed experiment is proposed in Figure 2.3a. Initially, in Figure 2.3a-I, the TENG is in contact status ($x = 0$), with the initial surface charge density σ_0. In Figure 2.3a-II, when the TENG is operated from $x = 0$ to $x = x_{max}$ with a large external load resistance, due to the obstructed charge transfer brought by the large resistance, most of positive charges remain in the top electrode and hence air breakdown may happen due to the HV induced by the positive charges. In Figure 2.3a-III, after motion stops at $x = x_{max}$, the residual charges gradually transfer to the bottom electrode. And then as shown in Figure 2.2a-IV, when the TENG is operated back to $x = 0$, the residual charge can be gradually transferred back to the top electrode. From III to IV, we can monitor the transfer amount of the residual charge and calculate the final charge density σ_1. Therefore, if $\sigma_0 > \sigma_1$ dramatically, the breakdown happens; if $\sigma_0 \approx \sigma_1$, there is no breakdown effect. The surface charge density can be recovered back to status I by triboelectrification or ionized-air injection [12].

Six CS mode TENGs with different initial charge densities achieved by different materials and treatments were measured [12, 17]. For TENG #2 with σ_0 of ~35.8 μC/m^2 (Figure 2.3b, measured in short-circuit condition, with the circuit shown in inset), σ_1 was measured as ~35.2 μC/m^2 (Figure 2.3c, with 100 GΩ resistance as the large load resistance, with the circuit shown in inset), which is almost the same as σ_0. However, for TENG #5 with σ_0 of ~82.3 μC/m^2 (Figure 2.3d), σ_1 was measured as only ~40.2 μC/m^2 (Figure 2.3e), which is much lower than σ_0. The measurement results for all the six TENGs are plotted in Figure 2.3f. We can notice that for TENGs #1 and #2 with relative low σ_0, $\sigma_0 \approx \sigma_1$; and for TENGs #3 ~ 6 with relative high σ_0, $\sigma_0 > \sigma_1$. These results indicate the existence of air breakdown for TENG operated in high-impedance-load condition with charge density much lower than the threshold for short-circuit condition (240 μC/m^2), and the corresponding threshold surface charge density σ_t is only around 40–50 μC/m^2.

2.3.2 Quantitative Measurement of Breakdown Discharge Points

However, in the V-Q plot of TENGs, the breakdown discharge points are more meaningful as they confine the breakdown and nonbreakdown areas, providing complete information for operations of TENGs. Toward quantitative measurement of the breakdown discharge points, based on the capacitive property, the assessment method is developed by charging the target TENG at different displacement x until the breakdown condition. Here, the schematic measurement circuit for the proposed method is shown in Figure 2.4a [18]. Another TENG (TENG1) is added as the high-voltage source to trigger the target TENG (TENG2) to approach the breakdown condition. Switch 1 (S1) and switch 2 (S2) are used to enable different measurement steps.

Detailed process flow of this method, including an experiment part and a data analysis part, is shown in Figure 2.4b. First of all, it is critical to keep the surface charge density identical as reflected by $Q_{SC,max}$ to ensure the consistency of measurement at different x. Thus, in Step 1, S1 was turned on and S2 was turned

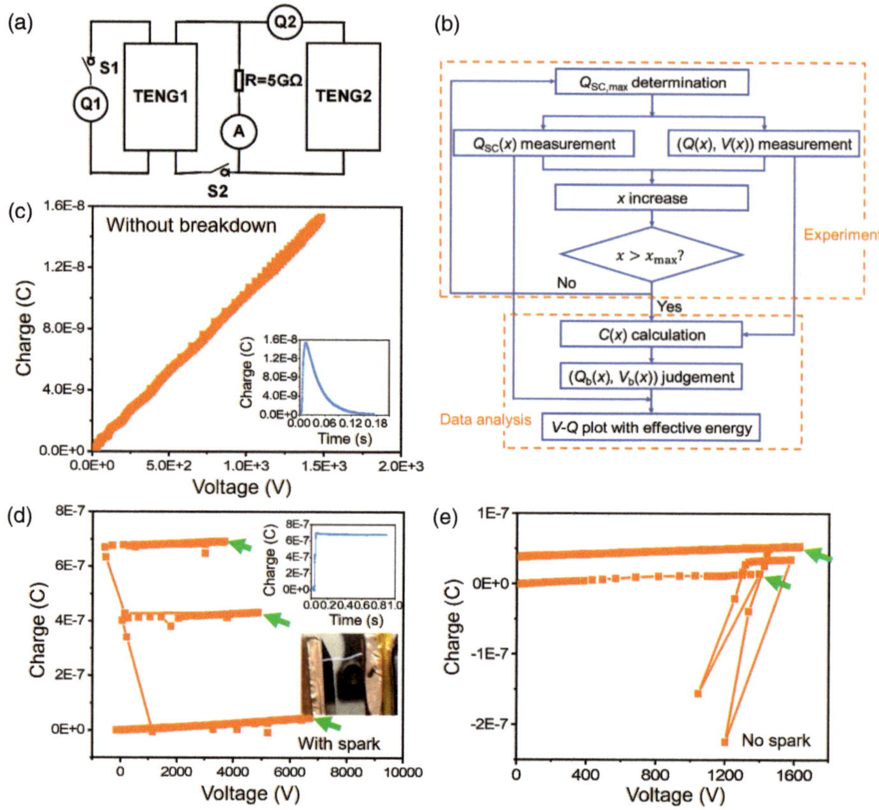

Figure 2.4 The methods and charge–voltage (Q-V) curve of the discharge breakdown points measurement. (a) The circuit and (b) process flow of the measurement method. (c–e) The Q-V curves. Source: [18] / Springer Nature / CC BY-4.0.

off to measure $Q_{SC,max}$; if $Q_{SC,max}$ is lower than the expected value, an additional triboelectrification process is conducted to approach that. And then in Step 2, x was set to a certain value, and the short-circuit charge transfer $Q_{SC}(x)$ at a certain x was measured by coulometer Q1. In step 3, S1 was turned off, S2 was turned on, and then the TENG1 was triggered to supply high-voltage output for TENG2. The charge flowing into TENG2 and the voltage across TENG were measured at the same time, in which the charge was measured by coulometer Q2, and the voltage was obtained by multiplying the resistance R with the current flowing through it as measured by current meter I, as detailed in Methods. The turning points obtained in this (Q,V) were considered the breakdown points. And then, if $x < x_{max}$, the process was repeated starting from step 1 with an increased x, until x_{max} was achieved to finish the experimental measurement part. For the data analysis part, first, $C(x)$ was calculated from the slope of the linear part in the measured (Q,V), by considering it the nonbreakdown part. And then, the first turning point $(Q_b(x), V_b(x))$ was determined at the variant R^2 value by linearly fitting $C(x)$, which was considered the threshold breakdown point. Finally, for any $x \in [0, x_{max}]$, all the $(Q_b(x), V_b(x))$

can be transferred into ($Q_{SC}(x) - Q_b(x)$, $V_b(x)$) as the breakdown points plotted in the V-Q cycle.

To verify the assessment method, a CS mode TENG was evaluated first since its theoretical simulations have been well conducted for comparison. Figure 2.4c–e illustrates typical measurement results of Q-V curves for CS mode, by the method described above. Figure 2.4c shows the Q-V curve when the voltage supply is not high enough to enable the breakdown. It can be noticed that this Q-V curve has good linearity, and the slope can reflect the capacitance. The transferred charge always returns to the initial condition after one charging process, as shown in the inset of Figure 2.4c. When the voltage is high enough, visible sparks between triboelectric surfaces can be easily observed, directly showing the existence of air breakdown, which corresponds to the glow or arc breakdown phenomena as shown in the inset of Figure 2.4d. As shown in Figure 2.4d, once the spark happens, the transferred charge experiences a sudden unrecoverable rise-up, and the voltage is suddenly decreased. The line before breakdown is in good linearity for capacitance extraction. The simultaneous observation of the sparks and the sudden changes in the Q-V curves is quite repeatable after numerous tests. Figure 2.4e is another typical Q-V curve for breakdown when there is no spark observed, in which unrecoverable sudden changes of both Q and V were measured. Thus, it can be considered a breakdown without sparks. These breakdown points might be due to the dark discharge phenomenon as the "Townsend region" of discharge [19–22], in which the voltage might be increased after breakdown, as consistent with the measured results. Both glow/arc and dark discharge points are considered the breakdown threshold points in this research since the significant discharge observed can induce energy loss in TENG. These breakdown points ($Q_b(x)$, $V_b(x)$) are indicated by green arrows in Figure 2.4d,e.

The measured breakdown points with various x can be summarized to define the nonbreakdown areas. Figure 2.5 shows results of a CS mode TENG with $Q_{SC,max} = 67\,\text{nC}$. To avoid fluctuations in the measurement, every single dot in Figure 2.5 was repeated at least three times to obtain an average value with a standard derivation. Figure 2.5a illustrates experimental setups of the CS mode TENG. The static part of the TENG is fixed on an XYZ-3-directional stage to control the x. The moving part is fixed on and triggered by a linear motor. Parameters of the TENG in the experiment are listed in Table 2.2. Breakdown voltage results of V_1 are plotted in blue dots with error bars in Figure 2.5b, as compared with theoretical results in the orange line as calculated by Paschen's law. The measured and calculated results fit quite well, demonstrating the effectiveness of our proposed method. The capacitances at various x were calculated, as shown in Figure 2.5c, with the average experimental results in blue and the calculated results from the nonideal parallel-board equation in orange (see Methods). Considering the parasitic capacitance from conductive substrates and connecting wires, and difficulties in reaching the full contact status at $x = 0$ point, the measured capacitances are usually a little larger than the calculated ones. Figure 2.5d is the V-Q plot of CMEO marked with the breakdown area of the CS mode TENG. The blue line with dots is the experimental results by our method, as averaged from multiple measurement

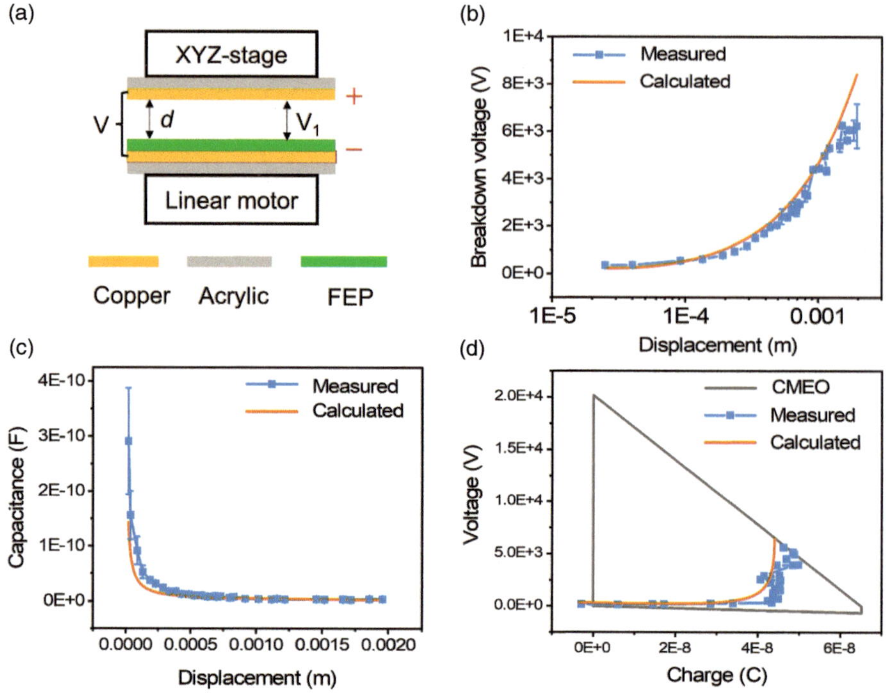

Figure 2.5 Breakdown discharge results for a CS-TENG. (a) Schematic diagram of CS-TENG. (b) Average breakdown voltage V_1 from experiment, compared with theoretical values calculated by Paschen's law. (c) Average capacitance C, compared with theoretical values calculated by the equation considering two-side edge effects. (d) Cycle for maximized energy output with the threshold breakdown curve, compared with theoretical curve by Paschen's law. $Q_{SC,max}$ = 67 nC. Source: [18]/Springer Nature/CC-BY-4.0.

Table 2.2 Parameters of the CS mode TENG.

Length (m)	0.02
Width (m)	0.02
$Q_{SC,max}$ (nC)	67
σ ($\mu C/m^2$)	168
d_{max} (m)	0.002
C_2 (F)	1.03×10^{-10}

results, while the orange line was calculated from Paschen's law. It is noticed that the measured breakdown points are usually with a little smaller voltage, which might be induced by imperfect parallel surfaces in TENG, and the parasitic capacitance.

A CFT mode TENG is also studied by this method as another well-modeled structure. The experiment results by the proposed method are illustrated in Figure 2.6, with $Q_{SC,max}$ of 25 nC and experiment setups as shown in Figure 2.6a. Detailed

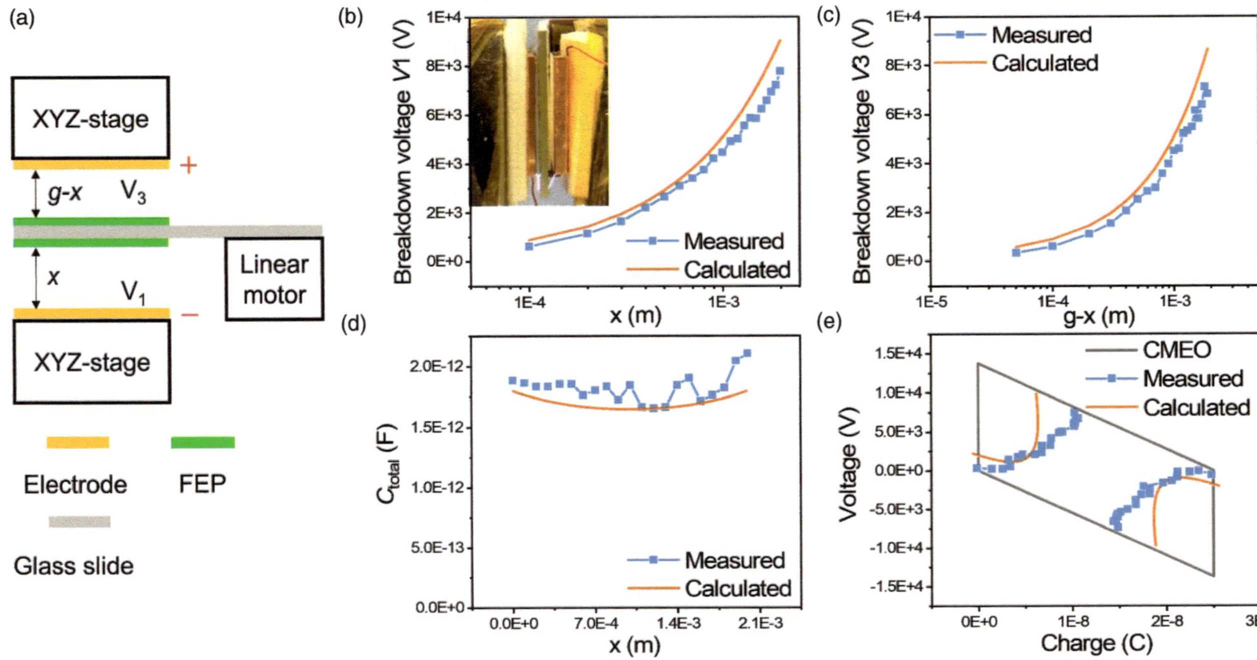

Figure 2.6 Breakdown discharge for CFT-TENG. (a) Schematic diagram of CFT mode TENG. Average breakdown voltage (b) V_1 and (c) V_3 from experiment, compared with theoretical values calculated by Paschen's law. Figure of visible spark insets in (b). (d) Average capacitance C_{total}, compared with theoretical values calculated by the equation considering two-side edge effects. (e) Cycle for maximized energy output with the threshold breakdown curve compared with theoretical curve by Paschen's law. $Q_{SC,max}$ = 25 nC. Source: [18] / Springer Nature / CC BY-4.0.

Table 2.3 Experiment parameters of the CFT mode TENG.

Gap (m)	0.002
Thickness of middle layer (m)	0.001435
Thickness of glass slide (m)	0.001
Equivalent thickness (m)	0.0004731
Q_{SC} (gap) (nC)	25
Q_{10} (nC)	2.39
σ (μC/m^2)	38.6
C_2 (F)	7.48×10^{-12}
$C_{total\text{-}exp\text{-}ave}$ (F)	1.82×10^{-12}

parameters in measurements are listed in Table 2.3. The middle dielectric layer of the CFT mode TENG is placed on a glass slide substrate and then fixed on a linear motor. Both two electrodes are connected on XYZ-stages to fix the maximum displacement. There are two air gaps between the middle layer and the two electrodes, and when either of them is with a voltage (V_1 or V_3) larger than the corresponding breakdown voltage [9], we can consider the breakdown of the TENG triggered since the tribocharge in the dielectric layer losses. The breakdown voltages of V_1 and V_3 from experiments are plotted in blue in Figure 2.6b,c, respectively, as consistent with that calculated by Paschen's law quite well. The inset figure in Figure 2.6b shows visible sparks between the middle layer and electrodes of each side. Figure 2.6d is the experimentally measured capacitance in blue, as compared with calculated results from the nonideal parallel-board equation in orange. Similar to results for the CS mode TENG, the capacitance extracted from experiments is always larger than theoretical ones, due to influences by parasitic capacitances. With the V-Q plot for CMEO, breakdown points of the CFT mode TENG as derived from the experiment are plotted in blue, as compared with that calculated from Paschen's law in orange, which shows the consistent trend as illustrated in Figure 2.6e. The breakdown voltages from experimental results are even lower than calculated ones as compared with that in CS mode TENG, which may be due to the impacts of the imperfect parallel surfaces from both gaps.

2.4 Photon Generation

2.4.1 Radio-Frequency EM-Wave Photon Generation

Discharge can generate photons in the wide EM-wave spectrum from the radio-frequency waves to the visible light. The EM-wave generation is related to the rapidly changed displacement current. The state-of-the-art EM-wave generation can be fundamentally explained by Faraday's law (Eq. 2.2) and Maxwell-Ampere's

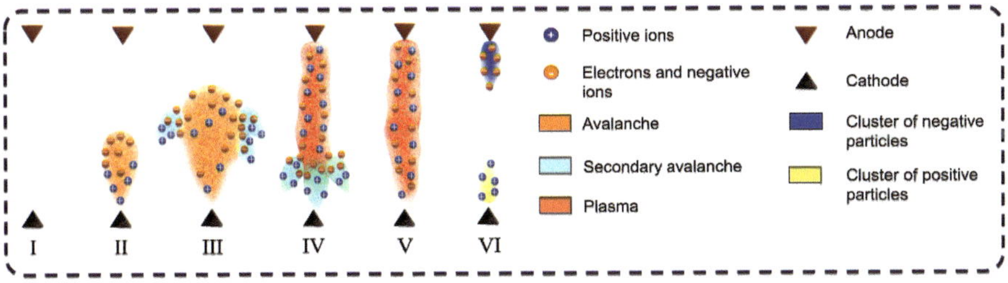

Figure 2.7 The whole process of the breakdown discharge.

law (Eq. 2.3):

$$\nabla \times \boldsymbol{E} = -\frac{\partial \boldsymbol{B}}{\partial t} \tag{2.2}$$

$$\nabla \times \boldsymbol{H} = \boldsymbol{J} + \frac{\partial \boldsymbol{D}}{\partial t} \tag{2.3}$$

Here, $\boldsymbol{E}$ and $\boldsymbol{D}$ are the electric field and electric displacement field, $\boldsymbol{B}$ and $\boldsymbol{H}$ are the magnetic flux density and the magnetic field strength, respectively, and $\boldsymbol{J}$ is the current density. The state-of-the-art wireless signals are mostly initiated by the varied conduction current density $\boldsymbol{J}$, followed by transformation between electric field ($\boldsymbol{E}$ or $\boldsymbol{D}$) and magnetic field ($\boldsymbol{B}$ or $\boldsymbol{H}$).

We put forward the paradigm-shift mechanism of EM-wave generation through the additional displacement current term $\partial \boldsymbol{P}_S/\partial t$ as proposed by Wang et al. [23–26] from the triboelectric effect [27]. And, thus, Eq. (2.3) becomes:

$$\nabla \times \boldsymbol{H} = \boldsymbol{J} + \varepsilon\frac{\partial \boldsymbol{E}}{\partial t} + \frac{\partial \boldsymbol{P}_S}{\partial t} \tag{2.4}$$

Therefore, to generate rapidly varied electric and magnetic fields for wireless transmission, the key is to obtain rapidly varied $\boldsymbol{P}_S$ at high frequency, which can be achieved through breakdown discharge effect. The breakdown discharge is induced by collisions between electric-field-driven high-speed moving electrons and air molecules, which generate ions and new electrons, forming electron avalanche and plasma. The whole process of the breakdown discharge is illustrated in Figure 2.7. This plasma can be considered collective oscillation of clustered charged particles, which can thus generate the rapidly varied displacement current and hence EM waves.

2.4.2 Visible Light Photon Generation

Besides the discharge-induced EM-wave generation, the ionization process of the gas molecules generates electrons and ions for conduction in the air. During the collisions with the ions or electrons, gas molecules or atoms can be excited to a higher energy level, and then quickly drop to a normal level, releasing a photon to carry the energy away. Such a photon can be either glow (plasma) discharge, or arc discharge [28]. With that, the concept of triboelectric plasma can be proposed by combining plasma sources with TENGs to realize atmospheric-pressure plasma with mechanical stimuli [29]. Four classical types of TENG-driven microplasma sources at atmospheric pressure, including dielectric barrier discharge (DBD), atmospheric-pressure nonequilibrium plasma jets (APNP-J), corona discharge and microspark discharge are demonstrated. The transient electric characteristics and optical emission spectra of typical triboelectric plasma are simultaneously analyzed and compared in detail, both experimentally and theoretically. The influence of design parameters of the microplasma devices, including the electrode gap, the diameter of wire electrode, and the flow rate, on the discharge performance are investigated, which can provide guidance for optimization. The applications of triboelectric plasma in patterned luminescence and plasma surface treatment are demonstrated successfully as a first-step evaluation as well as to prove the system's feasibility. This work may not only extend the applications of TENG to self-powered plasma but also enrich the diversity of plasma applications by providing a facile and portable supplement to traditional plasma sources.

2.5 Potential Applications

As a key fundamental phenomenon, studies on the discharge effects in TENG will have a significant impact on the potential applications. Depending on the types of applications, there are two potential ways. First, we may use methods to suppress the discharge to enhance the surface charge density for enhanced output performance. These methods may include but are not limited to environment, liquid–solid interface, three-phase design, etc., as detailed in the following chapters. Second, with the photons generated by the discharge effects, we may develop novel methods to achieve self-powered wireless sensing systems, with the triboelectric to provide enough energy for discharge, and the photons generated from the discharge for wireless transmission. This study will be further discussed in Chapter 10: tribophotonics as well. Other discharge-related applications include triboinduced electrospray for ultrasensitive mass spectrometry [30], triboinduced field emission of electrons [31], etc.

2.6 Summary

This chapter discussed the discovery, theoretical calculations, experimental verification, quantitative measurement, and photon generation of the discharge effect, as a key phenomenon in TENG. Discharge effect is in fact the key limiting factor of output performance of TENG, and thus studies on that have great significance in promoting the power generation. In the meanwhile, through the photon generation from the discharge effect, novel methods may be developed to achieve self-powered wireless sensing systems. These two aspects shape the potential applications of the discharge effect in TENGs, which will be further discussed in the following chapters.

References

1 Levatter, J.I. and Lin, S.C. (1980). Necessary conditions for the homogeneous formation of pulsed avalanche discharges at high gas pressures. *J. Appl. Phys.* 51 (1): 210–222.
2 Boyle, W. and Kisliuk, P. (1955). Departure from Paschen's law of breakdown in gases. *Phys. Rev.* 97: 255.
3 Townsend, J.S. (1915). *Electricity in Gases*, 337–391. москва: Рипол Классик.
4 Fu, J., Xu, G., Li, C. et al. (2020). Achieving ultrahigh output energy density of triboelectric nanogenerators in high-pressure gas environment. *Adv. Sci.* 7 (24): 2001757.
5 https://en.wikipedia.org/wiki/Gas-filled_tube
6 https://en.wikipedia.org/wiki/Corona_discharge
7 Gallo, C.F. (1975). Coronas and gas discharges in electrophotography: a review. *IEEE Trans. Ind. Appl.* IA-11 (6): 739–748.
8 Lieberman, M.A. and Lichtenberg, A.J. (2005). *Principles of Plasma Discharges and Materials Processing*. John Wiley & Sons.
9 Zi, Y., Wu, C., Ding, W., and Wang, Z.L. (2017). Maximized effective energy output of contact-separation-triggered triboelectric nanogenerators as limited by air breakdown. *Adv. Funct. Mater.* 27 (24): 1700049.
10 https://hypertextbook.com/facts/2000/Alice.Hong.shtml
11 Rumble, J. (ed.) (2021). *CRC Handbook of Chemistry and Physics*. Taylor & Francis.
12 Wang, S., Xie, Y., Niu, S. et al. (2014). Maximum surface charge density for triboelectric nanogenerators achieved by ionized-air injection: methodology and theoretical understanding. *Adv. Mater.* 26 (39): 6720–6728.
13 Wang, S., Lin, L., and Wang, Z.L. (2012). Nanoscale triboelectric-effect-enabled energy conversion for sustainably powering portable electronics. *Nano Lett.* 12 (12): 6339–6346.
14 Zi, Y., Niu, S., Wang, J. et al. (2015). Standards and figure-of-merits for quantifying the performance of triboelectric nanogenerators. *Nat. Commun.* 6: 8376.
15 Li, Y., Li, Y.H., Li, Q.X., and Zi, Y.Y. (2003). Computation of electrostatic forces with edge effects for non-parallel comb-actuators. *J. Tsinghua Univ. (Sci. & Tech.)* 43 (8): 1024–1026. 1030.
16 Niu, S., Liu, Y., Chen, X. et al. (2015). Theory of freestanding triboelectric-layer-based nanogenerators. *Nano Energy* 12: 760–774.
17 Wang, S., Zi, Y., Zhou, Y.S. et al. (2016). Molecular surface functionalization to enhance the power output of triboelectric nanogenerators. *J. Mater. Chem. A* 4 (10): 3728–3734.
18 Xia, X., Fu, J., and Zi, Y. (2019). A universal standardized method for output capability assessment of nanogenerators. *Nat. Commun.* 10 (1): 4428.
19 Gallo, C.F. (1977). Corona-a brief status report. *IEEE Trans. Ind. Appl.* 13 (6): 550–557.
20 Barnes, M.S., Keller, J.H., Forster, J.C. et al. (1992). Transport of dust particles in glow-discharge plasmas. *Phys. Rev. Lett.* 68 (3): 313–316.

21 Massines, F., Rabehi, A., Decomps, P. et al. (1998). Experimental and theoretical study of a glow discharge at atmospheric pressure controlled by dielectric barrier. *J. Appl. Phys.* 83 (6): 2950–2957.
22 Marcus, R.K. and Broekaert, J.A.C. (2003). *Glow Discharge Plasmas in Analytical Spectroscopy*. Wiley.
23 Xia, X., Wang, H., Guo, H. et al. (2020). On the material-dependant mechanism of contact electrification. *Nano Energy* 78: 105343.
24 Wang, Z.L. (2020). On the first principle theory of nanogenerators from Maxwell's equations. *Nano Energy* 68: 104272.
25 Shao, J., Willatzen, M., and Wang, Z.L. (2020). Theoretical modeling of triboelectric nanogenerators (TENGs). *J. Appl. Phys.* 128 (11): 111101.
26 Wang, Z.L. (2017). On Maxwell's displacement current for energy and sensors: the origin of nanogenerators. *Mater. Today* 20 (2): 74–82.
27 Wang, H., Wang, J., Yao, K. et al. (2021). A paradigm shift fully self-powered long-distance wireless sensing solution enabled by discharge-induced displacement current. *Sci. Adv.* 7 (39): eabi6751.
28 https://en.wikipedia.org/wiki/Glow_discharge
29 Cheng, J., Ding, W., Zi, Y. et al. (2018). Triboelectric microplasma powered by mechanical stimuli. *Nat. Commun.* 9 (1): 3733.
30 Li, A., Zi, Y., Guo, H. et al. (2017). Triboelectric nanogenerators for sensitive nano-coulomb molecular mass spectrometry. *Nat. Nanotechnol.* 12 (5): 481–487.
31 Zi, Y., Wu, C., Ding, W. et al. (2018). Field emission of electrons powered by a triboelectric nanogenerator. *Adv. Funct. Mater.* 28 (21): 1800610.

3

Figure-of-Merit of Triboelectric Nanogenerator

3.1 Introduction

With the rapid development of the Internet of Things (IoTs), sustainably powering widely distributed small electronics becomes a big issue. As a new energy harvesting technology, nanogenerators, based on pyroelectric [1, 2], piezoelectric [3–5], and triboelectric effects [6–8], have been invented to scavenge tiny energy from the ambient environment into electricity. As a new type of energy harvester that is based on Maxwell's displacement current [9], the unique capacitive model of nanogenerators makes traditional characterization methods unsuitable for unveiling the output capability. Here, triboelectric nanogenerator (TENG) is focused as an example, which has been intensively studied due to its high output and high energy conversion efficiency. It has been demonstrated that with traditional approaches, only a small portion of the maximized energy per cycle E_m can be outputted [10], which cannot reflect the real output capability. To address this issue, the performance figure-of-merits (FOMs) of TENG [11], including a structural FOM and a material FOM, have been developed as a standard to quantitatively evaluate the output performance through cycles of maximized energy output (CMEO) in the voltage-charge (V-Q) plot.

However, to develop a standardized assessment method, there are still several critical challenges to be addressed, as stated below [11]. First, there are still no systematic studies on the standardized method for the output capability assessment, especially in experimental ways. This method is extremely important to delicately characterize nanogenerators toward practical applications and standardization in industry. In comparison with its counterpart, one reason for the success of solar cells can be attributed to its well-established standard and standardized method [12–15]. Second, in the current FOM as the standard for TENG, the breakdown effect, which universally exists as a key limiting factor of the maximized effective energy output of TENG [16–19], has not been considered. This discharging phenomenon brings the urgent need to revise the current definition of FOM and raises a great challenge in terms of standardized assessment. Last but not least, developing standards and a standardized assessment method that can be universally applied in various structures of TENGs and other types of nanogenerators that are based on capacitive model

Triboelectric Nanogenerators: Technology, Applications, and Commercialization, First Edition.
Yunlong Zi, Hengyu Guo, Jie Wang, Chi Zhang, Xiangyu Chen, and Qing Zhao.

[20–22] is critical to promote the research and applications of nanogenerators and draw broad interest from research communities.

To develop such kind of universal standardized method, Prof. Yunlong Zi's team first developed the process flow of the measurement on the maximized effective energy output E_{em} per cycle, and then the revised FOM [23] and output energy density [24] were defined. Environmental life-cycle assessment (LCA) and techno-economic analysis (TEA) were also conducted on TENGs as an evaluation standard in another dimension [25]. These studies provide standardized methods to assess output capacity that can be universally applied for nanogenerators, considering the breakdown limits, which will contribute a lot to further applications and industrializations of nanogenerator technology in terms of standardization.

3.2 Effective Maximized Energy Output

3.2.1 Traditional Methods to Characterize the TENG

Since the invention of TENG in 2012, there are several major parameters frequently used to characterize the output performance of a TENG: the open-circuit voltage (V_{OC}), the peak short-circuit current (I_{SC}-peak), and the optimized peak power (*P*-peak). These characterization methods are based on Voltage–Resistance (V–R), Current–Resistance (I–R) and Power–Resistance (P–R) plots, which have some practical meanings in estimating the peak output capability of TENG devices [26]. However, since I_{SC}-peak and *P*-peak can only provide instantaneous-achieved highest values, they fail to reflect the average current/power output performance of TENG, which is more significant for applications. To address this issue, Zhu et al. developed characterization methods based on the average short-circuit current (I_{SC}-ave) and average power (*P*-ave), which provide a better estimation of the performance of TENG [27]. However, simply comparing the average current and the average power is still not suitable to evaluate the performance of TENGs due to the difference in surface area. Therefore, the power density (*P*-den) is raised and the power density is used to present the output performance per area. In addition, the power density *P*-den is an important index to reflect the sufficiency of contact between the triboelectric materials. The TENG with poor sufficiency of contact has a low power density inevitably. Therefore, the *P*-den which reflects the sufficiency of friction of the TENG becomes an important merit to judge the TENGs.

As mentioned before, the traditional approaches based on Voltage–Current (V–I) plot or the Power–Resistance (P–R) plot are not suitable to understand the performance of TENG because the current I is not an independent variable, which relies on the transferred charge *Q* and the time *t* (related to the maximum separation distance x_{max} and the velocity *v*) as illustrated below.

$$I_{sc} = \frac{dQ_{sc}}{dt} = \frac{dQ_{sc} \times v}{dx} \tag{3.1}$$

Even using the same TENG, researchers will get different current values and thus different power values due to the distinct charge transfers in different displacements, and the varied velocity. Therefore, the current-based power value is not an accurate parameter to measure the output performance of the TENG.

On the other hand, the integration of the current multiplying the voltage over time could be used to present the actual output energy, according to the following equation.

$$E_{out} = \int VI\mathrm{d}t = \oint V\mathrm{d}Q \tag{3.2}$$

The integration is very vital because it indicates the TENG's output energy per cycle accurately, and the integration is not affected by the motion parameters [28]. In addition, when the integration is divided by the surface area, the output energy density (E_{out}-den) could be an important index to judge the TENG's output property per square meter and to reflect the sufficiency of contact between the triboelectric materials.

3.2.2 *V–Q* Plot

As shown in Equation (3.2), the integration of voltage V multiplying transferred charge Q can accurately present the TENG's output energy. Meanwhile, the voltage and the transferred charge could both be directly measured by experiments. This is the original idea to propose the V–Q plot.

The relation between V and Q becomes critical to determine the operation of TENG, as proposed by Niu et al. [20]:

$$V_{OC}(x) - \frac{1}{C(x)}Q = V \tag{3.3}$$

$V_{OC}(x)$ is the open-circuit voltage depending on the displacement x, and $C(x)$ is the capacitance between two electrodes. The integration of V and Q is the energy output per cycle, and the V–Q plot can be directly used to illustrate the energy output performance of TENG:

$$E = P_{ave}T = \int_0^T VI\mathrm{d}t = \int_{t=0}^{t=T} V\mathrm{d}Q = \oint V\mathrm{d}Q \tag{3.4}$$

P_{ave} is the average power output and I is the current output. These features make V–Q plot a crucial method to illustrate the maximized energy output of TENG.

Figure 3.1a–e presents the method to achieve the maximized energy output for a lateral-sliding mode TENG simulated by the finite-element method (FEM), as demonstrated by Zi et al. [11]. In Figure 3.1a, it is defined as $Q = 0$ and $V = 0$ when the TENG is at the initial status ($x = 0$). At first, the external load is set as 100 MΩ in Figure 3.1b. The cycle of energy output (CEO) is simulated from (0, 0) point and achieves a steady state after about two periods. The encircled area of the steady-state V–Q plot is the output energy per cycle of the TENG. In Figure 3.1c, the output energy is calculated with various external loads from 4.4 MΩ to 4.4 GΩ. It is obvious that the output voltage will be higher when the external resistance is larger, while the transferred charge Q_{SC} will be smaller. In Figure 3.1d, a four-step strategy is designed to obtain the cycle of the maximized energy output (CMEO). A switch is connected in parallel with the resistor, and the switch will be turned on at 0 and

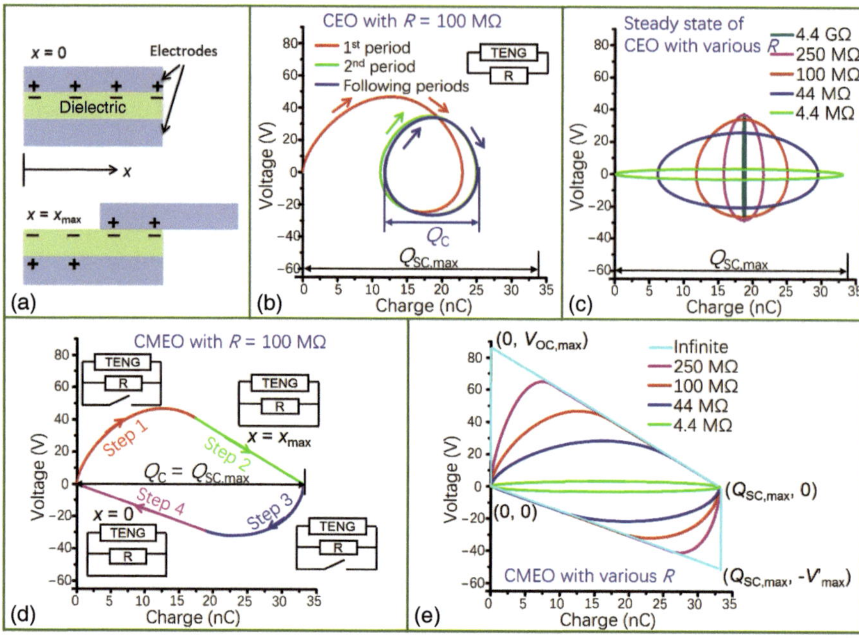

Figure 3.1 Operation cycles of TENG. (a) Schematic diagram of the LS mode TENG with displacement $x = 0$ and $x = x_{max}$; (b) The cycle for energy output (CEO) with load resistance $R = 100\,M\Omega$. The total cycling charge Q_C was marked and the inset shows the operation circuit; (c) The steady state of CEO with various load resistances; (d) The cycle for maximized energy output (CMEO) with load resistance $R = 100\,M\Omega$, with the maximum total cycling charge $Q_C = Q_{SC,max}$. The insets show the corresponding status of the switch in circuits during different steps; (e) The CMEO with various load resistances. The vertices of the CMEO with infinite load resistances are marked. Source: Reproduced with permission of [11]. 2015 ©Springer Nature.

x_{max} positions, which makes the transferred charge always equal to the maximum value $Q_{SC,max}$ obtained in the short-circuit condition. Then the energy output as the encircled area will be larger when the external resistance increases (Figure 3.1d). The maximized energy output can be achieved at CMEO with the infinite load resistance shown in Figure 3.6e. Therefore, the TENG's maximum output energy E_m can be defined as:

$$E_m = \frac{1}{2} Q_{SC,max} \left(V_{OC,max} + V'_{max} \right) = \frac{1}{2} Q_{SC,max}{}^2 \left[\frac{1}{C(x = 0)} + \frac{1}{C(x = x_{max})} \right] \quad (3.5)$$

$V_{OC,max}$ and V'_{max} are the maximum open-circuit voltages when $x = 0$ and $x = x_{max}$. The V–Q plot can also be measured experimentally. Two electrometers (for example, Keithley 6514) are connected to a TENG at the same time (one electrometer connected in parallel with TENG for the voltage and the other one connected in series with TENG for the transferred charge). The signals of the voltage and transferred charge will be collected by the Data Acquisition (DAQ) system and transmitted to the computer simultaneously. This is the first time utilizing the V–Q plot to maximize the TENG's output energy, which sets a foundation to propose the FOMs as the standards for quantitatively evaluating TENG.

3.2.3 Effective Maximized Energy Output Per Cycle

However, considering the breakdown discharge effect, not all energy in E_m can be output for applications. A large portion of the energy will be dissipated through the breakdown discharge effect induced by the high voltage from TENG. In sliding TENGs (LS and SFT modes), such breakdown discharge may involve air breakdown and dielectric breakdown [24]. While for contact TENGs (CS, CFT, SEC modes), the breakdown mainly happens in the air gap between triboelectric surfaces during separation process, while the threshold breakdown voltage V_b can be calculated using Paschen's law as shown in Equation (2.1) in Chapter 2. By comparing V_b and voltage between triboelectric surfaces, we can plot the breakdown regions (in red) in V–Q plots as shown in Figure 3.2, with parameters shown in Table 3.1.

As shown in Figure 3.2a–c, with the further increase of the charge density, the breakdown region for CS mode TENG appears at the charge density of 40–50 μC/m^2 and continues to increase until 240 μC/m^2 when the breakdown discharge happens even in the short-circuit condition. Here we can define the nonbreakdown region (while area) as the effective maximized energy output E_{em} per cycle. In the CS TENG with given parameters, the E_{em} saturates at around 0.02 J [18]. As for CFT and SEC modes, they may experience severe breakdown discharge when the charge densities are larger than ~20 μC/m^2 and ~45 μC/m^2, respectively, as shown in Figure 3.2d,e. As a result, their E_{em} will rapidly reduce to approach 0. The E_{em} v.s. charge density plots for contact TENGs are shown in Figure 3.2f.

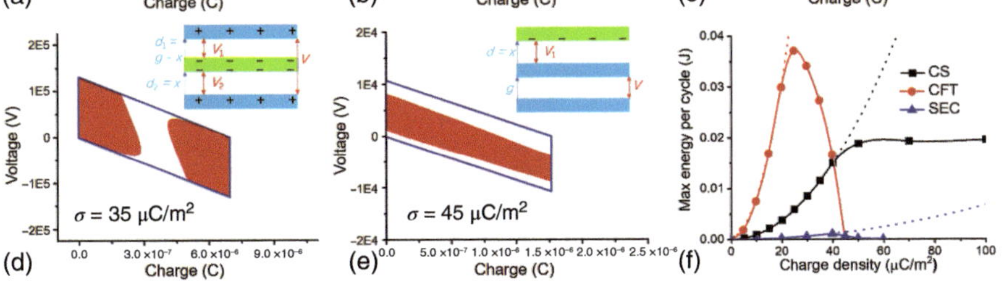

Figure 3.2 The breakdown areas and maximized energy output per cycle. (a–c) The breakdown and nonbreakdown areas in CS-TENG with different charge densities. (d) The breakdown and nonbreakdown areas in CFT-TENG. (e) The breakdown and nonbreakdown areas in SEC-TENG. (f) The relationship between the maximized energy output per cycle and the charge density in different TENGs. Source: Reproduced with permission of [29], 2017 © John Wiley & Sons.

Table 3.1 Parameters of TENGs for calculations.

TENG mode	CS	CFT	SEC
Dielectric layer	Thickness $d_r = 0.1$ mm, dielectric constant $\varepsilon_r = 2$, so effective thickness $d_0 = d_r/\varepsilon_r = 0.05$ mm [22]		
Area	Length $l \times$ Width w: 0.1 m × 0.1 m		
Gap g between electrodes (m)	N/A	0.03	0.01
Charge density (μC/m^2)	0–250	0–180	0–210
Displacement x (m)	0–0.03		

Since the mechanism of air breakdown effect is related to the avalanche effect induced by the electrons accelerated by electric field, this effect can be greatly suppressed by air with increased pressure or insulating gases. In the air with increased pressure, the mean-free-path of electrons is greatly reduced due to increased air molecule density, and then the energy of electrons may not be enough to break down air molecules to trigger avalanche effect. In insulating gases such as SF_6, the gas molecules require more energy to be broken down, and thus the required electric field increases as well [30]. Here Zi et al. calculated the breakdown threshold electric field for air with pressures of 1 atm, 2 atm, and 5 atm, and SF_6 gas under 1 atm as shown in Figure 3.3a, which clearly shows the

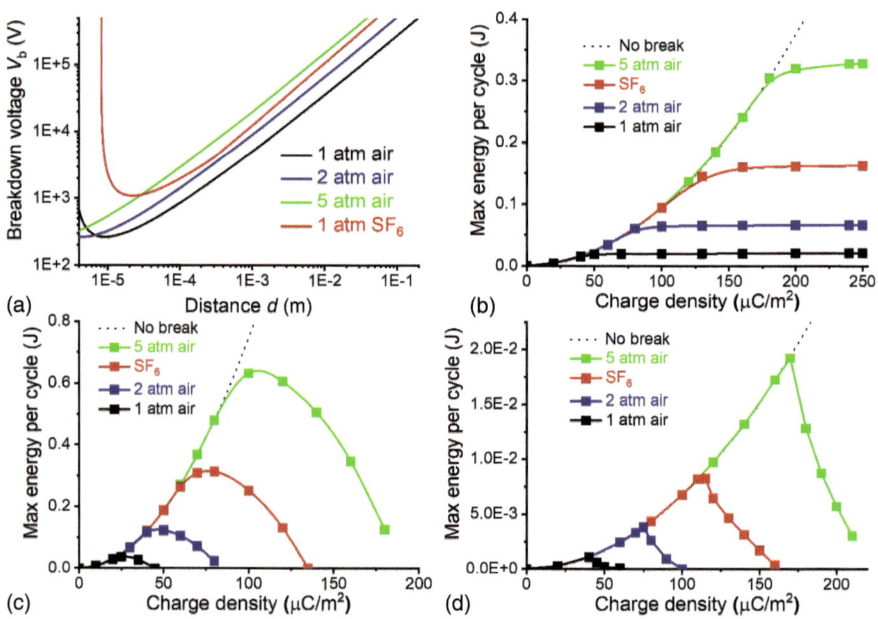

Figure 3.3 The (a) Paschen's curve and (b) maximized energy output per cycle under air with different pressures and insulating gas with 1 atm pressure. Source: Reproduced with permission of [29], 2017 © John Wiley & Sons.

increased threshold voltage induced by the high pressure or insulating gas. With that, we also calculated the maximized effective energy output E_{em} per cycle for each contact-separation-based TENGs, as shown in Figure 3.3b–d. This E_{em} can increase greatly with the increased pressure or insulating gases several times, with the increases in threshold short-circuit charge density as well. Such an effect was also demonstrated in further experimental studies [31].

3.3 Figure-of-Merit

3.3.1 Figure-of-Merits as the Common Standard

Although different modes of TENG devices have been proposed to satisfy special application in different environments, a satisfactory common standard, which can quantify the performance of TENG, remains elusive.

The V–Q plot which indicates the energy output of TENG sets a foundation for the common standard. Zi et al. developed the FOMs as the standard, which considers the materials, surface areas, maximum separation distances, surface charging densities, output powers, and so on, to normalize the parameters and quantitatively evaluate TENG [11]. A dimensionless structural FOM (FOM_S) of TENG was defined in the following equation, as the factor that only depends on the structural parameters and x_{max}.

$$\mathrm{FOM_S} = \frac{2\varepsilon_0}{\sigma^2}\frac{E_m}{Ax_{max}} \tag{3.6}$$

Here ε_0 is the vacuum permittivity. This FOM represents the TENG's merit from the structure. Figure 3.4 demonstrates the structural FOM of CS, LS, single electrode contact (SEC), sliding freestanding triboelectric layer (SFT) and the contact freestanding triboelectric layer (CFT) modes with the same charge density and area for comparison. The red square dots are the values simulated by FEM, and the black dash and solid lines are the values calculated by their corresponding analytical formulae while considering one-side and two-side edge effects, respectively. It is obvious that the calculated values fit the results from the FEM simulation very well, and CFT mode has the highest maximum value of the structural FOM.

Zi et al. make three general conclusions. First, the TENGs based on contact separation have better performances than the sliding-based TENGs due to their higher voltage at the same separation distance. Second, the freestanding structure can improve the TENG's performance due to the decreased capacitance between electrodes. Therefore, CFT mode has a higher structural FOM than the CS mode, and the structural FOM of SFT mode is better than that of LS mode. Third, the two-electrode-based TENG has a better performance than the single-electrode-based TENG with same materials and size due to the limited transferred charges and suppressed built-in voltages in SE TENG.

Zi also defined the performance FOM (FOM_P) of TENG as:

$$\mathrm{FOM_P} = \mathrm{FOM_S} \cdot \sigma^2 = 2\varepsilon_0\frac{E_m}{Ax_{max}} \tag{3.7}$$

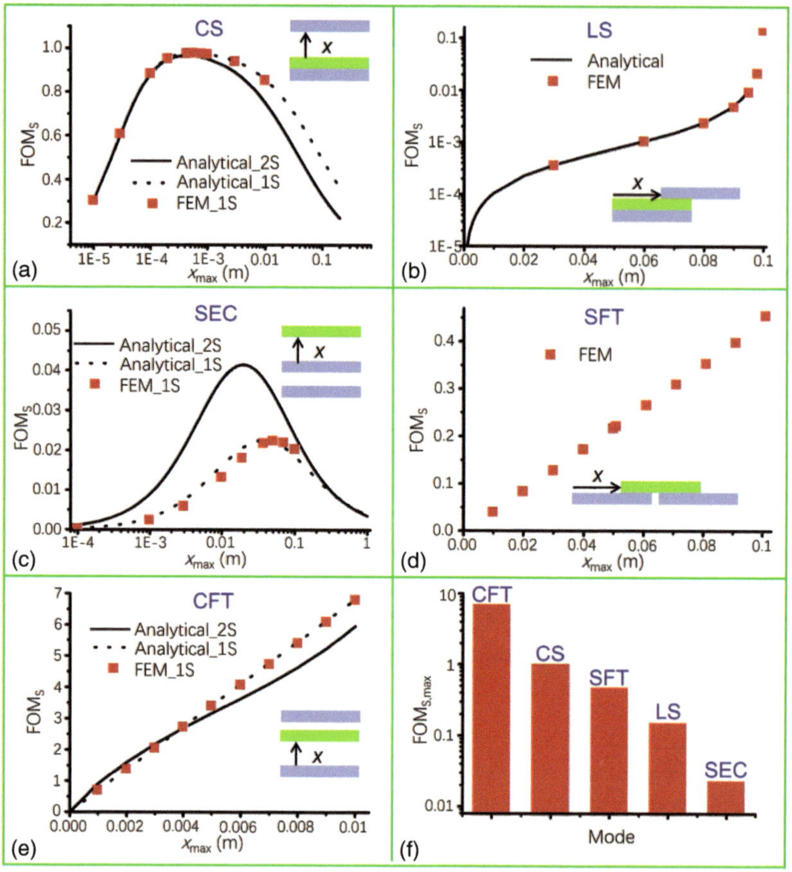

Figure 3.4 FOM_S versus x_{max} for different TENG structures. (a) ~ (e) FOM_S for vertical contact-separation (CS) mode, lateral sliding (LS) mode, single-electrode contact (SEC) structure, SFT structure and CFT structure calculated by analytical formulas and FEM simulation. The insets show the corresponding schematic diagrams of the corresponding structures. 1S and 2S stand for calculations and simulations considering 1-side and 2-side side effects of the nonideal parallel capacitor; (f) The maximum structural FOM ($FOM_{S,max}$) of different structures extracted from FEM simulations. Source: [11]/Springer Nature/CC-BY-4.0.

It is stated that the FOM_P can be considered the common standard to evaluate different TENGs since it is directly proportional to the maximum possible output power and closely related to the highest achievable energy-conversion efficiency, which is universally applicable for TENG systems with various device structures, external circuits, and sizes.

3.3.2 Revised Figure-of-Merits Based on E_{em} Considering Breakdown Discharge Effect

However, the FOM definition raised in 2015 did not consider breakdown discharge effect, which may not be quite different from the real output performance considering the discharge effect may waste a large portion of energy. Xia et al. developed an experimental method to evaluate the E_{em} by removing the energy in breakdown discharge region, which is more related to the real output performance of the TENG. Such an E_{em} can be considered the foundation for standardized evaluation of TENG through revised FOM_S and FOMP, which are defined as [23]:

$$FOM_S = \frac{2\varepsilon_0}{\sigma^2} \frac{E_{em}}{Ax_{max}} \tag{3.8}$$

$$FOM_P = FOM_S \cdot \sigma^2 = 2\varepsilon_0 \frac{E_{em}}{Ax_{max}} \tag{3.9}$$

Based on the experimental data, the example CS mode and CFT mode TENGs in that work were evaluated with FOM_S of 0.077754 and 2.40115, respectively. The high FOM_S of CFT TENG originates from the small capacitance and double-side triboelectric charges for enhanced electrostatic induction [11]. However, considering the severe breakdown effect, the low threshold charge density is still the key limiting factor for CFT TENG.

3.3.3 Standardized Assessment of a Piezoelectric Nanogenerator, Compared with TENGs

To demonstrate the universal applicability of the proposed standardized assessment method, Xia et al. also applied the method to evaluate a piezoelectric nanogenerator (PENG), as shown in Figure 3.5. The evaluated PENG was made by a polyvinylidene difluoride (PVDF) film with a thickness of 28 μm and electrodes on both sides. The original PENG was fully packaged, so the measured *Q-V* curve shows no breakdown discharge, with the capacitance of 565 pF as extracted from the slope, as shown in Figure 3.5a. This is because the dielectric materials usually have a very high threshold breakdown electric field, such as 340 MV/m for PVDF [32], which makes the breakdown voltage high even for a thin film (estimated as 9520 V for the tested sample). If we cut the film to remove the packaged structure, breakdown discharge may happen along the cut surface due to the low threshold electric field in the air. With that, the breakdown points can be measured as marked by green arrows in Figure 3.5b, with the visible sparks for breakdown discharge shown in inset of Figure 3.5c. The CMEO of the PENG is shown in Figure 3.5c, which shows the open-circuit voltage of this PENG is only about 170V, which is still far below the breakdown point (>400 V). The low voltage output of PENG is related to the large capacitance of the thin film structure, despite the high charge density achieved through piezoelectric effect. However, the breakdown discharge may still happen in PENG or piezoelectric materials, especially for those with power-promoting circuits used [32–36]. By using the identical standardized method, we can also calculate the FOM_S of the PENG as 0.000262077, which is relatively small due to the low voltage. The comparison of FOM_S between CS-TENG, CFT-TENG, and PENG is shown in Figure 3.5d. However, according to the relatively high charge density, the material FOM (FOM_M) of the PENG is much larger than that of TENGs, as shown in Figure 3.5e. The overall performance FOM (FOM_P) is plotted in Figure 3.5f

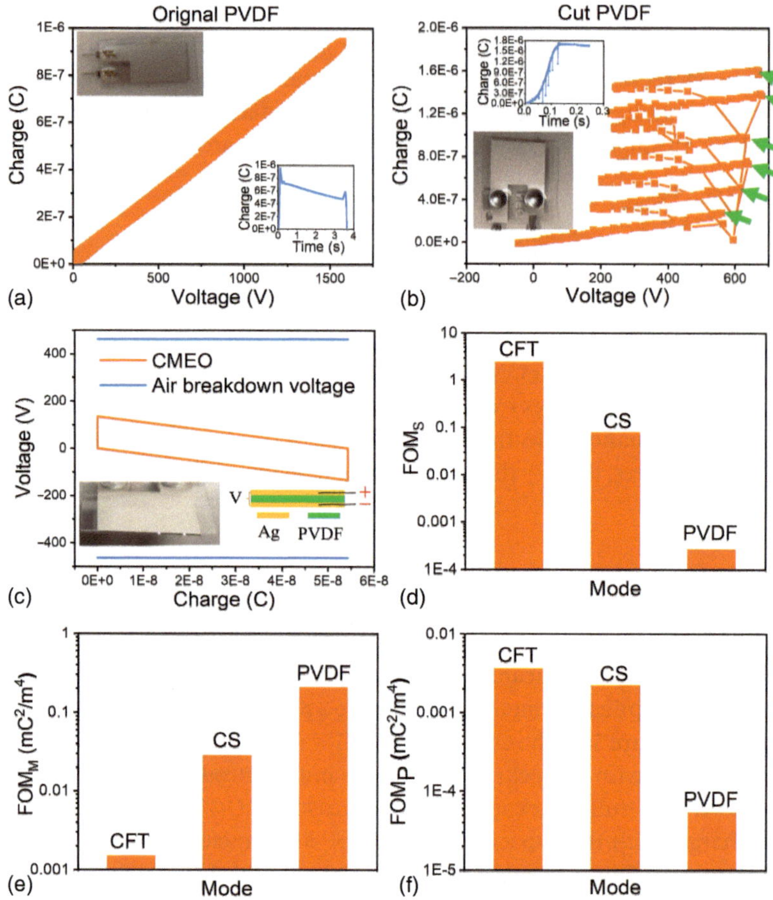

Figure 3.5 Further applications of the proposed method. $Q-V$ plots of (a) PVDF-film-based PENG without cutting, no breakdown happens; and (b) PVDF piezo-film after cutting, with blue lines of transfer charge and figures of experiment devices inset. Point targeted by green arrow in (b) is breakdown point. (c) Cycle for maximized energy with breakdown areas of PVDF film after cutting with a schematic diagram and figure with visible spark inset in. The measured (d) structural FOM; (e) material FOM and (f) performance FOM of different structures. Source: [23] / Springer Nature / CC BY-4.0.

as a final comparison as well. The comparison of figures-of-merit is summarized in Table 3.2. Such results demonstrate the proposed FOM can be the universal standard to evaluate all kinds of nanogenerators, which can facilitate a comparison of the performance with a well-aligned assessment method.

3.4 Output Energy Density

Energy density is a common key concept for energy storage, which measures the capability of the unit for energy storage. This concept is also widely used to measure

Table 3.2 Revised FOM of different nanogenerators.

Mode	FOM_S	FOM_M (mC^2/m^4)	FOM_P (mC^2/m^4)
CFT	2.40115	0.00148996	3.5776×10^{-3}
CS	0.07775	0.028224	2.19453×10^{-3}
PENG	2.62077×10^{-4}	0.204756	5.3662×10^{-5}

the potential of the energy source for energy supply. For energy harvesting, a more frequently used term is the power density. However, for TENG technology, considering its energy conversion process from kinetic energy to electrostatic energy and then to electricity [37], the overall energy density of the electrostatic energy will determine how much power density can be generated in electricity. Thus, it is important to investigate the output energy density per cycle for TENG. Besides, output energy density may provide a more "fair" evaluation standard for sliding-based TENGs compared to contact-separation-based TENGs, considering that the displacement x in sliding-based TENGs is usually large which makes the FOM_P low, while the advantage of the small volume taken by the sliding-based TENGs cannot be revealed in FOM_P. In the meanwhile, the sliding-based SFT mode TENG usually demonstrates ultrahigh power density and efficiency [38].

By using the maximized effective energy output per cycle, E_{em}, as the potential output energy of TENG, the output energy density can be expressed as [24]:

$$U = \frac{E_m}{V_L} \tag{3.10}$$

where the V_L represents the total operating volume of the TENG. The output energy density of contact-separation-based TENGs versus charge density σ can be easily calculated from the previous theoretical results in Section 3.2.3, with a similar trend to their E_{em}. For sliding-based TENGs, Fu et al. conducted numerical simulations through COMSOL Multiphysics software packages, which determine the breakdown discharge points including the charge transfer Q and the corresponding breakdown voltage V. Through these points, the nonbreakdown region in different charge density σ can be confined. For three typical sliding-based TENGs, the output energy densities of the SFT mode TENGs with dielectric sliding part (SFT-dielectric) and metal sliding part (SFT-metal), as well as the LS mode TENG were calculated, respectively, as shown in Figure 3.6a–c. These TENGs are all with electrode area of $w \times l$ = 25 mm × 25 mm, and the electrode gap g = 4 mm. It is noticed that breakdown discharge may happen not only in the gap between electrodes but also across the static dielectric layer. With that, the output energy density for all types of typical TENGs versus σ are plotted and compared in Figure 3.6d,e. It is clearly shown that the sliding-based TENGs may achieve much higher output energy density considering the obvious air gap during the contact-separation can be avoided. The maximum output energy density is almost ~10^4 J/m^3, which is still limited by the breakdown electric field 2.1×10^7 V/m of the dielectric PDMS

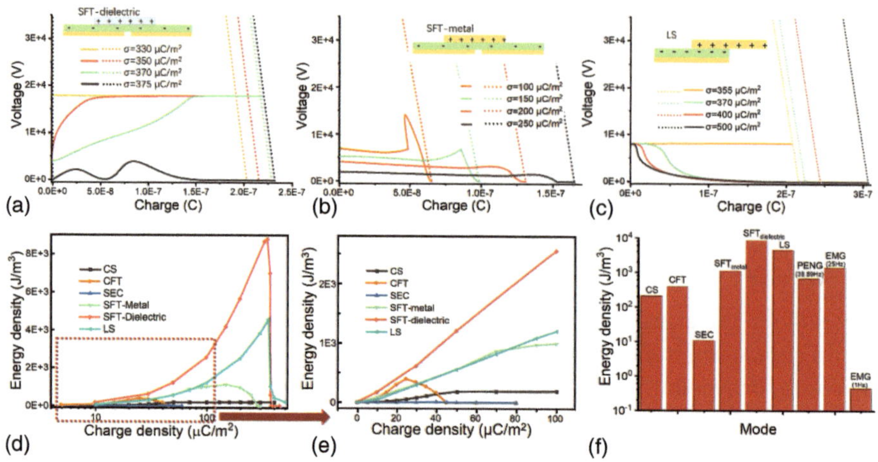

Figure 3.6 The output energy density of TENGs considering the breakdown. (a–c) The Q–V curve of SFT TENG with a dielectric or metal moving part and LS TENG under various charge densities, respectively, in which the external dotted wireframes are the CMEO without considering breakdown, and the areas below the internal line are the energy E_{em} that can be obtained when considering the breakdown effect. The insets are the working principle for each mode. In order to show the breakdown effect more clearly, we only show the local part of Q–V curve which is below 3.5×10^4 V. (d) The output energy density of different modes of TENG with considering the breakdown. In order to show the low charge density part clearly, we plotted the output energy density of all modes of TENG under a charge density below 100 μC/m² in Figure 3.6e. (f) The histogram of the energy density of different energy harvesting devices. Source: Reproduced with permission of [24], 2019 © American Chemical Society.

used in the simulation. If dielectric materials with a high breakdown electric field or high energy density limit can be used, the output energy density can be further enhanced.

Such results can be also compared to other mechanical energy harvesters, such as electromagnetic generators (EMGs) and PENGs. The study by Dong et al. [39] demonstrates a PENG with optimal parameters showing the output energy density of 691.7 J/m³, which is limited by the voltage output. This result is smaller than the maximum value achieved by SFT-TENG. For the commercial EMG from INFINITY SAV, considering its output performance may increase with the frequency [40], we made a comparison for EMG with low frequency (1 Hz) and high frequency (25 Hz), in which the output energy densities are both smaller than TENG. However, with increased rotational frequency, the EMG power output may eventually surpass that of TENG. The calculated output energy densities for TENGs compared with those for EMG and PENG were plotted in Figure 3.6f.

3.5 Environmental and Techno-Economic Analysis

LCA has become a universal new standard to evaluate environmental impacts and economic cost of a product or technology. In energy-harvesting devices, LCA can be adopted to evaluate environmental concerns such as CO_2 emissions from the device material preparation, manufacturing, and operation, as well as the levelized cost of energy (LCOE) of this technology. Such an evaluation can contribute a lot to achieving the worldwide goal of carbon neutrality in 2050–2060, as well as decision-making for investment.

The first LCA study of TENG was conducted by Ahmed et al. [25], based on two high-output TENGs: one is the micro-grating TENG (TENG A) [41] and the other is the radial fine array TENG (TENG B) [27]. Both of them demonstrate high output power density and efficiency. During manufacturing of these two TENGs, the cost of materials and manufacturing process are shown in Figure 3.7a, calculated through TEA, while the corresponding carbon footprints (CO_2 emission contribution) are shown in Figure 3.7b, and the detailed schematics for evaluation are shown in Figure 3.7c. The calculated CO_2 emission factor and LCOE can be compared to other existing technologies as shown in Figure 3.7d,e, showing some advantages of TENGs in environmental and TEA.

Xu et al. did an LCA analysis on the carbon footprint of TENG for ocean-wave energy harvesting [42]. This work proposes a method to inflate CO_2 gas into TENGs for energy harvesting of ocean waves to minimize the carbon footprint. LCA result

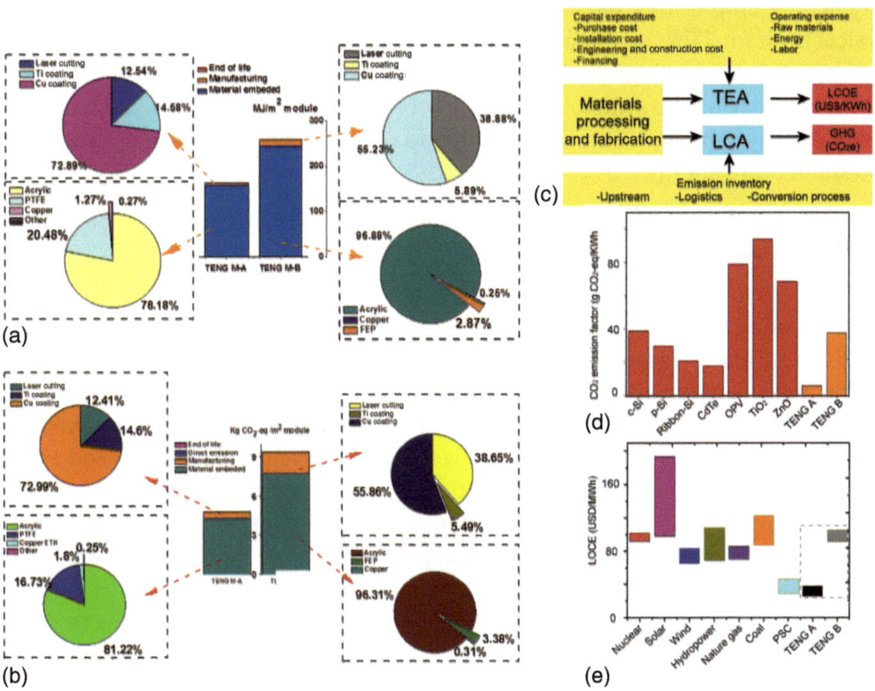

Figure 3.7 The cost and the CO_2 emission distributions of two TENGs. (a) The cost distributions of two TENGs. (b) The CO_2 emission distributions of two TENGs. (c) Schematic of the life-cycle assessment (LCA) and techno-economic analysis (TEA) of TENGs. (d) The LCOE of TENGs with different efficiencies. (e) The Comparison of LCOE between existing technologies. Source: Reproduced with permission of [25], 2017 © Royal Society of Chemistry.

showed that the CO_2 emissions generated by this scheme can be minimal, which can be compensated by the generated electricity, ultimately achieving negative carbon emissions. This research may provide a pathway for applying TENGs in green energy harvesting and attaining the sustainable development goals set by the United Nations and carbon neutrality.

3.6 Potential Applications

The proposed FOM and output energy density can be provided as a standard way to evaluate the output performance of TENG, with different emphasis:

Due to the definition, FOM is more related to the power output density of area and the efficiency, which mainly emphasizes when we have limited energy input, or when the displacement is limited by the energy input, what is the performance of the TENG. In this standard, the performance of TENG is considered to be related to the area, considering the triboelectricity is an interface effect. Therefore, the drawback of this standard is the occupied height, or the volume of the device, which was not emphasized.

Output energy density is more straightforwardly related to the energy output capability, which emphasizes more about the potential of the energy output. It sets a physical limit for TENG in energy output, with the main mechanism of breakdown discharge. It is more related to the volume, and thus it is closer to the practical applications considering the available volume for TENGs is usually limited. Such output energy density can also be easily compared to other energy technologies as a common standard in energy.

3.7 Summary

The standard evaluation of TENG is the key to this technology in output promotion and practical applications. The initial standard: FOM was proposed in 2015 from the maximized energy output, over the area and the displacement, as related to the power output density per area and the efficiency. Considering the breakdown discharge effect, the experimental standardized evaluation method and revised FOM were then proposed in 2019. The output energy density per volume of the TENG was further proposed. As another important standard evaluation method, the environmental and TEA based on LCA was also proposed to evaluate the environmental impact and the LCOE in TENG. These standards and evaluation methods provide pathways to objectively evaluate TENG units and compare them with other energy harvesting technologies [43].

References

1 Yang, Y., Guo, W., Pradel, K.C. et al. (2012). Pyroelectric nanogenerators for harvesting thermoelectric energy. *Nano Lett.* 12 (6): 2833–2838.

2 Yang, Y., Wang, S., Zhang, Y. et al. (2012). Pyroelectric nanogenerators for driving wireless sensors. *Nano Lett.* 12 (12): 6408–6413.

3 Wang, X. (2012). Piezoelectric nanogenerators—harvesting ambient mechanical energy at the nanometer scale. *Nano Energy* 1 (1): 13–24.

4 Hu, Y. and Wang, Z.L. (2015). Recent progress in piezoelectric nanogenerators as a sustainable power source in self-powered systems and active sensors. *Nano Energy* 14: 3–14.

5 Hu, D., Yao, M., Fan, Y. et al. (2019). Strategies to achieve high performance piezoelectric nanogenerators. *Nano Energy* 55: 288–304.

6 Fan, F.-R., Tian, Z.-Q., and Lin Wang, Z. (2012). Flexible triboelectric generator. *Nano Energy* 1 (2): 328–334.

7 Wang, S., Lin, L., and Wang, Z.L. (2015). Triboelectric nanogenerators as self-powered active sensors. *Nano Energy* 11: 436–462.

8 Wang, Z.L. (2013). Triboelectric nanogenerators as new energy technology for self-powered systems and as active mechanical and chemical sensors. *ACS Nano* 7 (11): 9533–9557.

9 Wang, Z.L. (2017). On Maxwell's displacement current for energy and sensors: the origin of nanogenerators. *Mater. Today* 20 (2): 74–82.

10 Zi, Y., Wang, J., Wang, S. et al. (2016). Effective energy storage from a triboelectric nanogenerator. *Nat. Commun.* 7: 10987.

11 Zi, Y., Niu, S., Wang, J. et al. (2015). Standards and figure-of-merits for quantifying the performance of triboelectric nanogenerators. *Nat. Commun.* 6: 8376.

12 Green, M.A., Emery, K., Hishikawa, Y. et al. (2015). Solar cell efficiency tables (Version 45). *Prog. Photovolt. Res. Appl.* 23 (1): 1–9.

13 Mousazadeh, H., Keyhani, A., Javadi, A. et al. (2009). A review of principle and sun-tracking methods for maximizing solar systems output. *Renew. Sustain. Energy Rev.* 13 (8): 1800–1818.

14 Ito, S., Nazeeruddin, M.K., Liska, P. et al. (2006). Photovoltaic characterization of dye-sensitized solar cells: effect of device masking on conversion efficiency. *Prog. Photovolt. Res. Appl.* 14 (7): 589–601.

15 Luque, A., Martí, A., López, N. et al. (2005). Experimental analysis of the quasi-Fermi level split in quantum dot intermediate-band solar cells. *Appl. Phys. Lett.* 87 (8): 083505.

16 Su, Z., Han, M., Cheng, X. et al. (2016). Asymmetrical triboelectric nanogenerator with controllable direct electrostatic discharge. *Adv. Funct. Mater.* 26 (30): 5524–5533.

17 Cui, N., Gu, L., Lei, Y. et al. (2016). Dynamic behavior of the triboelectric charges and structural optimization of the friction layer for a triboelectric nanogenerator. *ACS Nano* 10 (6): 6131–6138.

18 Wang, S., Xie, Y., Niu, S. et al. (2014). Maximum surface charge density for triboelectric nanogenerators achieved by ionized-air injection: methodology and theoretical understanding. *Adv. Mater.* 26 (39): 6720–6728.

19 Lee, K.Y., Chun, J., Lee, J.H. et al. (2014). Hydrophobic sponge structure-based triboelectric nanogenerator. *Adv. Mater.* 26 (29): 5037–5042.

20 Niu, S., Liu, Y., Wang, S. et al. (2013). Theory of sliding-mode triboelectric nanogenerators. *Adv. Mater.* 25 (43): 6184–6193.

21 Niu, S., Liu, Y., Chen, X. et al. (2015). Theory of freestanding triboelectric-layer-based nanogenerators. *Nano Energy* 12: 760–774.

22 Niu, S., Wang, S., Lin, L. et al. (2013). Theoretical study of contact-mode triboelectric nanogenerators as an effective power source. *Energy Environ. Sci.* 6 (12): 3576.

23 Xia, X., Fu, J., and Zi, Y. (2019). A universal standardized method for output capability assessment of nanogenerators. *Nat. Commun.* 10 (1): 4428.

24 Fu, J., Xia, X., Xu, G. et al. (2019). On the maximal output energy density of nanogenerators. *ACS Nano* 13 (11): 13257–13263.

25 Ahmed, A., Hassan, I., Ibn-Mohammed, T. et al. (2017). Environmental life cycle assessment and techno-economic analysis of triboelectric nanogenerators. *Energy Environ. Sci.* 10 (3): 653–671.

26 Chen, J., Zhu, G., Yang, W.Q. et al. (2013). Harmonic-resonator-based triboelectric nanogenerator as a sustainable power source and a self-powered active vibration sensor. *Adv. Mater.* 25 (42): 6094–6099.

27 Zhu, G., Chen, J., Zhang, T.J. et al. (2014). Radial-arrayed rotary electrification for high performance triboelectric generator. *Nat. Commun.* 5: 3426.

28 Li, X.Y., Tao, J., Wang, X.D. et al. (2018). Networks of high performance triboelectric nanogenerators based on liquid-solid interface contact electrification for harvesting low-frequency blue energy. *Adv. Energy Mater.* 8: 1800705.

29 Zi, Y., Wu, C., Ding, W. et al. (2017). Maximized effective energy output of contact-separation-triggered triboelectric nanogenerators as limited by air breakdown. *Adv. Funct. Mater.* 27 (24): 1700049.

30 Husain, E. and Nema, R.S. (1982). Analysis of Paschen curves for air, N_2 and SF_6 using the townsend breakdown equation. *IEEE Trans. Electr. Insul.* EI-17 (4): 350–353.

31 Fu, J., Xu, G., Li, C. et al. (2020). Achieving ultrahigh output energy density of triboelectric nanogenerators in high-pressure gas environment. *Adv. Sci.* 7 (24): 2001757.

32 Wang, Y., Cui, J., Wang, L. et al. (2017). Compositional tailoring effect on electric field distribution for significantly enhanced breakdown strength and restrained conductive loss in sandwich-structured ceramic/polymer nanocomposites. *J. Mater. Chem. A* 5 (9): 4710–4718.

33 Lipscomb, I.P., Weaver, P.M., Swingler, J. et al. (2009). The effect of relative humidity, temperature and electrical field on leakage currents in piezo-ceramic actuators under dc bias. *Sens. Actuators A* 151 (2): 179–186.

34 Saito, W., Kuraguchi, M., Takada, Y. et al. (2005). Design optimization of high breakdown voltage AlGaN–GaN power HEMT on an insulating substrate for $R_{ON}A$-V_B tradeoff characteristics. *IEEE Trans. Electron Devices* 52 (1): 106–111.

35 Wu, L., Do, X.-D., Lee, S.-G. et al. (2017). A self-powered and optimal SSHI circuit integrated with an active rectifier for piezoelectric energy harvesting. *IEEE Trans. Circuits and Syst. I: Regul. Pap.* 64 (3): 537–549.

36 Liang, J. and Liao, W.-H. (2012). Improved design and analysis of self-powered synchronized switch interface circuit for piezoelectric energy harvesting systems. *IEEE Trans. Ind. Electron.* 59 (4): 1950–1960.
37 Xu, G., Li, X., Xia, X. et al. (2019). On the force and energy conversion in triboelectric nanogenerators. *Nano Energy* 59: 154–161.
38 Wang, S., Xie, Y., Niu, S. et al. (2014). Freestanding triboelectric-layer-based nanogenerators for harvesting energy from a moving object or human motion in contact and non-contact modes. *Adv. Mater.* 26 (18): 2818–2824.
39 Dong, X., Yi, Z., Kong, L. et al. (2019). Design, fabrication, and characterization of bimorph micromachined harvester with asymmetrical PZT films. *J. Microelectromech. Syst.* 28 (4): 700–706.
40 Zi, Y., Guo, H., Wen, Z. et al. (2016). Harvesting low-frequency (<5 Hz) irregular mechanical energy: a possible killer application of triboelectric nanogenerator. *ACS Nano* 10 (4): 4797–4805.
41 Zhu, G., Zhou, Y.S., Bai, P. et al. (2014). A shape-adaptive thin-film-based approach for 50% high-efficiency energy generation through micro-grating sliding electrification. *Adv. Mater.* 26 (23): 3788–3796.
42 Xu, G., Li, X., Fu, J. et al. (2023). Environmental lifecycle assessment of CO_2-filled triboelectric nanogenerators to help achieve carbon neutrality. *Energy Environ. Sci.* 16 (5): 2112–2119.
43 Li, X., Xu, G., Xia, X. et al. (2019). Standardization of triboelectric nanogenerators: progress and perspectives. *Nano Energy* 56: 40–55.

4

Output Promotion by Environment

4.1 High Vacuum Environment

The breakdown effect in TENG can be expressed by Paschen's law. As shown in Figure 4.1a, the breakdown voltage has close relationships with the gas pressure and gap distance. For the contact-separation TENG (CS-TENG), the breakdown phenomenon is easier to occur in the air gap between two triboelectric layers, and the gap voltage is positively correlated with the surface charge density. To avoid the breakdown effect in the air gap, the gap voltage should be smaller than the breakdown voltage during the whole working range of the CS-TENG (the common working displacement is between 0–0.01 m) [2–6]. In other words, the curve of air gap voltage and gap distance has no intersection with the breakdown voltage curve. With the gas pressure decreasing to 10 Pa, we can find that the breakdown voltage curve is completely out of the curve of air gap and gap distance, even for the condition of surface charge density reaching to 1003 $\mu C/m^2$ (Figure 4.1a). Based on this principle, Wang et al. demonstrated that air breakdown can be fully eliminated in a high vacuum environment, and then the surface charge density can be improved from 140 $\mu C/m^2$ in atmosphere condition to 660 $\mu C/m^2$ in vacuum ($P \sim 10^{-6}$ torr) (Figure 4.1b) [1]. The corresponding output voltage also increased about five times (Figure 4.1c). To further increase the output performance of TENG, Wang et al. introduced a ferroelectric material (barium titanate material: BT) between the dielectric layer of PTFE and the back electrode. During the contact and separating process of CS-TENG, the ferroelectric material will be gradually polarized. Experimental results exhibit that the surface polarization from triboelectrification will induce built-in dielectric polarization inside the ferroelectric material, and the latter could enhance the former in subsequent contact-separating processes, finally increasing the triboelectric charge density of TENG (Figure 4.1d). Because there is no limit of air breakdown, the output charge density and output voltage are enhanced from 142 $\mu C/m^2$ and 26 V in atmospheric pressure to 1003 $\mu C/m^2$ and 180 V in high vacuum (10^{-6} torr), respectively (Figure 4.1e,f). More importantly, the output charge density can be further enhanced by regulating the dielectric breakdown strength of the dielectric material. As shown in Figure 4.1g, the instantaneous peak power density can reach 50 W/m^2. This work proposes a direction to optimize the output performance of TENG, and promotes the triboelectric charge

Triboelectric Nanogenerators: Technology, Applications, and Commercialization, First Edition.
Yunlong Zi, Hengyu Guo, Jie Wang, Chi Zhang, Xiangyu Chen, and Qing Zhao.

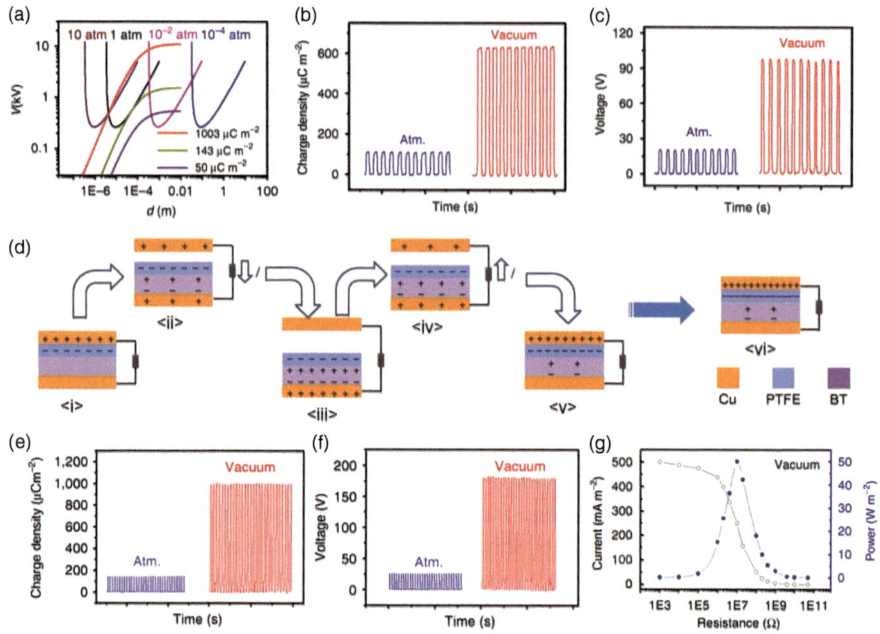

Figure 4.1 Output promotion of TENG by decreasing atmospheric pressure. (a) The breakdown voltage and gap voltage in TENG with different surface charge densities. (b) Output charge of CS-TENG in atmosphere and vacuum condition. (c) Output voltage of CS-TENG in atmosphere and vacuum condition. (d) Working mechanism of CS-TENG with the consideration of dielectric polarization inside the ferroelectric material. (e) Output charge of CS-TENG with BT material in atmosphere and vacuum condition. (f) Output voltage of CS-TENG with BT material in atmosphere and vacuum condition. (g) Output current and power of CS-TENG with BT material at various resistances in vacuum condition. Reproduced with permission of Ref. [1], 2017 © Springer Nature.

density to a new record-high value, which is much higher than previously expected. Moreover, it also aroused great interest in performance promotion of TENG.

4.2 High Atmospheric Pressure and High-Breakdown-Limit Gas Environments

Since air breakdown effect is closely related to the ambient environment, Zi et al. explored the air breakdown limits of TENG in air with different pressures and high-breakdown-limit gas environments [7]. In 2017, they first analyzed the breakdown effect in CS-TENG with theoretical calculation. Figure 4.2a plots the curves of breakdown voltage against gap distance depicted by Paschen's law in the conditions with different gas pressures and SF_6 gas. It is clear that Paschen's curve shifts toward left with the increase of gas pressure. The breakdown voltage also increases with the gas pressure increasing within most of gap distance because electrons could accumulate smaller kinetic energy in a lower mean-free path. In other words, at the same air gap, a larger voltage is needed to break down the air gap. Moreover, the breakdown voltage in SF_6 gas is always higher than the breakdown voltage in air condition because SF_6 has a higher molecular weight, which is difficult to be ionized by electron collision. The results indicate that TENG working in SF_6 gas condition could produce a higher output performance. Based on this theory, they also calculated the maximum output energy density of TENG during one working cycle under different air conditions. As depicted in Figure 4.2b, the maximum energy density of CS-TENG increases with the increase in gas pressure, and the maximum energy density in SF_6 gas is always higher than the value in air condition because of the higher critical breakdown threshold of SF_6 gas. These results propose a potential solution from the point of view of theoretical calculation to improve the output performance of TENG by fully packaging TENG in high-pressure gas or high-breakdown-limit gas environments.

To demonstrate this solution and further realize a high output performance of TENG, they also did a series of experiments in the high-pressure gas environment to suppress the breakdown effect [8]. As shown in Figure 4.2c, they built a high-pressure reactor to realize a reliable and stable working environment, and a small linear motor was set in the chamber to control the movement of TENG. The whole reactor was well sealed without the leakage issue. Townsend's electron avalanche model indicates the gas breakdown effect is inevitable in TENGs due to air gaps between two tribosurfaces or two electrodes. From the perspective of reducing the mean free path of electrons, increasing gas pressure of the working environment for TENG is an effective method, and thus the air breakdown effect is suppressed. For a CS-TENG, the Kapton was selected as the dielectric layer to contact with the copper. As depicted in Figure 4.2d, the output charge gradually increases with the gas pressure increasing, demonstrating the feasibility of enhancing TENG's performance by increasing gas pressure. They also measured the energy density of CS-TENG at different pressures, and the results indicate that the maximum energy density is increased 25 times at 10 atm (Figure 4.2e). Then, they also measured the output charge of sliding-freestanding mode TENG (SFT-TENG) at different atmospheric pressures. As shown in Figure 4.2f, the output charge increases with the atmospheric pressure and gradually reaches stability, which may be caused by the complex breakdown effect in SFT-TENG. The corresponding maximum energy density of SFT-TENG at 6.5 atm is five times that value at 1 atm (Figure 4.2g). In other words, this work provides a practical and effective solution to greatly enhance the output performance of all types of TENGs that are limited by breakdown effect.

Besides air breakdown effect, the relationship between contact electrification and ambient has also been demonstrated. Lin et al. installed the AFM equipment into a vacuum chamber to investigate atmospheric effects on contact electrification at nanoscale [9]. In order to control the atmosphere in contact electrification, the AFM equipment was installed into a vacuum chamber where pure gases (O_2, N_2, and Ar) can be refilled with gas pressure varying from 1×10^{-3} $Pa^{-1} \times 10^5$ Pa (Figure 4.2h). The transferred charge density on the dielectric surfaces (SiO_2, AlN, polyvinyl chloride (PVA), and polymethyl methacrylate (PMMA)) after being contacted by the Pt/Ir coated tip in different atmospheres at different gas pressures are shown in

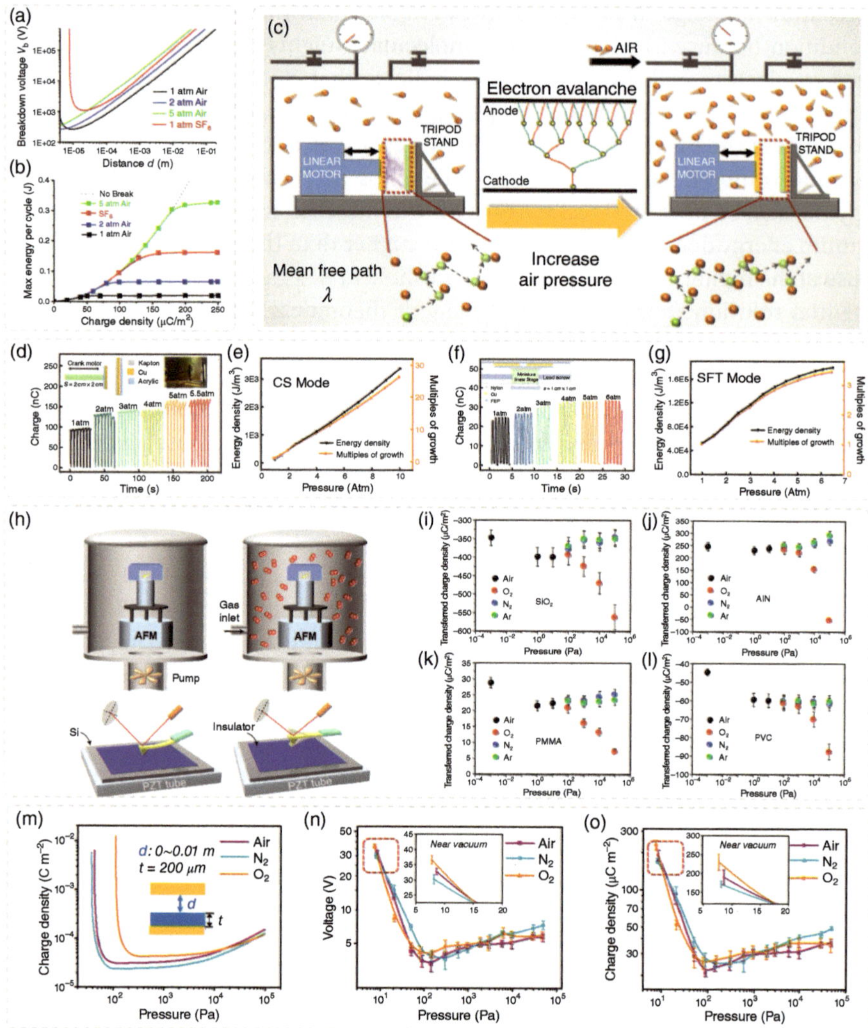

Figure 4.2 Output promotion of TENG by increasing atmospheric pressure and adopting breakdown-resistant gas. (a) Breakdown voltage in different gas conditions. (b) Maximum energy density of CS-TENG in different gas conditions. Reproduced with permission of Ref. [7], 2017 © John Wiley & Sons. (c) Schematic diagram shows the experimental facility and the working mechanism for suppressing breakdown effect in high-pressure gas conditions. (d) Output charge of CS-TENG in short-circuit conditions at different atmospheric pressures. (e) Output energy density of CS-TENG at various atmospheric pressures. (f) Output charge of SFT-TENG in short-circuit conditions at different atmospheric pressures. (g) Output energy density of SFT-TENG at various atmospheric pressures. Reproduced with permission of Ref. [8], 2020 © John wiley & Sons. (h) Schematic illustration of the AFM experiments in the vacuum chamber. (i–l) The transferred charge density on the (i) SiO_2, (j) AlN, (k) PMMA, and (l) PVC surfaces as induced by contacting with the tip in O_2, N_2, Ar, and air atmospheres at different gas pressures. Reproduced with permission of Ref. [9], 2019 © Elsevier. (m) Theoretical maximum surface charge density of CS-TENG in different gas environments. (n) Open-circuit voltage of the CS-TENG in different gas environments. (o) Output surface charge density of the CS-TENG in different gas environments. Reproduced with permission of Ref. [10], 2021 © Elsevier.

Figure 4.2i–l. Clearly, the triboelectric charge on all the dielectric surfaces becomes more negative when the air pressure changes from 1×10^{-3} Pa^{-10} Pa. Besides, the transferred charge on all the dielectric surfaces goes negative more significantly in O_2 atmosphere while is not significant in of N_2 and Ar. The negative charge density on the SiO_2 and PVC surfaces increases, and the positive charge density on the PMMA surface decreases with the increase of O_2 gas pressure. In particular, the transferred charges on the AlN surface are positive before the O_2 is refilled into the chamber, and the positive charge density decreases when the O_2 gas pressure increases. When the O_2 gas pressure reaches 1×10^5 Pa, the transferred charge polarity on the AlN surface reverses from positive to negative. Based on the experiment results, the O_2 in the air might play a key role in the effect of air pressure on the contact electrification because the transferred charge on the dielectric surfaces also becomes more negative when the air pressure increases from 1×10^{-3} Pa^{-10} Pa, which is consistent with the effect of O_2 on the contact electrification. The results suggest that N_2 and Ar cannot change the contact electrification properties of dielectrics significantly, while the adsorption of O_2 can make dielectrics more likely to be negatively charged in contact electrification.

The positive influence of O_2 on contact electrification was also verified at a large-scale TENG device. Yi et al. examined the output performance of TENG under a controlled atmosphere [10]. The curves of the allowed maximum surface charge density at various gas pressures for air, N_2, and O_2 are plotted in Figure 4.2m. To reduce the instability and measuring error as much as possible, the output performance of the CS-TENG was measured and recorded only when the gas pressure and the output values were stable. By using PTFE and Cu as trio-materials, the open-circuit voltage and output charge density of the CS-TENG are shown in Figure 4.2n,o, respectively. It can be seen that the surface charge density first declines gradually, reaches its minimum, and then increases rapidly with the decrease of gas pressure. As plotted in the insets of Figure 4.2n,o, in a near-vacuum environment where the negative effect of electrostatic breakdown is negligible, CS-TENG in O_2 atmosphere exhibits the highest output values among the three gases, indicating an improvement of oxygen on the contact electrification of TENG. Therefore, the output performance of TENG can also be improved by changing the O_2 concentration in the high atmospheric pressure.

4.3 Interfacial Liquid Lubrication

Interface liquid is the other promising approach to avoid air breakdown effect, thus improving the output performance. Shi and coworkers employed a liquid lubrication strategy to increase the service life of sliding-mode TENG (S-TENG) for the first time, which found that the output performance of lubricated TENG can also be improved compared to unlubricated TENG [11]. By using polyimide (PI) film and Al film as contact materials, the authors studied the effect of applied load on the output performance of S-TENG by adding lubricants (Figure 4.3a). The results displayed that the voltage output of S-TENG has 2.4–3.5 times improved with the lubrication

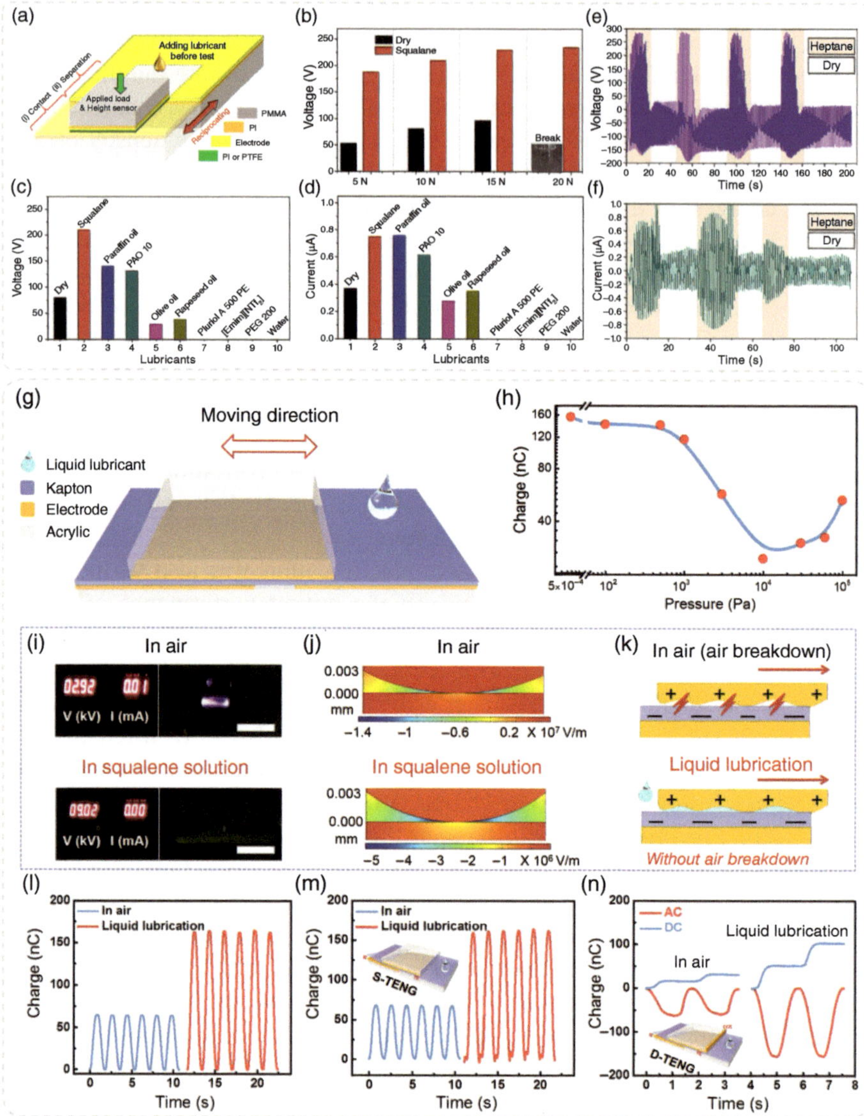

Figure 4.3 Output promotion of TENG by interface liquid lubrication. (a) Schematic diagram of the slide-mode TENG by adding lubricant. (b) Influence of the applied load on voltage for PI-Al contact under dry and squalane-lubricated conditions. (c) Voltage and (d) current for dry and different liquid-lubricated TENG under 10 N. (e,f) Influence of adding/removing heptane to/from PI-Al contact on (e) voltage and (f) current. Reproduced with permission of Ref. [11], 2021 © Elsevier. (g) Structural schematic of sliding FS-TENG with interface liquid lubrication. (h) Transferred charges of sliding FS-TENG at various atmospheric pressures. (i) Practical pictures of breakdown between two electrodes in air at 2.9 kV and without breakdown in liquid lubricant solution at 9 kV (scale bar, 0.4 cm). (j) Finite-element simulation of the electric field distribution in microgap between triboelectric electrode and dielectric film in air atmosphere or liquid lubricant solution (the surface charge density of each FEP film was 50 μC/m^2). (k) The schematic of interfacial breakdown in air and avoiding interface breakdown through interface liquid lubrication during sliding motion. (l–n) Transferred charge of the (l) FS-TENG, (m) S-TENG, and (n) dual mode AC/DC-TENG in air and with interface liquid lubrication. Reproduced with permission of Ref. [12], 2020 © John Wiley & sons.

of squalene than that with dry contact under different applied forces (Figure 4.3b). Moreover, the influence of different liquids on the output performance of S-TENG was also conducted, and the results were compared as shown in Figure 4.3c,d. Interestingly, the voltage and current output of S-TENG lubricated by squalane, paraffin oil, and PAO10 have been improved, and squalene-lubricated S-TENG has the highest enhancement. Other liquids such as Pluriol A 500 PE, PEG 200, [Emim][NTf_2] ionic liquid, and water exhibit little effect on the output performance of S-TENG.

To reveal the mechanism of liquid lubrication on TENG, Zhou et al. systematically studied the relationship between interface liquid lubrication and air breakdown effect in TENG [12]. Here, squalene was added to the interface of FS-TENG as a liquid lubricant, and Cu and Kapton film were employed as tribomaterials (Figure 4.3g). Through comparing the capacitance of fabricated TENG device and the capacitor with deposited electrode, it is found that the capacitance of practical device is only half as much as the latter, indicating part contacts of the two triboelectric layers in practical operation and the existence of large air voids at the triboelectric interface. Considering the presence of air voids, the breakdown effect in FS-TENG was investigated by using a vacuum control system. The output charge of FS-TENG at various atmospheric pressures is shown in Figure 4.3h. With the decline of atmospheric pressure, the transferred charge shows a first decreasing and then increasing tendency, finally closing to a stable value. This tendency with changes in atmospheric pressure highly matches with variation of air breakdown voltage based on Paschens' law, demonstrating the loss of electrostatic charges induced by interfacial air breakdown on triboelectric interface. The above results highly verify that interfacial breakdown induced by poor contact veritably occurs in sliding-mode TENG, causing a significant reduction in its output performance. It is worth noting that due to the suppression of air breakdown under a higher vacuum environment, the electrostatic charge loss would be significantly reduced, which enables charge accumulation on the surface of Kapton film during continuous triboelectrification process. So, a stable and enhanced output performance can be realized for the sliding-mode TENG. Then, the authors investigated the influence of squalane on breakdown effect due to its characteristics of low polarity and high dielectric constant than air. The comparison of breakdown effect between two electrodes in the media of air and liquid lubricant is shown in Figure 4.3i. At the same gap distance, obvious breakdown phenomenon occurs at 2.9 kV in atmospheric environment, while there is no breakdown in liquid lubricant even at 9 kV, which demonstrates that breakdown is harder to take place in liquid lubricant compared to in the atmosphere. Additionally, the simulated results at the same contact state indicate that the electrostatic field strength in microgap between triboelectric electrode and dielectric film can be weakened in one order of magnitude when liquid lubricant replaces the air (Figure 4.3j). These results indicate that as compared to air, liquid lubricant can suppress breakdown effect by improving breakdown field strength requirement and lowering electrostatic field strength in the microgap between the two triboelectric layers, which could enormously reduce electrostatic charge loss and improve charge accumulation on triboelectric layers at the same contact state (Figure 4.3k).

Based on the above results and analysis, the authors measured the suppression breakdown effect of liquid lubrication on performance of all-sliding TENG including FS-TENG, S-TENG, and dual-mode AC/DC-TENG (D-TENG) under 10 N. Compared with normal device, all the corresponding transferred charges of FS-TENG, S-TENG, and D-TENG of lubricated device increase more than 2.5 times under 10 N (Figure 4.3l–n), respectively. These results show that interface liquid lubrication can truly suppress interface breakdown, thus improving the electric outputs of sliding TENG. Therefore, this work verified that the output performance of sliding TENG can be largely improved by using interface liquid lubrication through suppressing breakdown effect in the microgap between the two triboelectric layers.

4.4 Humidity

As previous works reported, water always plays a negative role in triboelectrification for CS-TENG [13–18]. The possible reasons are as follows: (i) The water molecules adsorbed on the friction surface in a high humidity environment will dissipate the triboelectric charge through the conductive path formed by water [13–17]; (ii) Water molecules adsorbed between the metal/polymer interfaces, due to the existence of hydrogen bonds, replacing electron donors and reducing electron transfer, decreasing charge transferred under humid conditions [18]. Although humidity plays a negative role in triboelectrification in CS-TENG, its effect on S-TENG and FS-TENG is different. Hu et al. first reported that for S-TENG with Cu and PI as tribomaterials, the transferred charge initially increases with increasing relative humidity and then decreases beyond a certain value, indicating an optimal working humidity [17]. A similar result is further verified by investigating the open-circuit voltage and transferred charge of a capsule TENG with a PVC-Cu pair under different humidity (Figure 4.4a) [19]. The possible explanation is that in a low-humidity environment, electron transfer only occurs in the regions where the PVC and the Cu surfaces are convex because the PVC and the Cu can only touch each other in these regions (Figure 4.4ci), which limits the total number of electrons transferred between the PVC and the Cu film, thus resulting in a relatively low output at low RH values. As the relative humidity increases, the dispersed water bridge is formed between the friction materials, filling the grooves on the tribosurfaces under microscopic view (Figure 4.4cii,iii), providing more channels for transferred electrons (or negative charges) (Figure 4.4civ,v), thus resulting in higher output performance. When the relative humidity is further increased, a continuous water layer forms on the surface, resulting in a decrease in the number of transferred electrons.

Besides, the effect of humidity on triboelectrification also depends on the materials occupied. As previously reported, with the relative humidity increasing from 40% to 80%, the output voltage of a TENG based on a hydrophilic polyamide film decays logarithmically [22], while the voltage changes of a TENG based on a hydrophobic polytetrafluoroethylene film is negligible. Consequently, the changes in output performance under humid conditions are possibly influenced by the surface-free energy and hydrophobicity of the polymer. As the humidity increases, water

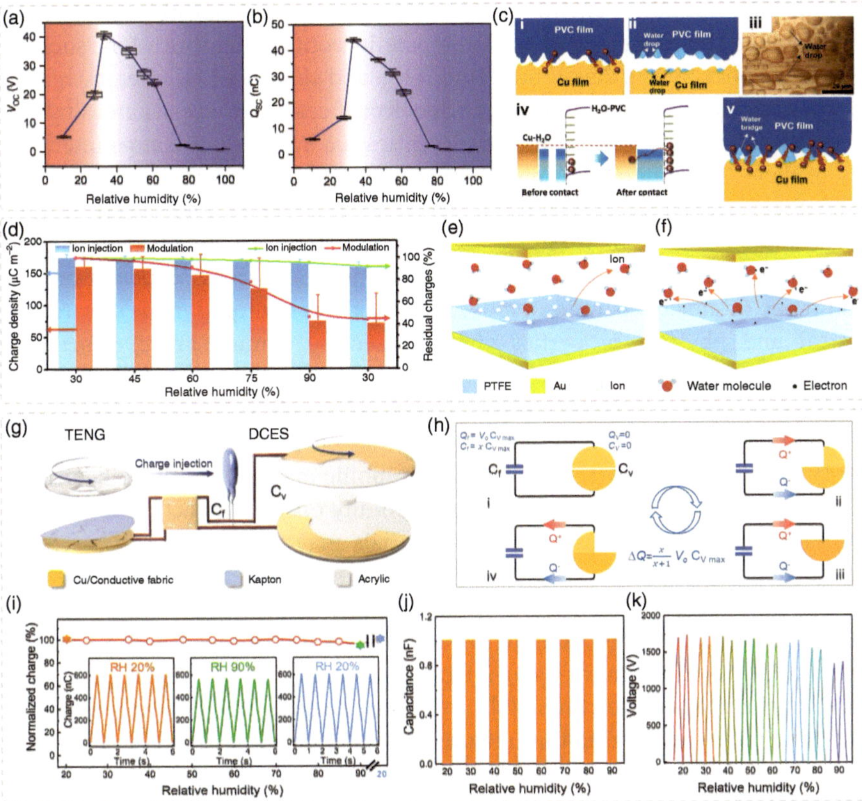

Figure 4.4 Output promotion of TENG by humidity. (a,b) Electrical output performances of a capsule TENG with a PVC-Cu pair including (a) V_{oc}-RH and (b) Q_{sc}-RH relationships. (c) Schematic mechanisms for the effect of humidity on triboelectrification: (i) Electron transfer from the Cu film to the PVC without the influence of a water film. Schematic showing electron transfer that only occurs in the regions where the Cu film is in direct contact with the PVC film. (ii) Very thin, discontinuous water layers in the concave regions of PVC and Cu film surfaces. (iii) Optical micrograph of the Cu film with a small number of water microdroplets on the concave surface. (iv) Schematic of charge transfer from the Cu to the PVC through the water film. (v) Electron transfer from the Cu film to the PVC with the addition of a "water bridge". Reproduced with permission of Ref. [19], 2020 © Elsevier. (d) Charge density and residual surface charges introduced through the ion injection and modulation approaches on the PTFE surface under different relative humidity conditions. (e,f) The schematic of the dissipation of (e) ionic charges and (f) electronic charges in a humid atmosphere. Reproduced with permission of Ref. [20], 2021 © Royal Society of Chemistry. (g) Structural schematic of DCE-TENG. (h) Charge transferred between the C_v and C_f. (i–k) Humidity resistance performance and durability of the DCE-TENG. (i) Transferred charge of DCE-TENG, (j) the capacitance of C_v, and (k) open-circuit voltage of TENG under different humidity. Reproduced with permission of Ref. [21], 2021 © John Wiley & Sons.

molecules adsorb onto the surface of hydrophilic materials, creating a conductive water layer that dissipates surface charge. Conversely, hydrophobic surfaces hinder water molecule adsorption and effectively inhibit water molecule spreading on the surface, maintaining stable output performance in high-humidity environments [23, 24]. Additionally, larger surface contact angles can suppress surface discharge effects in high-humidity environments [17]. Some water-adsorbing materials can immobilize water molecules to participate in contact electrification, such as polyvinyl alcohol [25–27], starch molecules [28], and cellulose [29], thereby reducing the negative impact of humidity on triboelectrification, or even increasing the output performance of TENG [16]. The electrification and surface charge retention ability of dielectric materials in high-humidity environments are determined not only by their hydrophobicity but also by their inherent contact electrification properties. In summary, the presence of adsorbed water is crucial for the charging processes in TENGs. Its role in the triboelectrification process depends on the motion mode (contact-separation mode, sliding or free-standing mode), materials (chemical and physical properties characterized by contact angle, functional group, roughness, and inherent triboelectrification ability), and other operational parameters. Generally, in contact-separation mode, humidity has a negative effect on triboelectrification, while the influence of the humidity on the sliding-mode TENG is less or even positive. The use of hydrophilic materials as the triboelectric layer may lead to a negative effect of high humidity on the contact electrification effect, whereas the use of hydrophobic materials may result in an ignorable or positive effect.

To avoid the negative influence of humidity, surface-charge engineering is verified as an alternative approach. Liu et al. make the dielectric surface acquire negative charge with ions or electrons by using ion injection or modulation, and then studied the effect of charge species on the stability of surface charges [20]. Under 90% humidity, surface charges introduced through the modulation approach largely decline to 43% of the initial value while surface charges obtained through ion injection exhibit excellent stability, maintaining 95% of the initial value (Figure 4.4d). The remarkable difference in the decay trend of charges preloaded through the above surface charge engineering under humid conditions indicates that ions are more stable compared to electrons in a high-humidity environment (Figure 4.4e,f). This may be due to the fact that ions forming covalent bonds with the surface atoms of dielectric materials are quite hard to dissipate through the conductive path of water molecules.

One of the other promising strategies to achieve high output perforce for TENG is using a dual-capacitor enhancement system reported by Zhou et al. [21]. The basic structure of dual-capacitor enhancement TENG (DCE-TENG) mainly consists of a rotary FS-TENG, a rectifier, and a dual-capacitor enhancement system including a designed variable capacitor (C_v) and a commercial ceramic capacitor as a fixed capacitor (C_f) (Figure 4.4g). Unlike the conventional TENG, the working mechanism of DCE-TENG is based on the couple processes of low charge injection from TENG and charge transferred between no-wear C_v and C_f (Figure 4.4h). Noteworthily, the value of C_v exhibits much less sensitivity to the influence of moisture compared to triboelectrification and electrostatic induction process. Then, the output performance of DCE-TENG under various relative humidity conditions was investigated (Figure 4.4i). As the relative humidity rises from 20% to 90%, the transferred charge of DCE-TENG only declines from 600 to 570 nC, indicating a high retention value of 95% compared to the output under 20% relative humidity. More importantly, the transferred charge quickly returns to its initial value when the relative humidity gets back to 20%. Additionally, the influence of relative humidity on the capacitance of C_v and output performance of TENG is also measured. Clearly, the capacitance of C_v maintains a stable value with the relative humidity varying from 20% to 90% (Figure 4.4j). Besides, although the open-circuit voltage of TENG exhibits a slight decline under the high humidity of 90% (Figure 4.4k), the voltage is still far more than 600 V, which is enough to ensure a stable output of DCE-TENG. The stable value of TENG under different relative humidity might be owing to the continuous contact state of two friction surfaces, and thus greatly reduces the negative effect from high relative humidity. All the above results suggest that due to the less influence of humidity on charge injection from TENG and charge transferring between C_v and C_f, DCE-TENG can work with a stable output even under a high-humidity environment.

4.5 Summary

In this chapter, the methods for output promotion of TENG by environment are presented including high vacuum environment, high-pressure gas environment and high-breakdown-limit gas environment, interfacial lubrication, and suppressing the negative influence of humidity. By regulating the working environment of TENG, the air breakdown process in TENG can be suppressed or even fully avoided, so the maximum surface charge density and energy density of TENG can be enhanced. Besides, by using hydrophobic materials, surface-charge engineering and a dual-capacitor enhancement system, the negative effect of humidity on TENG will be changed to an ignorable or positive effect, thus achieving a high performance of TENG in a humidity environment. Combination of advanced encapsulated technologies, these methods can greatly promote the practical applications of TENG.

References

1 Wang, J., Wu, C.S., Dai, Y.J. et al. (2017). Achieving ultrahigh triboelectric charge density for efficient energy harvesting. *Nat. Commun.* 8: 88.

2 Wang, S.H., Xie, Y.N., Niu, S.M. et al. (2014). Maximum surface charge density for triboelectric nanogenerators achieved by ionized-air injection: Methodology and theoretical understanding. *Adv. Mater.* 26: 6720.

3 Zhang, C.L., Zhou, L.L., Cheng, P. et al. (2020). Surface charge density of triboelectric nanogenerators: theoretical boundary and optimization methodology. *Appl. Mater. Today* 18: 100496.

4 Fu, J.J., Xia, X., Xu, G.Q. et al. (2019). On the maximal output energy density of nanogenerators. *ACS Nano* 13: 13257.

5 Zi, Y.L., Niu, S.M., Wang, J. et al. (2015). Standards and figure-of-merits for quantifying the performance of triboelectric nanogenerators. *Nat. Commun.* 6: 8376.

6 Xia, X., Fu, J.J., and Zi, Y.L. (2019). A universal standardized method for output capability assessment of nanogenerators. *Nat. Commun.* 10: 4428.

7 Zi, Y.L., Wu, C.S., Ding, W.B., and Wang, Z.L. (2017). Maximized effective energy output of contact-separationtriggered triboelectric nanogenerators as limited by air breakdown. *Adv. Funct. Mater.* 27: 1700049.

8 Fu, J.J., Xu, G.Q., Li, C.H. et al. (2020). Achieving ultrahigh output energy density of triboelectric nanogenerators in high-pressure gas environment. *Adv. Sci.* 7: 2001757.

9 Lin, S.Q., Xu, L., Tang, W. et al. (2019). Electron transfer in nano-scale contact electrification: Atmosphere effect on the surface states of dielectrics. *Nano Energy* 65: 103956.

10 Yi, Z., Liu, D., Zhou, L. et al. (2021). Enhancing output performance of direct-current triboelectric nanogenerator under controlled atmosphere. *Nano Energy* 84: 105864.

11 Wu, J., Xi, Y., and Shi, Y. (2020). Toward wear-resistive, highly durable and high performance triboelectric nanogenerator through interface liquid lubrication. *Nano Energy* 72: 104659.

12 Zhou, L., Liu, D., Zhao, Z. et al. (2020). Simultaneously enhancing power density and durability of sliding-mode triboelectric nanogenerator via interface liquid lubrication. *Adv. Energy Mater.* 10: 2002920.

13 Nguyen, V. and Yang, R. (2013). Effect of humidity and pressure on the triboelectric nanogenerator. *Nano Energy* 2: 604–608.

14 Zhang, X.S., Han, M.D., Wang, R.X. et al. (2014). High-performance triboelectric nanogenerator with enhanced energy density based on single-step fluorocarbon plasma treatment. *Nano Energy* 4: 123–131.

15 Nguyen, V., Zhu, R., and Yang, R. (2015). Environmental effects on nanogenerators. *Nano Energy* 14: 49–61.

16 Wen, R., Guo, J., Yu, A. et al. (2019). Humidity-resistive triboelectric nanogenerator fabricated using metal organic framework composite. *Adv. Funct. Mater.* 29: 1807655.

17 Hu, Y., Wang, X., Li, H. et al. (2020). Effect of humidity on tribological properties and electrification performance of sliding-mode triboelectric nanogenerator. *Nano Energy* 71: 104640.

18 Li, L., Wang, X., Zhu, P. et al. (2020). The electron transfer mechanism between metal and amorphous polymers in humidity environment for triboelectric nanogenerator. *Nano Energy* 70: 104476.

19 Wang, K., Qiu, Z., Wang, J. et al. (2022). Effect of relative humidity on the enhancement of the triboelectrification efficiency utilizing water bridges between triboelectric materials. *Nano Energy* 93: 106880.

20 Liu, L., Zhou, L., Zhang, C. et al. (2021). A high humidity-resistive triboelectric nanogenerator via coupling of dielectric material selection and surface-charge engineering. *J. Mater. Chem. A* 9: 21357–21365.

21 Zhou, L., Gao, Y., Liu, D. et al. (2021). Achieving ultrarobust and humidity-resistant triboelectric nanogenerator by dual-capacitor enhancement system. *Adv. Energy Mater.* 11: 2101958.

22 Zhang, H., Yang, Y., Su, Y. et al. (2013). Triboelectric nanogenerator as self-powered active sensors for detecting liquid/gaseous water/ethanol. *Nano Energy* 2: 693–701.

23 Lee, K.Y., Chun, J., Lee, J. et al. (2014). Hydrophobic sponge structure-based triboelectric nanogenerator. *Adv. Mater.* 26: 5037–5042.

24 Bui, V.T., Oh, J.H., Kim, J.N. et al. (2020). Nest-inspired nanosponge-Cu woven mesh hybrid for ultrastable and high-power triboelectric nanogenerator. *Nano Energy* 71: 104561.

25 Du, W., Nie, J., Ren, Z. et al. (2018). Inflammation-free and gas-permeable on-skin triboelectric nanogenerator using soluble nanofibers. *Nano Energy* 51: 260–269.

26 Liu, D., Liu, J., Yang, M. et al. (2021). Performance enhanced triboelectric nanogenerator by taking advantage of water in humid environments. *Nano Energy* 88: 106303.

27 Wang, N., Feng, Y., Zheng, Y. et al. (2021). New hydrogen bonding enhanced polyvinyl alcohol based self-charged medical mask with superior charge retention and moisture resistance performances. *Adv. Funct. Mater.* 31: 2009172.

28 Wang, N., Zheng, Y., Feng, Y. et al. (2020). Biofilm material based triboelectric nanogenerator with high output performance in 95% humidity environment. *Nano Energy* 77: 105088.

29 Zhang, R., Dahlström, C., Zou, H. et al. (2020). Cellulose-based fully green triboelectric nanogenerators with output power density of 300 W m^{-2}. *Adv. Mater.* 32: 2002824.

5

DC-TENG: A New Paradigm

5.1 Introduction

Electrostatic discharge is a ubiquitous phenomenon in our daily life, which is a sudden flow of electricity between two charged materials caused by contact or dielectric breakdown [1, 2]. For instance, during the dry winters, the body accumulates static electricity easily, and it is very prone to discharge when it contacts with conductors or other people. Generally, electrical discharge is considered a negative effect in the electronic industry and often causes electrostatic discharge failure of electronic components and integrated circuits. Intensive work has been dedicated to avoiding electrostatic discharge to protect safety of electronic components and instruments [3].

In conventional triboelectric nanogenerators (TENGs), electrostatic breakdown existing between the two triboelectric layers still has a negative effect on the output performance of TENG because the triboelectric charges will be released by electrostatic breakdown and then reduce the induced charges in external circuit. This phenomenon has been demonstrated in many previous works [4, 5]. Wang et al. fabricated a flexible TENG with silicone rubber as a triboelectric layer and a mixture of silicon rubber, carbon black, and carbon nanotubes as a triboelectric electrode [6]. With optimized structural design, the triboelectric charge density is up to 250 $\mu C/m^2$. In addition, by designing a three-layer TENG, the triboelectric charge density still increases to around 250 $\mu C/m^2$. However, in high-vacuum environment, the triboelectric charge density of the polytetrafluoroethylene (PTFE) film can be up to 660 $\mu C/m^2$, while the measured charge density of the same dielectric material is only around 120 $\mu C/m^2$ in atmospheric condition [7]. By further coupling surface polarization from triboelectrification (TE) and hysteretic dielectric polarization from ferroelectric material in vacuum, the triboelectric charge density boosts to 1003 $\mu C/m^2$. The charge quantity loss can be roughly estimated with its charge density gap in air and vacuum conditions, and we can find that a dominant part is wasted because of air breakdown.

To harvest this part of energy during electrostatic breakdown, a next-generation direct-current TENG (DC-TENG) is designed via the TE effect and by inducing artificial lightning [8, 9]. This novel DC-TENG, arising from electrostatic breakdown, can produce a constant current output, which has been demonstrated

Triboelectric Nanogenerators: Technology, Applications, and Commercialization, First Edition.
Yunlong Zi, Hengyu Guo, Jie Wang, Chi Zhang, Xiangyu Chen, and Qing Zhao.

to directly power electronic devices without any rectifier and energy storage unit. The DC-TENG not only can promote the miniaturization of self-powered systems used in wearable electronics and the Internet of Things but also can provide a paradigm-shifting technique for TENG to effectively convert mechanical energy into electric energy. Furthermore, it can be considered a prototype of lighting energy harvesting. In addition, it has been demonstrated for mechanical energy harvesting combined with the conventional TENG to enhance the final energy output and self-powered sensor for measuring various motion parameters, showing great potential for application in energy and sensing.

5.2 Basic Principle

The basic model of this DC-TENG is illustrated in Figure 5.1a. Its working mechanism is based on TE and electrostatic breakdown [8]. When the frictional electrode (FE) contacts with the triboelectric layer (material: PTFE), electrons will transfer between two materials due to the different abilities of losing electrons. As the PTFE is an electret, it can hold a quasi-permanent electric charge. When the slider moves on the dielectric layer and the charge-collecting electrode (CCE) reaches above the surface of the negatively charged PTFE, a high electric field will be built in the gap between the CCE and the PTFE layer. If the electrostatic field exceeds the dielectric strength of the atmosphere between them, air will be ionized and become conductive. Then, electrons will transfer from the surface of PTFE to the CCE and then to the FE in external circuit forming an electric loop. Thus, a continuous DC output can be obtained if a continuous external force is applied on the slider. Looking at the DC-TENG from a circuit perspective, it can be considered an electric charge source and a broken-down capacitor in which the CCE/gap/PTFE serves as the capacitor, as shown in Figure 5.1b. The TE process can be considered the charged process of PTFE surface. The CCE is rationally designed to induce a high voltage drop across the air gap, and the process of air breakdown is equivalent to a "switch" to timely release the voltage potential, and the "switch" is controlled by the avalanche breakdown process. It should be emphasized that the constant current output strongly relies on

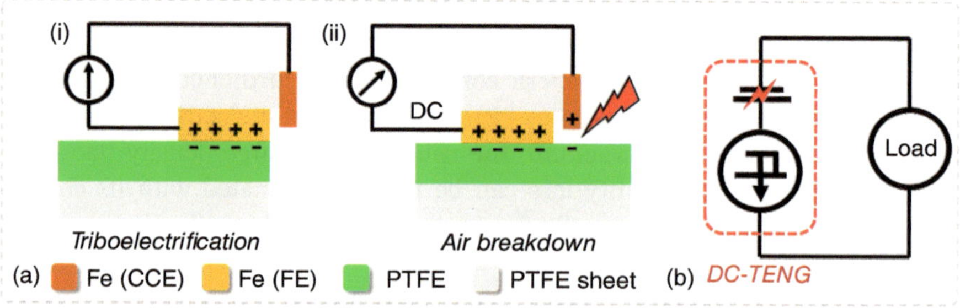

Figure 5.1 Working principle of DC-TENG. (a) Working mechanism of DC-TENG based on triboelectrification and air breakdown. (b) The simple circuit diagram of DC-TENG. Reproduced with permission of Ref. [10], 2020 © John Wiley & Sons.

whether the electric charge source can provide enough charges for air breakdown. Therefore, the constant current TENG is built based on sliding-mode TENG due to the high triboelectric efficiency of sliding electrification.

5.3 Physical Model

Based on the basic principle of DC-TENG, several physical models have been proposed to reveal the working mechanism of DC-TENG and optimize its output performance. In 2020, Xu et al. proposed an equivalent circuit model of DC-TENG based on theoretical research and simulation model [11]. Figure 5.2a shows one of the most exemplary structures of air breakdown DC-TENG to investigate the theory of air breakdown DC-TENG. To overcome the randomness of the breakdown peaks, an automatic statistical analysis and fitting method are proposed, serving as an effective tool in investigating the theory of the air breakdown DC-TENG. As shown in Figure 5.2b, the breakdown process was divided into three parts

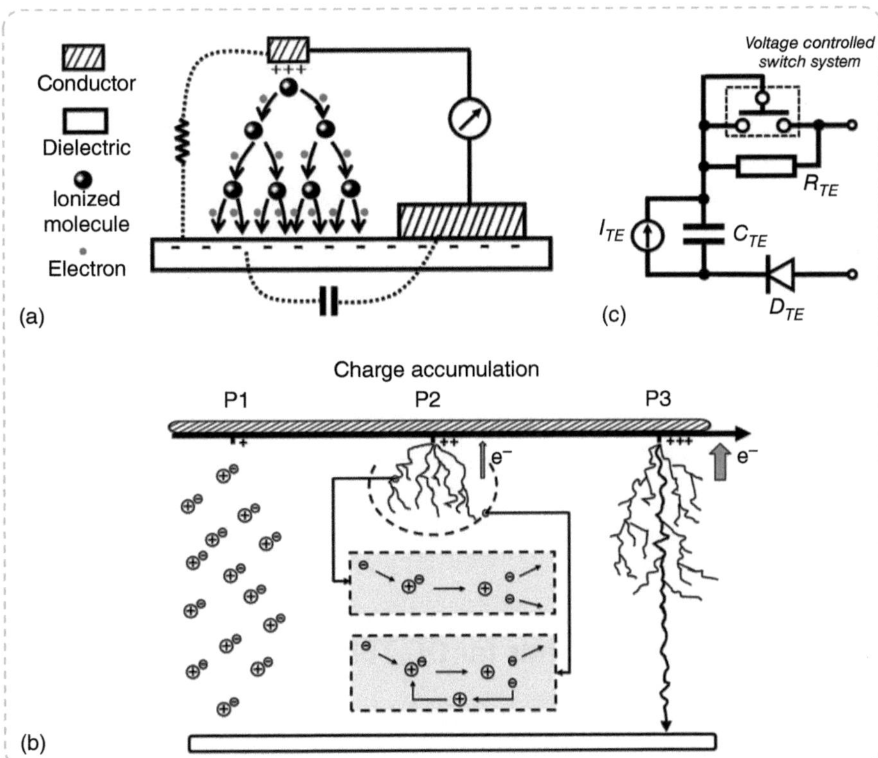

Figure 5.2 An equivalent circuit model of DC-TENG. (a) The simplified structure of DC-TENG. (b) The breakdown process in three states over time: turn-off state (P1); corona discharge state (P2); breakdown state (P3). (c) The equivalent circuit diagram of DC-TENG. Reproduced with permission of Ref. [11], 2020 © AIP Publishing.

over time: turn-off state (P1), corona discharge state (P2), and breakdown state (P3). According to the proposed method, an equivalent circuit of air breakdown DC-TENG is derived, containing a current source, an inner capacitor, a diode, and a voltage-controlled gate. This work provides a theoretical research and simulation model for the air breakdown DC-TENG, which is suitable for in-depth understanding and further advancement of DC-TENG.

Then, Dai et al. proposed a detailed theoretical transient physical-field model for air-breakdown DC-TENG [12]. As shown in Figure 5.3a, they considered that the detailed working mechanism of DC-TENG could be two breakdown domains (the breakdown around CCE and the breakdown around FE) rather than only one breakdown domain in DC-TENG through theoretical analysis. The simulated distribution of the electric field of DC-TENG in Figure 5.3b revealed that the strong local electric field around the FE and CCE could cause air breakdown. As shown in Figure 5.3c, under the combined breakdown effect in the harmful breakdown domain (HBD, the breakdown phenomenon around FE) and beneficial breakdown domain (BBD, the breakdown phenomenon around CCE), when the sliding displacement of the air-breakdown DC-TENG is large enough, the remaining charge density distribution on the surface of the triboelectric layer can be divided into three areas. As the slider moves forward, the charged area on the surface of PTFE on the left of FE will gradually increase and further intensify the breakdown effect in BBD. The remaining charge density will gradually decrease and tend to converge, as shown in the middle area. As the slider moves forward again, the breakdown phenomenon in the HBD occurs, which will decrease the surface charge density. Finally, the effective charge density of DC-TENG from BBD will decrease.

The physical field model and its numerical simulation proposed above were also of great significance for improving the equivalent circuit model of the air breakdown DC-TENG. The equivalent circuit model based on this work is shown in Figure 5.3d, indicating that both the breakdown effect at BBD and HBD will influence the output characteristics, and their breakdown behaviors are coupled. Therefore, to enhance the output performance of DC-TENG, increasing triboelectric charge density until a very weak breakdown in HBD, restraining the breakdown effect in HBD and strengthening the breakdown effect in BBD are crucial directions. This work reveals the physical essence of the air-breakdown DC-TENG and is of great significance in guiding their optimal design in various application scenarios.

5.4 Optimization Methods for DC-TENG

DC-TENG based on the TE and electrostatic breakdown can directly convert mechanical energy into a DC output without any rectifier or energy storage devices. Therefore, boosting the output performance of DC-TENG is important to promote the practical application of TENG. Up to now, great efforts have been made to optimize the performance of TENG. The optimization methods of air breakdown DC-TENG can be briefly divided into three parts from its basic principle: improving

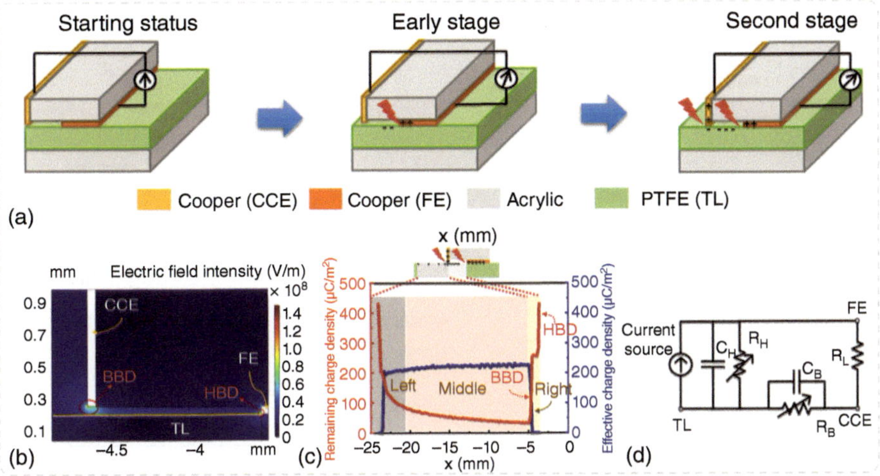

Figure 5.3 The transient physical-field model of air-breakdown DC-TENG. (a) The detailed working mechanism of DC-TENG considering two breakdown domains. (b) The simulated distribution of electric field of DC-TENG. (c) Remaining surface charge density and effective charge density of DC-TENG considering the two breakdown domains. (d) Equivalent circuit model of DC-TENG considering the two breakdown domains. Reproduced with permission of Ref. [12], 2020 © Elsevier.

the surface charge density, which is similar to the conventional AC-TENG, enhancing the process of electrostatic breakdown, and rational structural design.

5.4.1 Improving Triboelectrification

To build a guideline for selecting appropriate triboelectric materials for high-performance DC-TENG, Zhao et al. proposed a set of selection rules to screen whether a triboelectric material is suitable to be used in DC-TENG [13]. As shown in Figure 5.4a, they provide a primary model based on friction coefficient, surface charge density, and dielectric polarization to formulate the primary selection rules. Then, the relationship between surface charge density and effective charge density of DC-TENG of different triboelectric films was explored as shown in Figure 5.4b. The results indicate that dielectric polarization and surface charge density of different triboelectric films codetermine the effective charge density of DC-TENG. To satisfy the application requirements of DC-TENG, a comprehensive model of the performance selection of triboelectric materials for DC-TENG is also given, which includes surface charge density, friction coefficient, strength of polarization, the utilization rate of charges and stability. Finally, the polyvinyl chloride (PVC) and PTFE films show outstanding comprehensive properties, where the PVC shows a larger effective charge density and PTFE shows a better friction property (Figure 5.4c). This work offers a basic guideline for triboelectric materials selection and will promote the practical applications of DC-TENG.

To enhance the output performance of all kinds of TENGs, Cui et al. found that nitrile contacting with PTFE has superior triboelectric charge density in

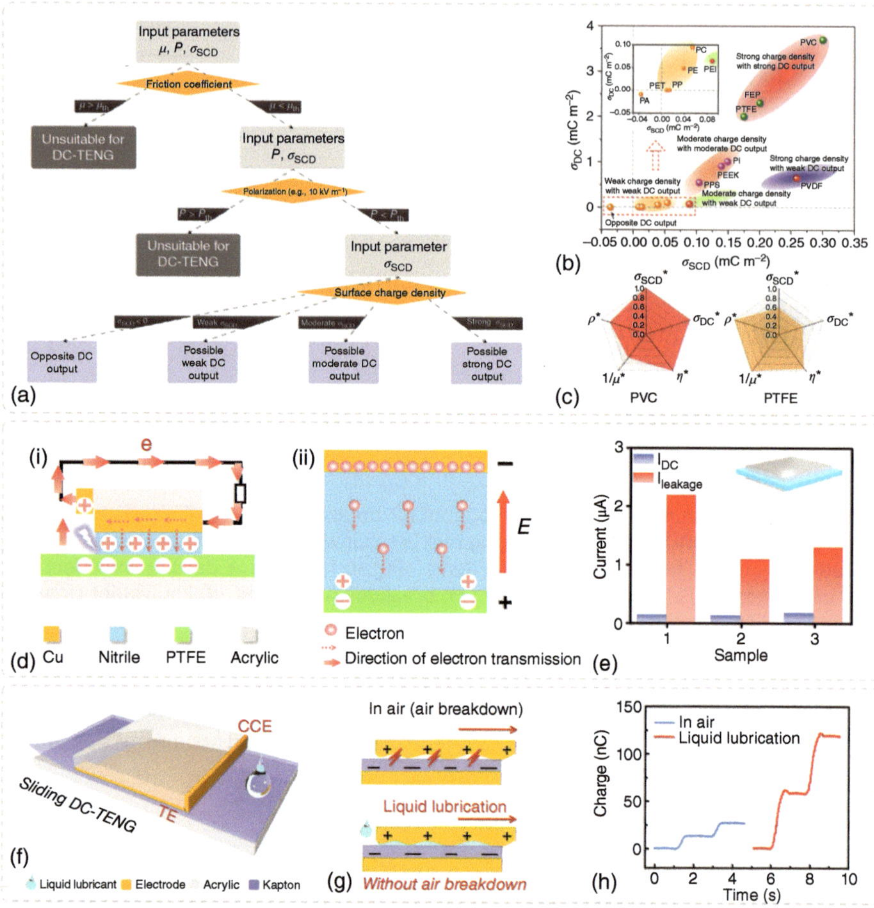

Figure 5.4 Improving triboelectrification. (a) Primary selection rules of triboelectric materials for DC-TENG. (b) The relationship between surface charge density and effective charge density of DC-TENG of different triboelectric layers. (c) Comprehensive selection rules of triboelectric materials for DC-TENG. Reproduced with permission of Ref. [13], 2021 © Springer Nature. (d) Working mechanism of DEDC-TENG. (e) Comparison of leakage of nitrile and current of DEDC-TENG. Reproduced with permission of Ref. [14], 2022 © Elsevier. (f) The schematic diagram of liquid lubricated sliding DC-TENG. (g) The schematic of interfacial breakdown in air and avoiding interface breakdown through interface liquid lubrication during sliding motion. (h) Transferred charge of the sliding DC-TENG with and without interface liquid lubrication. Reproduced with permission of Ref. [15] 2020 © John Wiley & Sons.

comparison to many conventional materials such as copper, polyethylene, and Kapton [14]. Therefore, by incorporating a dielectric enhancement layer on the friction electrode, the dielectric enhanced DC-TENG (DEDC-TENG) can increase the TE and thus achieve a higher output value of DC-TENG. The detailed working principle of the DEDC-TENG is shown in Figure 5.4d. To produce a continuous DC output, there should be a complete electric loop in the device. Here, electrons can transfer from the nitrile to the PTFE layer based on TE, from the PTFE layer to CCE based on electrostatic breakdown, and then from CCE to FE in external circuit. How can the charge transfer from the FE to the nitrile? They demonstrate that electrons passing through the nitrile film to the dielectric layer is in the form of leakage current (Figure 5.4d<ii>). As shown in Figure 5.4e, the leakage current of nitrile exceeds 1 μA with a voltage of 60 V applied on the two sides of the nitrile layer, which is higher than the output current of the DEDC-TENG, indicating that electrons can pass through the nitrile in the form of leakage current. This work provides a simple and universal strategy for elevating the performance of DC-TENG and proposes a new working mechanism and structural design of DC-TENG.

Given that the energy loss induced by air breakdown and the inferior durability seriously limit practical applications of TENG, Zhou et al. proposed a universal method to improve the power outputs and durability of both sliding-mode AC-TENG and sliding-mode DC-TENG, by adding appropriate liquid lubrication into the sliding interface (Figure 5.4f) [15]. It is demonstrated that interfacial breakdown between triboelectric layer and electrode can be largely restricted by the liquid lubricant as shown in Figure 5.4g. Therefore, more triboelectric charges will be reserved on the dielectric surface rather than released by air breakdown, and then the charges collected by CCE in sliding DC-TENG via electrostatic breakdown will also be enhanced, as the results are shown in Figure 5.4h. This work provides a promising universal strategy to enhance the electric outputs and durability of DC-TENG, which could greatly promote its practical applications.

5.4.2 Enhancing Electrostatic Breakdown

Enhancing the electrostatic breakdown process is another effective strategy to improve the output performance of DC-TENG. Paschen's law, which describes the gaseous breakdown voltage (V_b) as a function of gas pressure (P) and gap distance (d), can be given by

$$V_b = \frac{BPd}{\ln(APd) - \ln\left[\ln\left(1 + \frac{1}{\gamma_{se}}\right)\right]} \tag{5.1}$$

where A represents the saturation ionization of the gas at a specific electric field and pressure, B is a constant related to gas excitation and ionization energies, and γ_{se} is the secondary electron emission coefficient. Hence, the constants A, B, and γ_{se} are dependent on the gas's intrinsic properties.

The Paschen's curves within a wide gas pressure range indicate that the oxygen atmosphere is more prone to breakdown under atmospheric pressure compared with air and nitrogen, as shown in Figure 5.5a. Yi et al. further investigated the effects of atmospheric pressure and composition on the output performance of DC-TENG [16]. In different atmospheric conditions, the output charge density and open-circuit voltage first increases and then decreases with the rising of atmospheric pressure (Figure 5.5b, c), which is consistent with the theoretically calculated results based on Paschen's law. They found the oxygen atmosphere is beneficial for both contact electrification and electrostatic breakdown, which contribute to more charges collected from gas breakdown and the improvement of output performance. Due to the lower threshold voltage and easy breakdown characteristics, the output power of DC-TENG working in oxygen atmosphere realizes eightfold increase compared with standard atmospheric condition.

With the sliding AC-TENG and DC-TENG as measuring tools, respectively, Liu et al. systematically studied the influence of air pressure and temperature on the process of TE and electrostatic breakdown [10]. The effective charge of DC-TENG first increases with the decrease of air pressure and reaches the maximum value at around 300 Pa, which corresponds to the minimum breakdown voltage in air described by Paschen's law. The results indicate that appropriate low atmospheric pressure is favorable for the process of electrostatic breakdown and improves the output performance of DC-TENG. Besides, the output charge of DC-TENG exponentially increases with the increase of temperature, as shown in Figure 5.5d. The enhanced output performance of DC-TENG at high temperature can be explained from three aspects: (i) the enhanced sliding electrification through the increased contact intimacy between the electrode and triboelectric layer in high temperature; (ii) the enhanced electrostatic breakdown through avalanche breakdown; (iii) the reduced residual surface charges on the triboelectric layer through thermionic emission with the increasing of temperature (Figure 5.5e). Therefore, the output power density of the DC-TENG at 473 K and 300 Pa is improved by three orders of magnitude compared with 298 K and standard atmosphere.

Given that there are still residual surface charges on the triboelectric layer after electrostatic breakdown in DC-TENG, the inferior charge utilization limits the further improvement of its output performance. Inspired by the fable "The Crow and the Pitcher", Chen et al. designed an electric field-enhanced DC-TENG to enhance the output performance of DC-TENG under atmospheric conditions [17]. As shown in Figure 5.5f, a field-enhancing layer is added under the triboelectric layer to improve the electric field intensity between the CCE and triboelectric layer, which is similar to the ability of stones to increase the water level. The electrostatic breakdown efficiency dramatically increases from 22.4% to 46% through the external field-enhanced breakdown effect, resulting in more triboelectric charges participating in the breakdown process. Compared with the conventional DC-TENG, the output charge and short-circuit current of electric field-enhanced DC-TENG realize nearly twofold increase (Figure 5.5g, h).

In addition, Liu et al. investigated the influence of environmental humidity on the output performance of DC-TENG [18]. With increasing humidity, the breakdown voltage presents a declining tendency, which is consistent with the theoretical simulation results, as shown in Figure 5.5i. Moreover, as the relative humidity (RH) increases from 30% to 90%, the output performance of a sliding mode AC-TENG with the hydrophobic materials as the friction layer also increases, indicating the enhanced output performance of sliding electrification under high humidity. Benefitting from the synergistic enhancement of TE effect and electrostatic breakdown, the effective charge density of microstructure-designed DC-TENG based on PVC film reaches up to ~2.97 mC/m^2 even under 90% RH, which is almost 1.42-fold larger than that under 30% RH (Figure 5.5j). These results demonstrate that the DC-TENG based on TE effect and electrostatic breakdown provides an effective

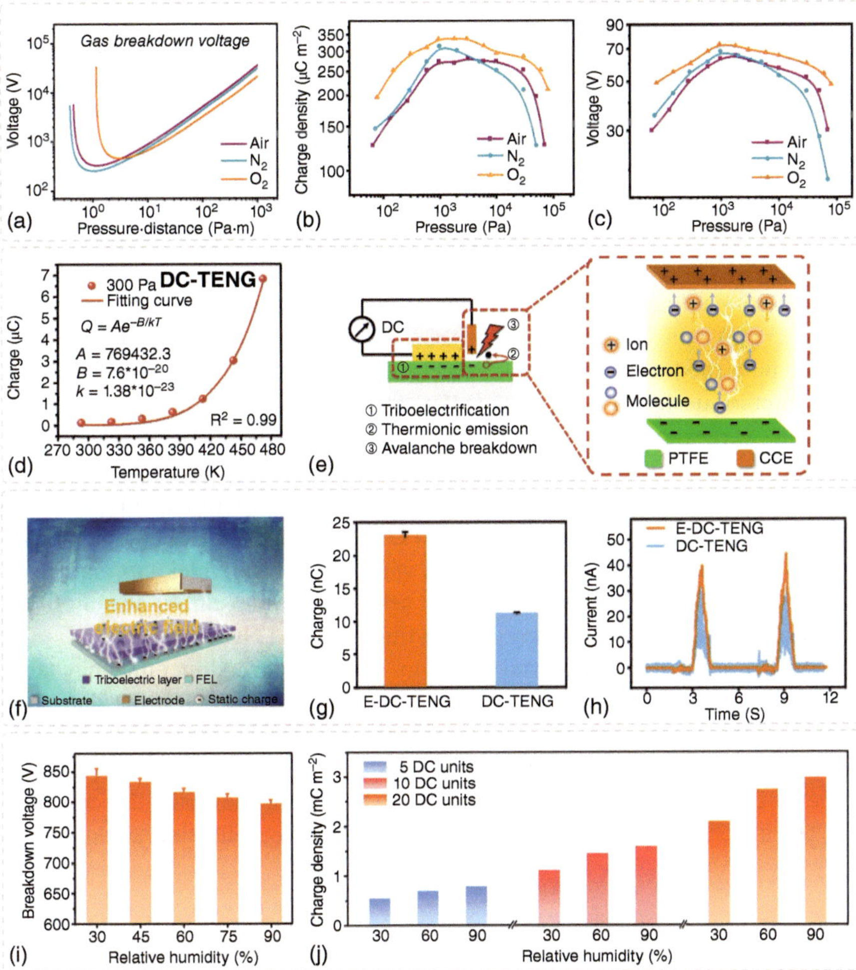

Figure 5.5 Enhancing electrostatic breakdown. (a) Paschen's curves of gas breakdown voltage for air, nitrogen, and oxygen. (b) Effective charge density and (c) open-circuit voltage of DC-TENG under different atmospheric conditions. Reproduced with permission of Ref. [16], 2021 © Elsevier. (d) Output charge of DC-TENG under different temperatures with the pressure of 300 Pa. (e) The mechanism of enhanced performance of DC-TENG in high temperature. Reproduced with permission of Ref. [10], 2020 © John Wiley & Sons. (f) Schematic diagram of the electric field-enhanced DC-TENG. (g) Output charge and (h) short-circuit current of conventional DC-TENG (blue line) and electric field-enhanced DC-TENG (yellow line). Reproduced with permission of Ref. [17], 2021 © John Wiley & Sons. (i) Breakdown voltage under different environmental humidity. (j) Effective charge density of microstructure-designed DC-TENG based on PVC film under different environmental humidity. Reproduced from Ref. [18], 2022 © John Wiley & Sons.

solution to the widespread output attenuation problem of conventional TENGs in high-humidity environment.

5.4.3 Advanced Structure Design

In addition to improving the TE and enhancing electrostatic breakdown, advanced structure design is also an effective way to improve the output performance and stability of DC-TENG. Due to lack of charge accumulation process compared with conventional AC-TENG, it is still limited for DC-TENG to achieve higher output performance only by enhancing TE and electrostatic breakdown. Zhao et al. provided a method to improve the charge density of DC-TENG by microstructural design with a rationally patterned electrode structure (Figure 5.6a) [19]. The working principle of microstructure-designed DC-TENG (MDC-TENG) is shown in Figure 5.6b. The triboelectric charges on the friction-charged dielectric surface can be released by electrostatic breakdown and collected by CCEs repeatedly. The output performance of MDC-TENG can be improved by tailoring the electrode structure to 50 units, and the effective surface charge density of MDC-TENG with the size of 1 cm × 5 cm can be improved to 5.4 mC/m^2 (Figure 5.6c). This work not only provides a miniaturization strategy of TENG as an energy supply or self-powered sensor but also presents a paradigm shift for large-scale energy harvesting by DC-TENG.

To solve the damage of material wear to energy conversion efficiency and robustness, Gao et al. designed a rolling DC-TENG (R-DC-TENG) by changing the sliding friction electrode to a rolling friction electrode (Figure 5.6d) [20]. The working principle of R-DC-TENG is similar to the traditional DC-TENG. The experimental results show that R-DC-TENG can still maintain the constant current output characteristics by substituting sliding electrification with rolling

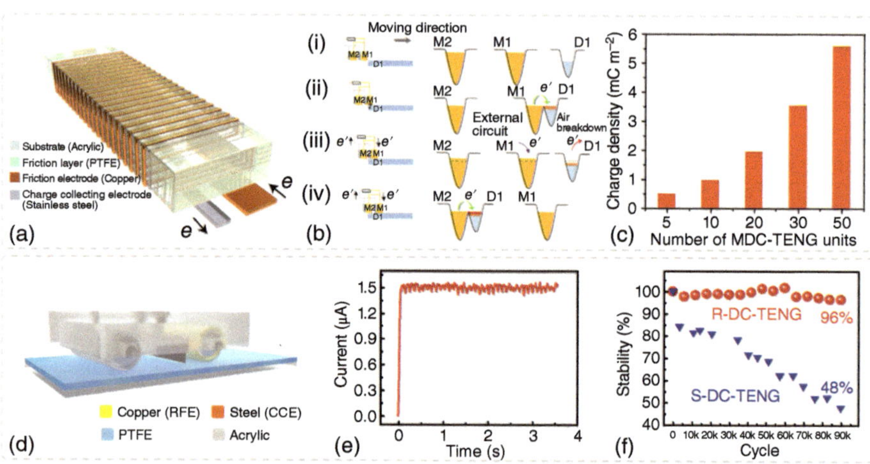

Figure 5.6 The advanced structure design of DC-TENG. (a) The structural schematic of MDC-TENG. (b) The schematic diagram of electron transition process. (c) Charge density of MDC-TENG with different numbers of MDC-TENG units. Reproduced with permission of Ref. [19], 2022 © Springer Nature. (d) The structural schematic of R-DC-TENG. (e) The constant current output of R-DC-TENG. (f) The stability test of R-DC-TENG. Reproduced with permission of Ref. [20], 2021 © Elsevier.

electrification (Figure 5.6e), and its stability has also been greatly improved. After nearly 100 000 cycles of continuous operation, its output performance can still maintain 96% of the initial value (Figure 5.6f).

5.5 DC-TENG for Energy Harvesting

The DC-TENG can be applied to harvest mechanical energy and convert it into electric energy, which can be potentially utilized as micro-/nanopower source to directly power electronic devices. Compared with the conventional AC-TENG, DC-TENG can avoid the utilization of rectifiers. As an electronic device, the rectifiers will inevitably consume electric power. Meanwhile, the volume of the rectifier cannot be ignored, which may restrict the application of the TENGs in micro-/nanodevices. Moreover, this structure that does not contain a rigid rectifier is very convenient to be utilized in flexible wearable electronics. Chen et al. reported a DC fabric TENG (DC F-TENG) that could harvest biomotion energy by taking advantage of the harmful and annoying electrostatic breakdown phenomenon of clothes [21]. The DC F-TENG is interwoven by warp yarns and weft yarns as a common plain structure, as shown in Figure 5.7a. The warp yarn is a three-ply twisted nonconductive PA yarn, and the weft yarns are composed of two kinds of yarns, the nonconductive PA yarns (frictional yarns) and conductive PA yarns coated with Ag (electrostatic breakdown yarns), as shown in Figure 5.7a. When the DC F-TENG moves forward and backward on the surface of PTFE film, the nonconductive PA yarns have TE effect with PTFE and lose electrons (Figure 5.7b). Thus, the PTFE film is negatively charged. Once the electric potential difference exceeds the breakdown voltage, the air will begin conducting (Figure 5.7b). A discharge between the PTFE film and the electrostatic breakdown electrode occurs due to the electrical breakdown of air. If the DC F-TENG continues to move toward the left, it will produce continuous DC outputs. A small DC F-TENG (1.5 cm × 3.5 cm) can easily light up 416 serially connected LEDs. Furthermore, the yarn supercapacitor can be fabricated and woven into the DC F-TENG to harvest and store energy, which can be used to power electronic devices, such as the hygrothermograph or calculator, which shows great convenience and high efficiency in practice. Cheng et al. provided a wearable fabric-based direct current TENG (FDC-TENG) with high power output, light-weight and highly flexible by simply coating two electrodes on the top side (breakdown electrode) and bottom side (friction electrode) of a polyester–cotton fabric (Figure 5.7c) [22]. The top breakdown electrode harvests triboelectric charges through the conductive plasma channel created by air breakdown. A polyester–cotton fabric is selected as the substrate due to the high strength of polyester and the high air permeability of cotton. The silver paste is chosen to fabricate the breakdown electrode and the friction electrode owing to its excellent conductivity and the strong tendency to lose electrons. When the FDC-TENG contacts with the friction layer (FEP in this article), the FEP is negatively charged and friction electrode is positively charged due to the TE effect. Afterward, when the FDC-TENG moves from the right end to the left (Figure 5.7d), there is no output current in the external circuit because the surface charge density of the FEP film on the left-hand side of the breakdown electrode is not high enough to break down

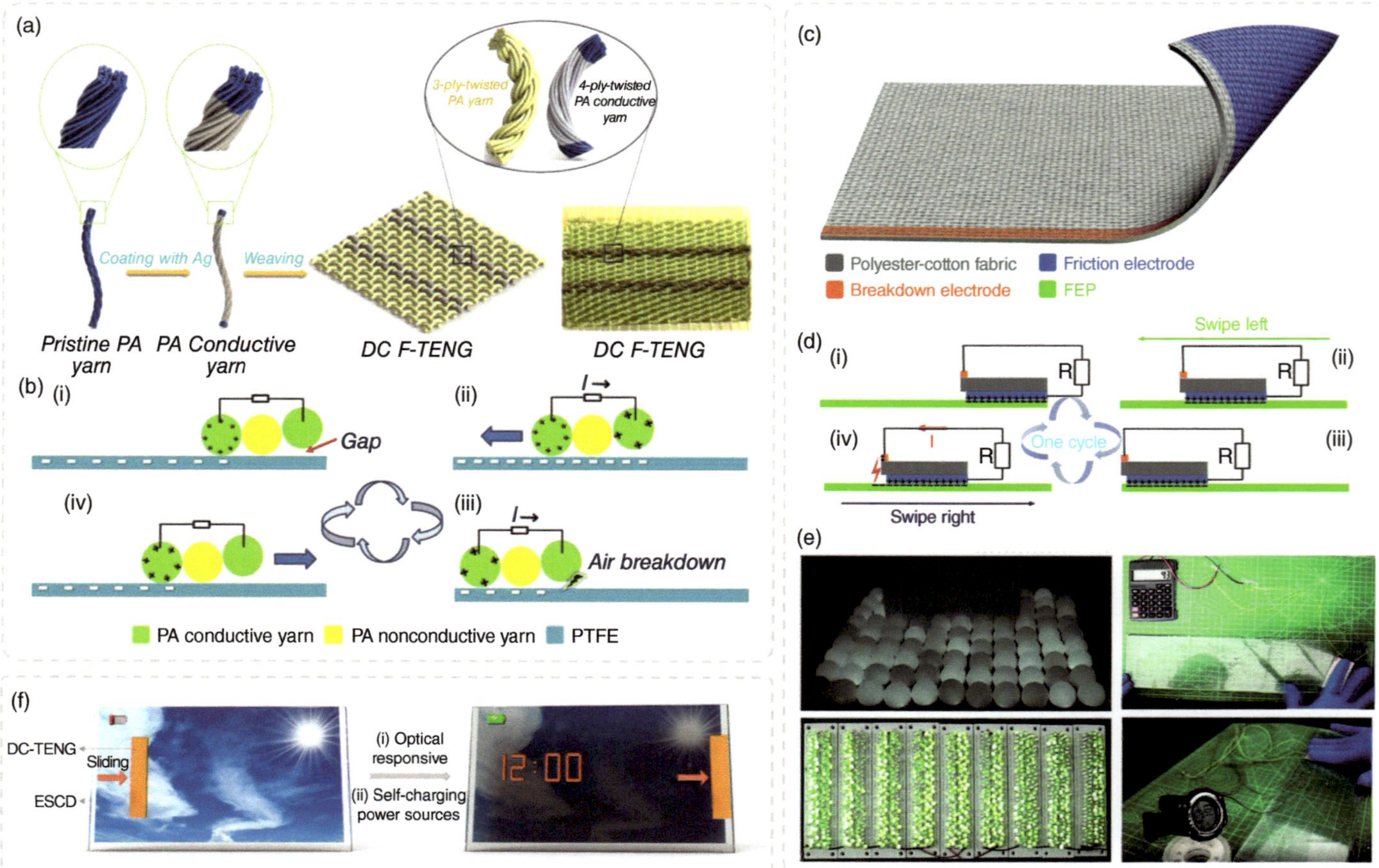
(a)
Coating with Ag
Weaving
3-ply-twisted PA yarn
4-ply-twisted PA conductive yarn
Pristine PA yarn
PA Conductive yarn
DC F-TENG
DC F-TENG
(b) (i)
(ii)
(iii)
(iv)
Gap
Air breakdown
PA conductive yarn
PA nonconductive yarn
PTFE
(c)
Polyester-cotton fabric
Friction electrode
Breakdown electrode
FEP
(d)
Swipe left
Swipe right
One cycle
R
I
(e)
(f)
DC-TENG
ESCD
Sliding
(i) Optical responsive
(ii) Self-charging power sources
12:00

Figure 5.7 DC-TENG for energy harvesting. (a) Schematic fabrication process and photograph of the DC F-TENG. (b) Working mechanism of the DC F-TENG. Reproduced with permission of Ref. [21], 2020 © American Chemical Society. (c) Schematic illustration of the FDC-TENG. (d) Working principle of the FDC-TENG. (e) Photographs of 99 bulbs, 1053 LEDs, a calculator, and a watch directly driven by FDC-TENG. Reproduced with permission of Ref. [22], 2021 © Royal Society of Chemistry. (f) Schematic diagram of the SC-ESCD, which integrates an electrochromic supercapacitor with the optical responding and energy storage characteristics, and a DC-TENG as an energy-harvesting device. Reproduced with permission of Ref. [23], 2021 © John Wiley & Sons.

air (Figure 5.7d). When the FDC-TENG moves from the left end to the right, a high electrostatic field will be built on the surface of the FEP film on the left-hand side of the breakdown electrode, and the air breakdown occurs as the electrostatic field is high enough, resulting in a DC output in the external circuit. As shown in Figure 5.7e, the finger-sized FDC-TENG can light up 99 bulbs or 1053 LEDs, as well as drive watches or calculators directly without rectifying or capacitor charging. These works not only provide novel design concepts for DC-TENG-based fabrics but also promote human biomechanical energy harvesting for wearable electronics.

Taking advantage of the high output voltage of DC-TENG, Guo et al. proposed a self-charging electrochromic supercapacitor device (SC-ESCD) [23], which has two parts, the DC-TENG as energy harvesting and conversion unit and electrochromic supercapacitor devices as energy store and optical responding unit, as shown in Figure 5.7f. A polyaniline-based ESCD was fabricated, which can show an obvious coloring and discoloring phenomenon with the external voltage switch on and off. Thus, when the DC-TENG slides on the ESCD window, the DC will directly charge the ESCD to change window's color without extra power supply, which is promising in smart information display screens or transmittance modulation windows.

5.6 DC-TENG for Self-Powered Sensing

DC-TENG also shows good characteristics in self-powered sensors. Yin et al. proposed a triboelectric vector sensor (TVS) based on DC-TENG, with four CCEs on the four sides of the FE (Figure 5.8a) [24]. When the FE slides forward, the corresponding back CCE will produce an electric signal; when the FE slides backward, the corresponding front CCE will produce an electric signal. According to the experimental and theory analysis, they found that the TAS can realize probing multiple physical quantities in a self-powered manner, such as displacement, velocity, acceleration, angle, and angular velocity as shown in Figure 5.8b. Given the aforementioned analysis, this DC-TENG shows great potential for self-powered sensors including but not limited to motion monitoring and tracking. With four CCEs connected to the negative pole of four different directional arrow indicator lights, and one common FE connected to their positive pole, when the FE moves in any direction, the LEDs in different directions, such as west, east, north, and south, can be directly lightened by TVS according to actual movement direction, achieving self-powered direction recognition in the dark environment (Figure 5.8c).

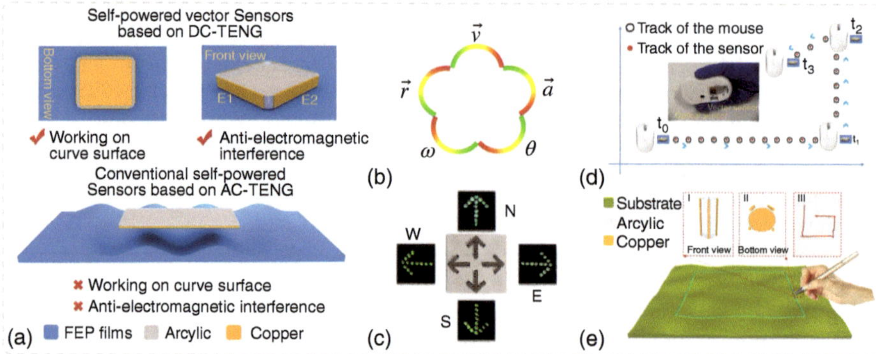

Figure 5.8 DC-TENG for self-powered sensing. (a) The schematic structure of the self-powered motion vector sensor based on DC-TENG compared with the conventional self-powered sensors based on AC-TENG. (b) Physical quantities obtained from the self-powered motion vector sensor. (c) A self-powered directional sensor composed of the TVS system with some LEDs as indicator lights. (d) The tracks of the commercial mouse equipped with an optical sensor and a self-powered vector sensor. (e) The TVS works on a curved surface. Reproduced with permission of Ref. [24], 2020 © John Wiley Sons.

More importantly, the orientation of other angles also can be monitored by the good linear fit relationship between the rotation angle and voltage response, indicating high precision of motion orientation recognition. To achieve the continued motion trajectory tracking, a self-powered intelligent mouse integrated with TVS has been designed. Combining the mathematical relationship between the motion displacement of intelligent mouse and the output signal of TVS, a template-matching algorithm is introduced to realize the real-time control of computer interface operations (Figure 5.8d). It is noted that the sensitivity of intelligent mouse with TVS is comparable to that of the commercial photoelectric mouse, indicating its reliability and potential commercial value. In addition to the plane motion trajectory tracking, the trajectory tracking on curve surface can also be achieved by the miniaturized TVS. Through integrating a miniaturized TVS (the area of FE is about 1 mm^2 and the thin copper wire is used as CCE) into a pen tip, a self-powered smart pen has been fabricated, and its displacement sensitivity of intelligent pen can reach 103.4 mV/mm (Figure 5.8e). These results demonstrate that TVS not only can realize handwriting recognition on flat even curved surfaces but also have the reliable ability to visually translate mechanical motion.

5.7 Hybrid of AC-TENG and DC-TENG

The relationship of TE, electrostatic induction, and electrostatic breakdown is the basis of the new energy harvesting technology of AC-TENG arising from electrostatic induction and DC-TENG arising from electrostatic breakdown. From the point of view of charge utilization, the electrostatic charges created by TE are used for inducing charges flowing in external circuit by electrostatic induction or inducing

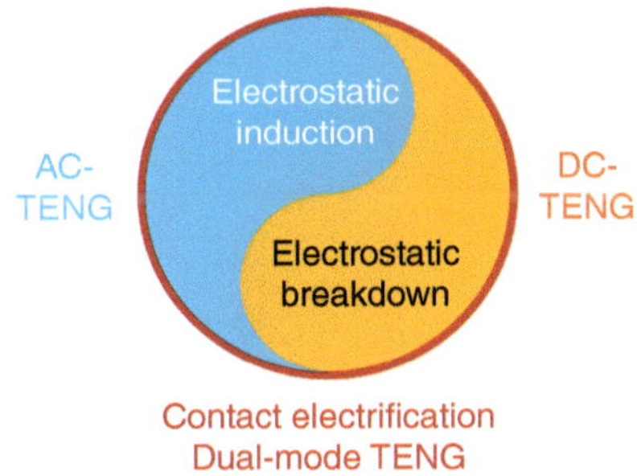

Figure 5.9 The relationship of triboelectrification, electrostatic induction, and electrostatic breakdown. Reproduced with permission of Ref. [25], 2020 © John Wiley & Sons.

ionization of gas to form a conductive path by electrostatic breakdown. It is thus natural to study the relationship of TE, electrostatic induction, and electrostatic breakdown. Generally, the charges created by TE in TENG can be used by electrostatic induction or electrostatic breakdown, namely the charges from electrostatic induction plus the charges from electrostatic breakdown are the whole charge created by TE, as shown in Figure 5.9. By rationally designing the structure of TENG and controlling the working environment of TENG, the amount of output charges can be regulated by these two principles (electrostatic induction and electrostatic breakdown), which is similar to the Yin and Yang diagrams in traditional Chinese culture.

As the AC-TENG or DC-TENG can only harvest the electrostatic energy by electrostatic induction or breakdown separately, the invented AC/DC-TENG (or named dual-mode TENG) shows particular superiority in mechanical energy harvesting and self-powered systems compared with the single-mode TENG. The AC/DC-TENG is defined as the simultaneous harvesting of the electrostatic energy from TE by coupling the two physical effects of electrostatic induction and electrostatic breakdown, which generally contains three types of electrodes in its structure: the FE, CCE, and induction electrode (IE). Zhou et al. reported a novel AC/DC-TENG to simultaneously harvest mechanical energy by electrostatic induction and dielectric breakdown in a single device (Figure 5.10a) [25]. Thanks to the coupling of the two principles, the whole energy output can be greatly enhanced. The working mechanism of the dual-mode TENG is shown in Figure 5.10b. Figure 5.10c is the output charge density of the dual-mode TENG, indicating the simultaneous production of AC and DC output.

Furthermore, Li et al. reported a self-powered vibration sensor system and achieved real-time and continuous detection of the vibration characteristics from an AC/DC-TENG [26]. The detailed structure is shown in Figure 5.10d. When the slider moves beyond the safe zone and into the dangerous zone, a high electrostatic field will be built between the CCE and the charged triboelectric layer, and the air will be ionized with a DC output in external circuit and driving the alarm light directly. Alike with the conventional AC-TENG, the output AC signal contains the moving information, such as amplitude, frequency, velocity, and acceleration, which can be used for a motion sensor. More importantly, instead of judging the position by the signal amplitude, the AC/DC-TENG can precisely judge the vibration amplitude by using the signal type changed from AC to DC (Figure 5.10e). In addition, this self-powered vibration sensor system can realize real-time feedback

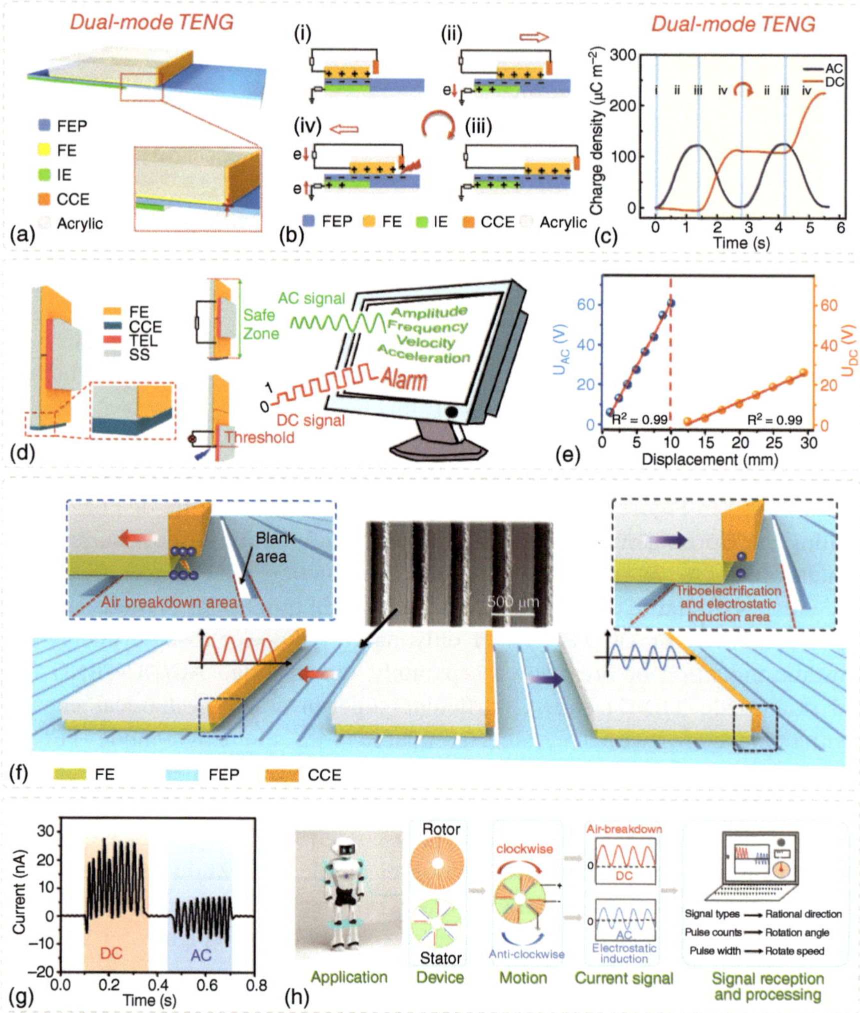

Figure 5.10 The dual-mode TENG for energy harvesting and self-powered sensors. (a) The schematic diagram of sliding dual-mode TENG composed of an AC-TENG and a DC-TENG. (b) The working mechanism of dual-mode TENG. (c) Output charge density of the dual-mode TENG. Reproduced with permission of Ref. [25], 2020 © John Wiley Sons. (d) The schematic illustration of AC/DC-TENG as a fully self-powered vibration monitoring sensor. (e) The typical output voltage of the AC/DC-TENG in monitoring vibration motion by using the signal type changed from AC to DC. Reproduced with permission of Ref. [26], 2020 © American Chemical Society. (f) Schematic diagram of the dual-signal vector sensor. (g) The DC and AC output signal of dual-signal vector sensor in different moving directions. (h) The dual-signal sensor for application in the human-robotic interaction. Reproduced with permission of Ref. [27], 2022 © John Wiley and Sons.

by detecting the signal type changes from AC to DC, and using DC output drives an alarm system directly, which provides a convenient and effective avenue for monitoring the structural health of construction.

In addition, Li et al. proposed a dual-signal vector sensor based on the effect of TE, electrostatic induction and electrostatic breakdown (Figure 5.10f) [27]. This vector can produce DC signal based on TE and electrostatic breakdown in one moving direction, and produce AC signal based on TE and electrostatic induction in another moving direction (Figure 5.10g). Due to the grid structure, the period of step-shaped DC voltage and pulse AC voltage signal in opposite motion are both corresponded to the period of grid. Therefore, the accuracy of dual-signal vector sensor is not affected by the changed temperature and humidity working environment because of the method of pulse signal counts. Benefiting from the fast switching between pulsed AC and DC signals of dual-signal vector sensor in opposite moving directions, a new motion direction recognition technology has been proposed. Moreover, a dual-signal angular sensor has been designed for the real-time monitoring of mechanical arm joint motion. As shown in Figure 5.10h, the rotational motion direction, angle, and speed can be obtained according to the signal types, pulse signal counts, and width, respectively, which greatly promotes the application of TENG-based vector sensor in human-robotic interaction. In addition, He et al. reported a dual-mode TENG composed of an AC-TENG and a DC-TENG in a coaxial structure [28]. The AC-TENG in outer ring can produce high AC output realizing efficient wind energy harvesting, while the DC-TENG in inner ring can produce DC output for wind speed sensing.

5.8 Summary

This chapter presents the DC-TENG arising from electrostatic breakdown, which is considered a new paradigm of TENG and shows potential applications for efficient energy harvesting and self-powered sensor. The author systematically analyses this DC-TENG from basic principle, physical model, and optimization methods to applications, and points out further research directions. This DC-TENG can produce constant current output with merits of driving electronics directly without any rectifiers, antielectromagnetic interference of output signal, and no limitation of dielectric breakdown, which shows competitive potentials in miniaturization of self-powered systems, efficient energy harvesting and energy storage, and self-powered sensing. In addition, the DC-TENG can be coupled with the conventional AC-TENG building a dual-mode TENG, and has been successfully applied for displacement monitoring and angle sensing in smart robotics and automatic vehicles. Recently, some new structures of DC-TENG have been proposed such as double-channel air breakdown and bidirectional DC-TENG [29–31], which further promotes the materials selection of DC-TENG. Compared with the relatively matured research field of AC-TENG from electrostatic induction, the research of DC-TENG arising from electrostatic breakdown is just beginning. At this infant development stage, there are still many challenges and problems, and they require more efforts and further in-depth studies to promote this research field.

References

1 Shashoua, V.E. (1958). Static electricity in polymers. 1. Theory and measurement. *J. Polym. Sci.* 33: 65.

2 Shashoua, V.E. (1963). Static electricity in polymers. 2. Chemical structure and antistatic behavior. *J. Polym. Sci. Part A Gen. Pap.* 1: 169.

3 Wang, J., Liu, D., Zhou, L.L., and Wang, Z.L. (2019). Nanogenerators from electrical discharge. In: *Electrostatic Discharge-From Electrical Breakdown in Micro-Gaps to Nano-Generators* (ed. S.H. Voldman). London: IntechOpen.

4 Wang, S.H., Xie, Y.N., Niu, S.M. et al. (2014). Maximum surface charge density for triboelectric nanogenerators achieved by ionized-air injection: methodology and theoretical understanding. *Adv. Mater.* 26: 6720.

5 Zhang, C.L., Zhou, L.L., Cheng, P. et al. (2020). Surface charge density of triboelectric nanogenerators: theoretical boundary and optimization methodology. *Appl. Mater. Today* 18: 100496.

6 Wang, J., Li, S.M., Yi, F. et al. (2016). Sustainably powering wearable electronics solely by biomechanical energy. *Nat. Commun.* 7: 12744.

7 Wang, J., Wu, C.S., Dai, Y.J. et al. (2017). Achieving ultrahigh triboelectric charge density for efficient energy harvesting. *Nat. Commun.* 8: 88.

8 Liu, D., Yin, X., Guo, H.Y. et al. (2019). A constant current triboelectric nanogenerator arising from electrostatic breakdown. *Sci. Adv.* 5: eaav6437.

9 Liu, D., Zhou, L.L., Wang, Z.L., and Wang, J. (2021). Triboelectric nanogenerator: from alternating current to direct current. *iScience* 24: 102018.

10 Liu, D., Zhou, L.L., Li, S.X. et al. (2020). Hugely enhanced output power of direct-current triboelectric nanogenerators by using electrostatic breakdown effect. *Adv. Mater. Technol.* 5: 2000289.

11 Xu, S., Guo, H., Zhang, S.L. et al. (2020). Theoretical investigation of air breakdown direct current triboelectric nanogenerator. *Appl. Phys. Lett.* 116: 263901.

12 Dai, K.R., Liu, D., Yin, Y.J. et al. (2022). Transient physical modeling and comprehensive optimal design of air-breakdown direct-current triboelectric nanogenerators. *Nano Energy* 92: 106742.

13 Zhao, Z.H., Zhou, L.L., Li, S.X. et al. (2021). Selection rules of triboelectric materials for direct-current triboelectric nanogenerator. *Nat. Commun.* 12: 4686.

14 Cui, S.N., Zhou, L.L., Liu, D. et al. (2022). Improving performance of triboelectric nanogenerators by dielectric enhancement effect. *Matter* 5: 180.

15 Zhou, L.L., Liu, D., Zhao, Z.H. et al. (2020). Simultaneously enhancing power density and durability of sliding-mode triboelectric nanogenerator via interface liquid lubrication. *Adv. Energy Mater.* 10: 2002920.

16 Yi, Z.Y., Liu, D., Zhou, L.L. et al. (2021). Enhancing output performance of direct-current triboelectric nanogenerator under controlled atmosphere. *Nano Energy* 84: 105864.

17 Chen, S.Y., Liu, D., Zhou, L.L. et al. (2021). Improved output performance of direct-current triboelectric nanogenerator through field enhancing breakdown effect. *Adv. Mater. Technol.* 6: 2100195.

18 Liu, L., Zhao, Z.H., Li, Y.H. et al. (2022). Achieving ultrahigh effective surface charge density of direct-current triboelectric nanogenerator in high humidity. *Small* 18: 2201402.

19 Zhao, Z.H., Dai, Y.J., Liu, D. et al. (2020). Rationally patterned electrode of direct-current triboelectric nanogenerators for ultrahigh effective surface charge density. *Nat. Commun.* 11: 6186.

20 Gao, Y.K., Liu, D., Zhou, L.L. et al. (2021). A robust rolling-mode direct-current triboelectric nanogenerator arising from electrostatic breakdown effect. *Nano Energy* 85: 106014.

21 Chen, C., Guo, H., Chen, L. et al. (2020). Direct current fabric triboelectric nanogenerator for biomotion energy harvesting. *ACS Nano* 14: 4585.

22 Cheng, R.W., Dong, K., Chen, P.F. et al. (2021). High output direct-current power fabrics based on the air breakdown effect. *Energy Environ. Sci.* 14: 2460.

23 Guo, W.B., Cong, Z.F., Guo, Z.H. et al. (2021). Multifunctional self-charging electrochromic supercapacitors driven by direct-current triboelectric nanogenerators. *Adv. Funct. Mater.* 31: 2104348.

24 Yin, X., Liu, D., Zhou, L. et al. (2020). A motion vector sensor via direct-current triboelectric nanogenerator. *Adv. Funct. Mater.* 30: 2002547.

25 Zhou, L., Liu, D., Li, S. et al. (2020). Rationally designed dual-mode triboelectric nanogenerator for harvesting mechanical energy by both electrostatic induction and dielectric breakdown effects. *Adv. Energy Mater.* 10: 2000965.

26 Li, S., Liu, D., Zhao, Z. et al. (2020). A fully self-powered vibration monitoring system driven by dual-mode triboelectric nanogenerators. *ACS Nano* 14: 2475.

27 Li, S.X., Zhao, Z.H., Liu, D. et al. (2022). A self-powered dual-type signal vector sensor for smart robotics and automatic vehicles. *Adv. Mater.* 34: 2110363.

28 He, L., Zhang, C., Zhang, B. et al. (2022). A dual-mode triboelectric nanogenerator for wind energy harvesting and self-powered wind speed monitoring. *ACS Nano* 16: 6244.

29 Shan, C.C., Liu, W.L., Wang, Z. et al. (2021). An inverting TENG to realize the AC mode based on the coupling of triboelectrification and air-breakdown. *Energy Environ. Sci.* 14: 5395.

30 Li, G., Wu, S.Y., Sha, Z. et al. (2022). Dual-breakdown direct-current triboelectric nanogenerator with synergistically enhanced performance. *Nano Energy* 99: 107355.

31 Shan, C.C., He, W.C., Wu, H.Y. et al. (2022). A high-performance bidirectional direct current TENG by triboelectrification of two dielectrics and local corona discharge. *Adv. Energy Mater.* 12: 2200963.

6

Promotion of Contact Electrification at Liquid–Solid Interface

6.1 Introduction

Decay of the tribocharges on the insulator surface induced by thermionic emission was demonstrated as an approach that can be used to distinguish electron transfer from ion transfer in solid–solid CE [1, 2]. Based on the thermionic emission theory, Lin et al. designed temperature-dependent charge decay experiments at nanoscale by using Kelvin probe force microscopy (KPFM) to quantify electron transfers and ion transfers in the CE between aqueous solutions and insulators [3]. In the experiments, the initial surface charge density of the insulator samples, such as SiO_2, Si_3N_4, and MgO, was measured by using KPFM. Then, an aqueous liquid droplet is slid over the insulator surface, and the charges will transfer from the liquid to the insulator surface in this process. As shown in Figure 6.1a, the charges on the insulator surface generated in the liquid–solid CE could be electrons or ions. The ions are produced by ionization reaction on the oxide and nitride surfaces [4–8]. After the CE, the insulator sample was heated to a certain temperature (such as 513 K), and the electrons were thermally excited and emitted from the insulator surface according to the thermionic emission theory in CE, as shown in Figure 6.1b. But for the ions, they formed covalent bonds with the atoms on the insulator surface, which correspond to chemical adsorptions instead of physical adsorption. For the ions' chemical adsorption on the solid surface, such as OH^- and H^+ on the SiO_2 surface, the energy threshold for removing the OH^- from SiO_2 surface is about 8.5 eV [9], and it is about 20 eV for removing an H^+ [10]. Therefore, the ions' chemical adsorbed on the surfaces is rather hard to be removed from the surfaces when the temperature is not too high.

Figure 6.1c gives the results of the charging and heating cycle tests. The initial surface charge density of the SiO_2 was close to zero. When it was contacted by DI water, negative charges transferred from the DI water to the SiO_2 surface. In the heating process (at 513 K and maintained for 10 minutes), the surface charge density decreased due to the thermionic emission of electrons. In the first cycle, it is noticed that some charges cannot be removed from the surface ("sticky" charges), which can be identified as ions generated on the surface in the liquid–solid CE. In the flowing cycles, the "sticky" charge density increased and reached saturation eventually. This is because the available charge positions on the insulator surface are limited

Triboelectric Nanogenerators: Technology, Applications, and Commercialization, First Edition.
Yunlong Zi, Hengyu Guo, Jie Wang, Chi Zhang, Xiangyu Chen, and Qing Zhao.

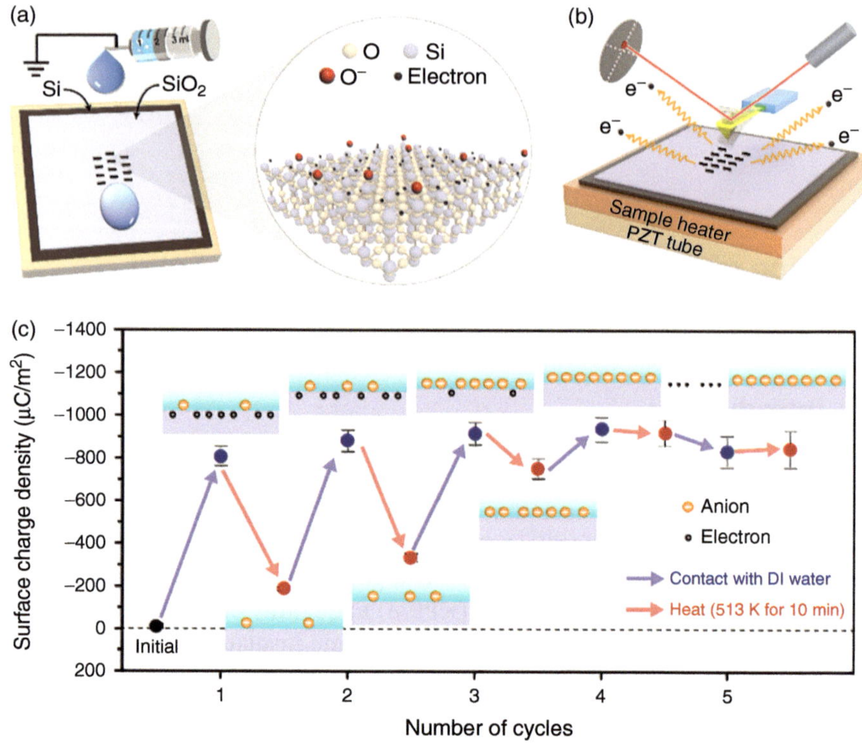

Figure 6.1 Temperature effect on the contact electrification between DI water and SiO_2. (a) The setup of the contact charge experiment. (b) The setup of AFM platform for the thermionic emission experiments. (c) The charge density on the SiO_2 surface in the charging and heating cycle tests. [3] / Springer Nature / CC BY-4.0.

with considering the breakdown field. When an ion is attached on the surface, an available charge position will be permanently occupied. In the next cycle of contact charging with DI water, both electrons and ions will transfer to the surface, but the number of the transferred electrons will decrease due to the decrease in the available charge position. Therefore, the removable charges on the insulator surface decreased with the increasing of test cycles. Though the ions are accumulating on the surface, the electron transfer was demonstrated to play a dominant role in the CE between a virgin SiO_2 surface and DI water in the very first contact. The ratio of electron transfers to ion transfers was more than 3.4.

Figure 6.2a–f shows the temperature effect on the CE between DI water and different insulators. The charge density on the insulator surface was found to decay exponentially, which was consistent with the thermionic emission theory. For some materials, the surface charge density became more positive, which implies that the solid received electrons in the CE. And some solid received holes, such as Si_3N_4, HfO_2, Al_2O_3, and AlN. It is easy to understand that the negative charge density (SiO_2) or positive charge density (Si_3N_4, HfO_2) decreased with time at high temperatures. This is because the negative charges on the solid surface are electrons and negative

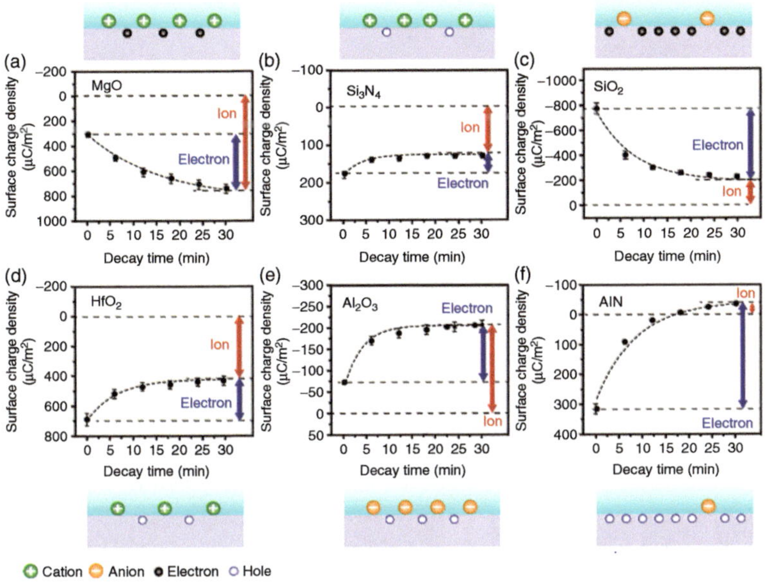

Figure 6.2 Temperature effect on the contact electrification between DI water and different insulator surfaces. (a) MgO, (b) Si_3N_4, (c) SiO_2, (d) HfO_2, (e) Al_2O_3, and (f) AlN. As for ion adsorption, both physical adsorption and chemical adsorption are possible, and they can be quantified using the method introduced by Lin et al. [3] / Springer Nature / CC BY-4.0.

ions, and the positive charges are holes and positive ions. The electrons and holes were emitted at high temperatures and resulted in a decay of the surface charge density. It is surprising that the charge density on some materials increased with decay time, such as MgO and Al_2O_3. It implies that the solids can receive both electrons and positive ions (MgO) when they contact with the DI water, or both holes and negative ions (Al_2O_3). These results imply that electron transfer and ion transfer in liquid–solid CE are independent of each other. Moreover, the electrons and ions generated in the liquid–solid CE may be segregated at different surface areas.

In order to promote the contact-electrification at solid–liquid interface, functional groups play a key role in determining the chemical properties of organic compounds. Moreover, they also have a significant impact on the electronegativity of electrically frictional materials [11]. The selection of monomers containing suitable functional groups [12] and the control of monomer ratios [13] are the effective means of controlling the electronegativity of polymeric electrically charged materials. As shown in Figure 6.3a, Tao et al. synthesized fluorinated poly(phthalazinone ether)s (FPPEs) by introducing phthalazinone moieties with strong electron-donating properties [14]. The materials containing 25% phthalazinone moieties exhibit a special crystalline behavior that allows them to trap and store more electrons. Fan et al. employed a physical etching process to prepare three regular polymer pattern arrays in the form of lines, cubes, and pyramids (Figure 6.3b) [15] Chemical modifications can cause direct changes in chemical bonding and are more commonly used in

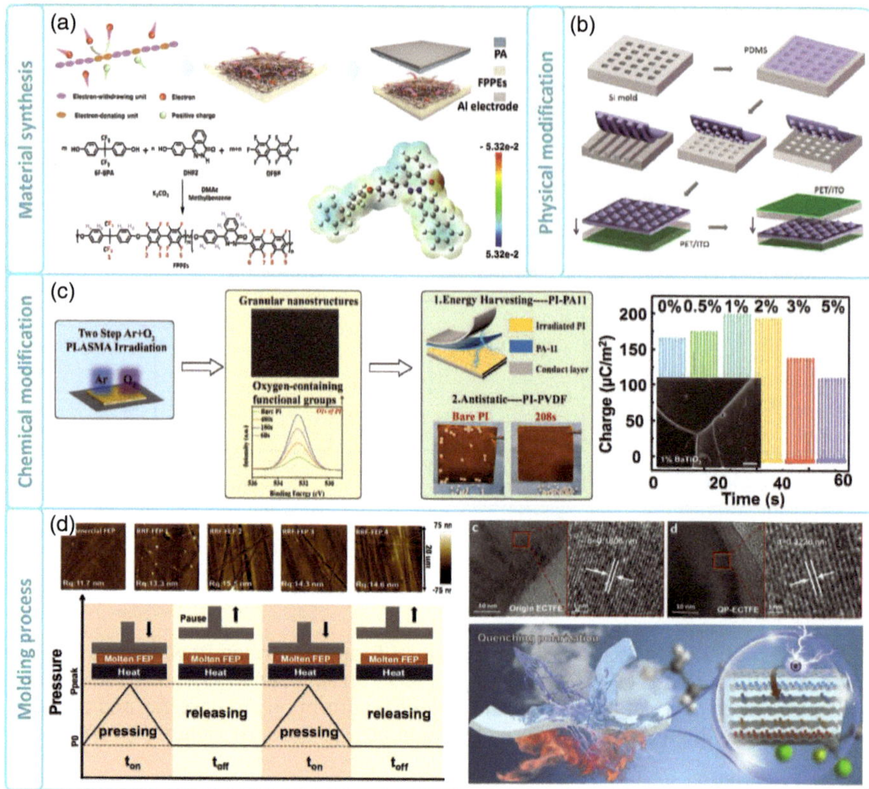

Figure 6.3 Modification for CE at solid–liquid interface. (a) Molecular structure design and synthesis of materials. Reproduced with permission of Ref. [14], 2023 © John Wiley & Sons. (b) Physical modification. Reproduced with permission of Ref. [15], 2022 © American Chemical Society. (c) Chemical modification including ion irradiation and inorganic doping. Reproduced with permission of Ref. [16], 2022 © Elsevier. (d) Special molding process to improve the CE properties of the material. Reproduced with permission of Ref. [58], 2022 © Springer Nature.

material modifications (Figure 6.3c). Wang and coworkers used the surface treatment technique of plasma irradiation to increase the C–O and C–O content on the PI surface [16]. This was achieved by utilizing in-situ two-step Ar + O_2 reactive ion etching (RIE), which resulted in the formation of granular nanostructures on the PI surface. By doping the F-PI films with $BaTiO_3$ nanoparticles, Wang et al. achieved the creation of electron deep traps and interfacial polarizations at interface [17].

In addition, well-designed processing can also enhance the charge density for contact initiation. As shown in Figure 6.3d, Liu et al. used repeated rheological forging to effectively modulate the surface functional group composition, crystallinity, and dielectric constant of fluorinated ethylene propylene. By utilizing a forged 30 μm-thick film, an air breakdown mode TENG was capable of achieving an ultra-high charge density of 510 $\mu C/m^2$ [18]. The team has also developed a quenched polarization (QP) method to generate ultra-high and long-lasting frictional charges on frictional electropolymers with weak dipolarity. The QP films can achieve a charge density of 391 μC/m^2, which is 200% higher than that of corona polarization. This method enables the use of charge storage materials that are not limited to highly insulating polymers [19].

Further research summarizes the factors at the microscopic level that influence the frictional starting properties of the polymer itself [18]. At the atomic level, the electronegativity of the atoms determines the electron-capturing capacity of the functional groups. This, in turn, influences the polarity and density of charges induced through friction [20, 21]. At the chain level, the molecular chain orientation plays a significant role [22]. It determines the probability of functional groups exposed in the surface region, coming into contact with the counter material, as well as the probability of the electron clouds overlapping [23]. Furthermore, higher crystallinity leads to an ordered arrangement of molecules, resulting in larger dipole moments and higher dipole polarization. The deep traps formed at the interface between the crystalline and amorphous regions enhance the storage of charge. In addition, material characteristics such as hardness [24], surface roughness [25], and temperature difference [26] play a significant role in charge generation. It has been shown that from 0 to 145 K, the output voltage, current, surface charge density, and output power increase by 2.7 times, 2.2 times, 3.0 times, and 2.9 times with the temperature difference between the hot and cold friction layers, respectively and decrease as the temperature difference continues to increase. And also an increase in the hardness of the material usually reduces the contact radius, thus indirectly affecting the charge output.

Contact electrification on liquid–solid interface is actually a long-standing unsolved issue in chemical–physical field, and this topic has recently drawn notable attention from researchers in many fields. One of the reasons for this situation is related to the fast development of L-S TENG. Ambient water motions, including the river flows, ocean waves, and even rain drops, contain tremendous mechanical energy and TENG that can collect energy from these ambient water motions is a desirable technique for on-site energy demand. Traditional energy harvesting techniques rely on electromagnetic generators [27, 28], or piezoelectric generators both have their limitations for collecting energy from liquid–solid motions, such as size limitation, complicated structure or low conversion efficiency. Therefore, the TENG devices, which have the advantages of small size, simple system, low cost, and direct interaction with water bodies, can be a key technique to solve the problems in this direction. The operation mechanism of L-S TENG, which is based on the combination of electrification effect and electrostatic screening effect, has been thoroughly studied by many previous works. The electrification between moving liquid and solid surfaces is the fundamental element, while the design of the TENG structure is either a single-electrode type or a free-standing type. In 2013, Lin et al. reported the first L-S TENG [29, 30], which is based on a single-electrode-type structure. Then, many other researchers developed a free-standing-type TENG for collecting energy mainly relying on screen effect of the liquid [31, 32]. Last year, Wang and coworkers reported a new transistor-inspired L-S TENG device based on the bridge phenomenon between an electrification surface and an Al electrode [33].

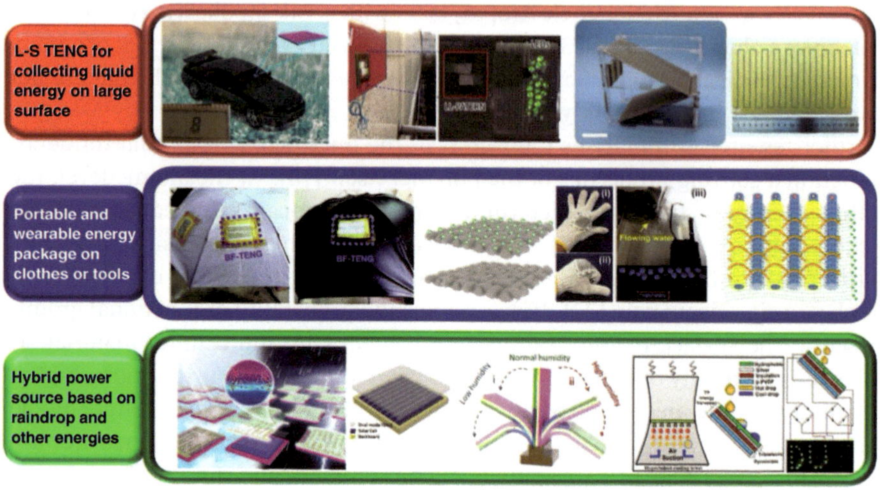

Figure 6.4 The application of L-S TENG for energy harvesting. (a) energy collection on large surface. Reproduced with permission of Ref. [35], 2017 © Elsevier. (b) portable and wearable energy package. Reproduced with permission of Ref. [41], 2020 © Royal Society of Chemistry. (c) hybrid power source. Reproduced with permission of Ref. [45], 2021 © Elsevier.

In this device, the spreading of water droplets on the electrification surface can bridge conventional interfacial effect into a bulk effect, leading to a very high instantaneous power density. One hundred commercial LEDs can be powered by one droplet (100.0 μl), which is so far the highest performance of L-S TENG. Hence, with the development of the fundamental study of S–L electrification, a series of breakthroughs can also be expected in the near future. Moreover, Wang et al. also designed a novel slippery lubricant-impregnated porous surface-based TENG (SLIPS-TENG) to overcome fast degradation of the triboelectric properties of interfacial materials under harsh environments. The SLIPS-TENG is now one of the emerging areas of liquid–solid contact electrification, which eliminates many inherent challenges encountered in current TENG.

Currently, the design of L-S TENG for energy harvesting can be briefly divided into three categories, as shown in Figure 6.4. First, the L-S TENG can be applied for energy collection on large surfaces (Figure 6.4a), such as windows [33, 34], building walls and so on. For example, Liang et al. developed a multiunit transparent TENG that can be directly applied to the window of a car [33, 34], where instantaneous power density can reach 12.66 mW/m^2. Meanwhile, L-S TENG investigated by Choi et al. [35] prepared the surface nanostructure based on the natural surface of lotus leaves. Relying on the fluorinated electrification surface with high hydrophobicity, this TENG can work for 28 days without performance degradation [34, 35]. Zhong et al. proposed a multilayer L-S TENG with fine electrode arrays targeting droplet or stream energy [36], which can fully convert the gravitational potential energy into electricity. In addition to the raindrop, the similar L-S TENG can also be applied to collect energy from ocean waves and streams. The design of L-S TENG for water wave energy is almost the same as that for raindrop energy, where the free-standing electrode array should be utilized to generate continuous displacement current.

For the electrification surface of L-S TENG, the charge density and the hydrophobicity are two major parameters to determine the performance of energy generation. Actually, the increase of the hydrophobicity does not always lead to the increase of the charge density. Many researchers have reported that the excessive increase of the roughness and the contact angle results in the suppression of transferred charges [36–38]. For hydrophobic materials, the increase of surface roughness leads to high hydrophobicity, while it may also decrease the effective contact area between the droplet and the surface (the liquid cannot completely infiltrate the uneven surface in microscale). On the other hand, the high hydrophobicity allows the liquids to smoothly move on the solid surface, while the current output from TENG is directly related to the motion velocity, suggesting that the high hydrophobicity is good for producing a high current signal. Hence, a suitable hydrophobicity (neither too high nor too low) is quite important for L-S TENG and this value needs to be specially designed for different devices. Moreover, the charge injection treatment is another commonly used method to modify the electrification surface of L-S TENG [38]. The excessive charges are injected onto the solid surface by ion guns or corona polarization, while the energy generation relies on the electrostatic screening effect caused by moving droplets (the solid electrification surface is always at a saturation state) [39]. For the charge injection treatment, the solid material should be the electret material, otherwise the surface charge density cannot be maintained during the operation. So far, a series of electret polymers have been applied for the study, including PTFE, FEP, CYTOP, and so on. Here, the surface charge density and its persistence after the injection treatment are decided by the polarity of the polymer molecules (the dipole moment, the density of the polar molecule, and so on). According to the previous reports, CYTOP film with high-temperature polarization is for now the best option for this kind of L-S TENG [40].

Secondly, the portable and wearable energy package is also one of the major research directions of L-S TENG, as can be seen in Figure 6.4b. Zhao et al. demonstrated a bi-electrode freestanding that can be attached on the surface of an umbrella [41] and this is a typical application of L-S TENG as a portable energy package. The generated output power density from this device can reach 1.838 W/m^2, which is enough to light up 30 commercial LEDs and this performance results from the device on the real umbrella. Besides umbrellas, raincoats and shoes are also suitable targets for integrating TENG. Xiong et al. reported a wearable all-fabric-based TENG, which can work as a smart cloth to harvest energy from water flow [42]. Moreover, this fiber-based L-S TENG can achieve additional function of self-cleaning, where hydrophobic nanoparticles are coated on various fabrics to serve as the electrification materials. The device can generate an instantaneous output power density of 0.14 W/m^2, while both the electrostatic energy and mechanical energy of water can be collected by this L-S TENG. On the other hand, the electrification on S–L interface is usually deficient in comparison with the case of solid–solid, especially when the device is completely immersed in the liquid. Accordingly, for the wearable L-S TENG devices, researchers need to further enhance the liquid/solid interactions by inducing more turbulences via designing microturbulators on electrification surface [43] or by designing some tiny drainage channel on the electrification surface to increase contact-separation motions between solid and liquid [44].

Third, hybrid energy generator is an interesting and intensively studied research direction of TENG, which is aiming at collecting multiple types of energy from surrounding environments. L-S TENG has excellent performance for collecting energy from raindrops and it can effectively work for the hybrid energy system (see Figure 6.4c). The advantage of L-S TENG is its universal applicability, which allows it to work parallelly with many other types of energy harvesters. For example, Zheng et al. reported the first hybrid energy system based on solar cells and L-S TENG [45, 46], which can harvest the energy from both sun and raindrops. The transparent TENG is attached on the top surface of a solar cell, while the falling raindrop can induce energy output by contacting the TENG surface. Considering the low efficiency of solar cells during rainy days, this hybrid system can well compensate for the energy loss of solar cells. Then, many progresses have been made for this solar-raindrop energy system. For example, different nanopatterns have been applied on the TENG surface, which can further enhance the light absorption of solar cells [47, 48]. Self-healing and self-cleaning materials have been employed for fabricating L-S TENG, while the light transparency and the long-term operation of the solar cells can be further protected [45]. In addition to solar energy, the L-S TENG can also work with wind energy harvester. Ren et al. reported a self-adaptive TENG, which can show spontaneous deformation with different humidity [49] (see Figure 6.4c). The humidity-sensitive material is integrated on the joint site of TENG, while the humidity change leads to the banding of TENG to different positions. In windy days with normal humidity, the TENG arrays all stand up and the vibration of these TENG arrays in the wind can generate electrical energy. Then, in rainy days, the high humidity in the air drives the TENG arrays to lay down and the electrification surface can harvest the raindrop energy based on CE effect. Moreover, Jiang et al. proposed a hybrid generator consisting of a free-standing-type L-S TENG and a pyroelectric generator (PENG) for collecting both the thermal energy and kinetic energy of the thermal fluids [50]. The output power of the TENG decreases with the increase of droplet temperature since the CE effect is related to ambient temperature. This hybrid energy harvester can achieve a peak power density of 2.6 $\mu W/cm^2$ and a maximum energy increment of 238% in comparison with a pure PENG device is also obtained, where 28 commercial LEDs can be lit up by this device. For the hybrid energy system, different energy generators can be combined with the help of rectifier circuits or storage units, where the unified energy density is higher than a single unit. Despite its diversity based on different physical effects and mechanisms, in order to simultaneously harvest different environmental energy and compensate for the drawbacks of each unit, the cooperation method between different generator units still needs further exploration.

References

1 Lin, S., Xu, L., Xu, C. et al. (2019). Electron transfer in nanoscale contact electrification: effect of temperature in the metal-dielectric case. *Adv. Mater.* 31: 1808197.

2 Xu, C., Zi, Y., Wang, A. et al. (2018). On the electron-transfer mechanism in the contact-electrification effect. *Adv. Mater.* 30: 1706790.

3 Lin, S., Xu, L., Wang, A., and Wang, Z.L. (2020). Quantifying electron-transfer in liquid-solid contact electrification and the formation of electric double-layer. *Nat. Commun.* 11: 399.

4 Raiteri, R., Martinoia, S., and Grattarola, M. (1996). pH-dependent charge density at the insulator-electrolyte interface probed by a scanning force microscope. *Biosens. Bioelectron.* 11: 1009–1017.

5 Davis, J., James, R., and Leckie, J. (1978). Surface ionization and complexation at oxide-water interface. 1. Computation of electrical double-layer properties in simple electrolytes. *J. Colloid Interf. Sci.* 63: 480–499.

6 Lagstrom, T., Gmur, T., Quaroni, L. et al. (2015). Surface vibrational structure of colloidal silica and its direct correlation with surface charge density. *Langmuir* 31: 3621–3626.

7 Bousse, L., Rooij, N., and Bergveld, P. (1983). Operation of chemically sensitive field-effect sensors as a function of the insulator-electrolyte interface. *IEEE T. Electron Dev.* 30: 1263–1270.

8 Gmur, T., Goel, A., and Brown, M. (2016). Quantifying specific ion effects on the surface potential and charge density at silica nanoparticle-aqueous electrolyte interfaces. *J. Phys. Chem. C* 120: 16617–16625.

9 Burg, K. and Delahay, P. (1981). Photoelectron emission spectroscopy of inorganic anions in aqueous solution. *Chem. Phys. Lett.* 78: 287–290.

10 Takakuwa, Y., Niwano, M., Nogawa, M. et al. (1989). Photon-stimulated desorption of H^+ ions from oxidized Si(111) surfaces. *Jap. J. Appl. Phys.* 28: 2581–2586.

11 Cao, W.T., Ouyang, H., Xin, W. et al. (2020). A stretchable highoutput triboelectric nanogenerator improved by mxene liquid electrode with high electronegativity. *Adv. Funct. Mater.* 30 (50): 2004181. https://doi.org/10.1002/adfm.202004181.

12 Wen, R.M., Zhao, B., Fan, L.M. et al. (2023). Controlling the output performance of triboelectric nanogenerator through filling isostructural metal-organic frameworks with varying functional groups. *Adv. Mater. Technol.* 8: 2201330. https://doi.org/10.1002/admt.202201330.

13 Wu, Y., Cuthbert, T.J., Luo, Y. et al. (2023). Cross-link-dependent ionogel-based triboelectric nanogenerators with slippery and antireflective properties. *Small* 19 (24): e2301381. https://doi.org/10.1002/smll.202301381.

14 Tao, X., Fu, S., Li, S. et al. (2023). Large and tunable ranking shift in triboelectric series of polymers by introducing phthalazinone moieties. *Small Methods* 7 (6): e2201593. https://doi.org/10.1002/smtd.202201593.

15 Fan, F.R., Lin, L., Zhu, G. et al. (2012). Transparent triboelectric nanogenerators and self-powered pressure sensors based on micropatterned plastic films. *Nano Lett.* 12 (6): 3109–3114. https://doi.org/10.1021/nl300988z.

16 Sun, X., Liu, Y., Luo, N. et al. (2022). Controlling the triboelectric properties and tribological behavior of polyimide materials via plasma treatment. *Nano Energy* 102: 107691. https://doi.org/10.1016/j.nanoen.2022.107691.

17 Tao, X., Li, S., Shi, Y. et al. (2021). Triboelectric polymer with high thermal charge stability for harvesting energy from 200 °C flowing air. *Adv. Funct. Mater.* 31 (49): 2106082. https://doi.org/10.1002/adfm.202106082.

18 Liu, Z., Huang, Y., Shi, Y. et al. (2022). Fabrication of triboelectric polymer films via repeated rheological forging for ultrahigh surface charge density. *Nat. Commun.* 13 (1): 4083. https://doi.org/10.1038/s41467-022-31822-2.

19 Liu, Z.Q., Huang, Y.Z., Shi, Y.X. et al. (2023). Creating ultrahigh and long-persistent triboelectric charge density on weak polar polymer via quenching polarization. *Adv. Funct. Mater.* 33: 2302164. https://doi.org/10.1002/adfm.202302164.

20 Li, S.Y., Fan, Y., Chen, H.Q. et al. (2020). Manipulating the triboelectric surface charge density of polymers by low-energy helium ion irradiation/implantation. *Energy Environ. Sci.* 13 (3): 896–907. https://doi.org/10.1039/c9ee03307f.

21 Wang, J., Wu, H.Y., Fu, S.K. et al. (2022). Enhancement of output charge density of teng in high humidity by water molecules induced self-polarization effect on dielectric polymers. *Nano Energy* 104: https://doi.org/10.1016/j.nanoen.2022.107916.

22 Calzolai, L., Gorst, C.M., Zhao, Z.H. et al. (1995). ^{1}H NMR investigation of the electronic and molecular structure of the four-iron cluster ferredoxin from the hyperthermophile Pyrococcus furiosus. Identification of Asp 14 as a cluster ligand in each of the four redox states. *Biochemistry* 34 (36): 11373–11384. https://doi.org/10.1021/bi00036a010.

23 Maggiore, J.C., Burrell, J.C., Browne, K.D. et al. (2020). Tissue engineered axon-based "living scaffolds" promote survival of spinal cord motor neurons following peripheral nerve repair. *J. Tissue Eng. Regen. Med.* 14 (12): 1892–1907. https://doi.org/10.1002/term.3145.

24 Verners, O., Lapčinskis, L., Ģermane, L. et al. (2022). Smooth polymers charge negatively: controlling contact electrification polarity in polymers. *Nano Energy* 104: 107914. https://doi.org/10.1016/j.nanoen.2022.107914.

25 Guan, D., Cong, X., Li, J. et al. (2021). Quantitative characterization of the energy harvesting performance of soft-contact sphere triboelectric nanogenerator. *Nano Energy* 87: 106186. https://doi.org/10.1016/j.nanoen.2021.106186.

26 Cheng, B., Xu, Q., Ding, Y. et al. (2021). High performance temperature difference triboelectric nanogenerator. *Nat. Commun.* 12 (1): 4782. https://doi.org/10.1038/s41467-021-25043-2.

27 Snyder, G.J. and Toberer, E.S. (2008). Complex thermoelectric materials. *Nat. Mater.* 7: 105–114.

28 Dudem, B., Huynh, N.D., Kim, W. et al. (2017). Nanopillatr-array architectured PDMS-based triboelectric nanogenerator integrated with a windmill model for effective wind energy harvesting. *Nano Energy* 42: 269–281.

29 Lin, Z., Cheng, G., Lee, S. et al. (2014). Harvesting water drop energy by a sequential contact-electrification and electrostatic-induction process. *Adv. Mater.* 26: 4690–4696.

30 Lin, Z., Cheng, G., Lin, L. et al. (2013). Water-solid surface contact electrification and its use for harvesting liquid-wave energy. *Angew. Chem. Int. Ed.* 52: 12545–12549.

31 Lee, J.H., Kim, S., Kim, T.Y. et al. (2019). Water droplet-driven triboelectric nanogenerator with superhydrophobic surface. *Nano Energy* 58: 579–584.

32 Kwak, S.S., Lin, S., Lee, J.H. et al. (2016). Triboelectrification-induced large electric power generation from a single moving droplet on graphene/polytetrafluoroethylene. *ACS Nano* 10: 7297–7302.

33 Xu, W., Zheng, H., Liu, Y. et al. (2020). A droplet-based electricity generator with high instantaneous power density. *Nature* 578: 392–396.

34 Liang, Q., Yan, X., Liao, X., and Zhang, Y. (2016). Integrated multi-unit transparent triboelectric nanogenerator harvesting rain power for driving electronics. *Nano Energy* 25: 18–25.

35 Choi, D., Kim, D.W., Yoo, D. et al. (2017). Spontaneous occurrence of liquid-solid contact electrification in nature: toward a robust triboelectric nanogenerator inspired by the natural lotus leaf. *Nano Energy* 36: 250–259.

36 Zhong, W., Xu, L., Zhan, F. et al. (2020). Dripping channel based liquid triboelectric nanogenerators for energy harvesting and sensing. *ACS Nano* 14: 10510–10517.

37 Lee, J. and Hwang, W. (2018). Theoretical study of micro/nano roughness effect on water-solid triboelectrification with experimental approach. *Nano Energy* 52: 315–322.

38 Helseth, L.E. (2019). The influence of microscale surface roughness on water-droplet contact electrification. *Langmuir* 35: 8268–8275.

39 Wang, S., Xie, Y., Niu, S. et al. (2014). Maximum surface charge density for triboelectric nanogenerators achieved by ionized-air injection: methodology and theoretical understanding. *Adv. Mater.* 26: 6720–6728.

40 Jang, S., La, M., Cho, S. et al. (2020). Monocharged electret based liquid-solid interacting triboelectric nanogenerator for its boosted electrical output performance. *Nano Energy* 70: 104541.

41 Zhao, L., Liu, L., Yang, X. et al. (2020). Cumulative charging behavior of water droplet driven freestanding triboelectric nanogenerators toward hydrodynamic energy harvesting. *J. Mater. Chem. A* 8: 7880–7888.

42 Xiong, J., Kin, M., Wang, J. et al. (2017). Wearable all-fabric-based triboelectric generator for water energy harvesting. *Adv. Energy Mater.* 7: 1701243.

43 Zhang, L., Zhang, N., Yang, Y. et al. (2018). Self-powered all-in-one fluid sensor textile with enhanced triboelectric effect on all-immersed dendritic liquid-solid interface. *ACS Appl. Mater. Interfaces* 10: 30819–30826.

44 Qian, Y., Nie, J., Ma, X. et al. (2019). Octopus tentacles inspired triboelectric nanogenerators for harvesting mechanical energy from highly wetted surface. *Nano Energy* 60: 493–502.

45 Yang, D., Ni, Y., Su, H. et al. (2021). Hybrid energy system based on solar cell and self-healing/self-cleaning triboelectric nanogenerator. *Nano Energy* 79: 105394.

46 Zheng, L., Cheng, G., Chen, J. et al. (2015). A hybridized power panel to simultaneously generate electricity from sunlight, raindrops, and wind around the clock. *Adv. Energy Mater.* 5: 1501152.

47 Yoo, D., Park, S.-C., Lee, S. et al. (2019). Biomimetic anti-reflective triboelectric nanogenerator for concurrent harvesting of solar and raindrop energies. *Nano Energy* 57: 424–431.

48 Zhao, L., Duan, J., Liu, L. et al. (2021). Boosting power conversion efficiency by hybrid triboelectric nanogenerator/silicon tandem solar cell toward rain energy harvesting. *Nano Energy* 82: 105773.

49 Ren, Z., Ding, Y., Nie, J. et al. (2019). Environmental energy harvesting adapting to different weather conditions and self-powered vapor sensor based on humidity-responsive triboelectric nanogenerators. *ACS Appl. Mater. Interfaces* 11: 6143–6153.

50 Jiang, D., Su, Y., Wang, K. et al. (2020). A triboelectric and pyroelectric hybrid energy harvester for recovering energy from low-grade waste fluids. *Nano Energy* 70: 104459.

7

Output Promotion of Triboelectric Nanogenerator by Electromechanical Structures

7.1 Introduction

Triboelectric nanogenerator (TENG) emerges as a promising energy harvesting technology originating from mechanodriven Maxwell's displacement current. Through triboelectrification and electrostatic induction, TENG efficiently transforms mechanical energy into electrical energy [1–6]. With merits encompassing effective energy conversion, versatility across multiple energy sources, eco-friendliness, facile fabrication, portability, safety, and high integration [7–12], TENG assumes a pivotal role in domains spanning energy harvesting [13–18], self-powered devices [19–23], wearable technology [24–27], environmental monitoring [28–31], and medical therapy [32–35]. This technology is poised to drive scientific and technological innovation, fostering sustainable development [36–40].

The output performance of TENG determines the process of its practicability and application diversity. To assess and compare the performance of various TENG devices, Zi et al. have proposed a comprehensive evaluation standard consisting of both structural figure of merits (FOM_S) and material figure of merits (FOM_M), clarified that the surface charge density is the key to determining performance [41]. They used LS-TENG as an example, employing finite element method simulations to generate the V-Q curve under an external 100 MΩ resistance, as shown in Figure 7.1a, b. The area enclosed by the V-Q curve represents the energy output of the TENG per cycle, which is referred to as the cycles for energy output (CEO). The cyclic charge Q_C, defined as the difference between the maximum and minimum charge in the steady state, is calculated for each CEO. The V-Q curves with varying external resistances are illustrated in Figure 7.1c, revealing that the circulating charge Q_C consistently falls below the maximum transferred charge $Q_{SC,max}$. In the pursuit of increasing the output capacity by maximizing the circulating charge, Zi et al. utilized a switch with external resistors connected in parallel to induce an instantaneous short-circuit state, as illustrated in Figure 7.1d. The resulting V-Q curve obtained under this condition is termed the cycle for maximum energy output (CMEO). Figure 7.1e presents the CMEO with different external resistances, indicating that larger external resistances correspond to higher output energy per

Triboelectric Nanogenerators: Technology, Applications, and Commercialization, First Edition.
Yunlong Zi, Hengyu Guo, Jie Wang, Chi Zhang, Xiangyu Chen, and Qing Zhao.

cycle. Hence, the maximum energy output E_m can be achieved when the external resistance approaches infinity.

$$E_m = \frac{1}{2} Q_{SC,max} \left(V_{OC,max} + V'_{max} \right) \tag{7.1}$$

where $Q_{SC,max}$, $V_{OC,max}$, and V'_{max} are all proportional to the tribocharge density σ. So, E_m should be proportional to the square of σ. Moreover, considering that the average output power and energy conversion efficiency of CMEO are proportional to $\frac{E_m}{x_{max}}$, the dimensionless structure of FOM is proposed as

$$\mathrm{FOM_s} = \frac{2\varepsilon_0}{\sigma^2} \frac{E_m}{Ax_{max}} \tag{7.2}$$

The $\mathrm{FOM_S}$ of TENG with different modes/structures, including vertical contact separation (CS) mode, lateral sliding (LS) mode, single-electrode contact (SEC) mode, sliding freestanding triboelectric-layer (SFT) mode, and contact freestanding triboelectric-layer (CFT) mode, are displayed in Figure 7.1f. Their relationship is arranged in descending order as CFT > CS > SFT > LS > SEC. The properties of the material are only related to σ^2 that can be defined as $\mathrm{FOM_M}$. Among the materials commonly used in CE, fluorinated ethylene propylene (FEP) is acknowledged for its significant electronegativity and thus utilized as a standard reference. Combing liquid Galinstan, Zi et al. quantified the charge density of different materials, thereby establishing the dimensionless figure of merits of materials ($\mathrm{FOM_{DM}}$).

$$\mathrm{FOM_{DM}} = \frac{\sigma^2_{\frac{\text{Material}}{\text{Galinstan}}}}{\sigma^2_{\frac{\text{FEP}}{\text{Galinstan}}}} \tag{7.3}$$

Finally, combining the FOM of the structure and the material, the FOM of the total performance of TENG is defined as

$$\mathrm{FOM_P} = \mathrm{FOM_s} \cdot \sigma^2 = 2\varepsilon_0 \frac{E_m}{Ax_{max}} \tag{7.4}$$

The proposed standards and evaluation methods offer solid theoretical guidance to effectively enhance the output performance of TENG and pave the way for the practical applications and industrialization of this technology.

Through progressive refinement and enhancement of TENG technology, researchers have unveiled the emergence of air breakdown phenomena within the device, as evidenced in Figure 7.2. In the realm of contact-separated TENGs, as charge accumulates on the tribolayer, an excessive potential difference precipitates direct air breakdown, as depicted in Figure 7.2a. In the case of the free-standing tribolayer TENG, air breakdown exists between the friction materials and between the electrodes, as shown in Figure 7.2b.

In a recent study by Wang et al., an innovative experiment was conducted involving the application of an air ionization gun to introduce static surface charge onto an FEP film [42]. The outcome revealed a pivotal surface charge density, marked as critical, that incited air breakdown, as exemplified in Figure 7.3a. The air breakdown phenomenon is pronounced when a surplus of electrons, surpassing the threshold of 240 μC/m^2, is introduced into a 50 μm PTFE film, as graphically depicted in

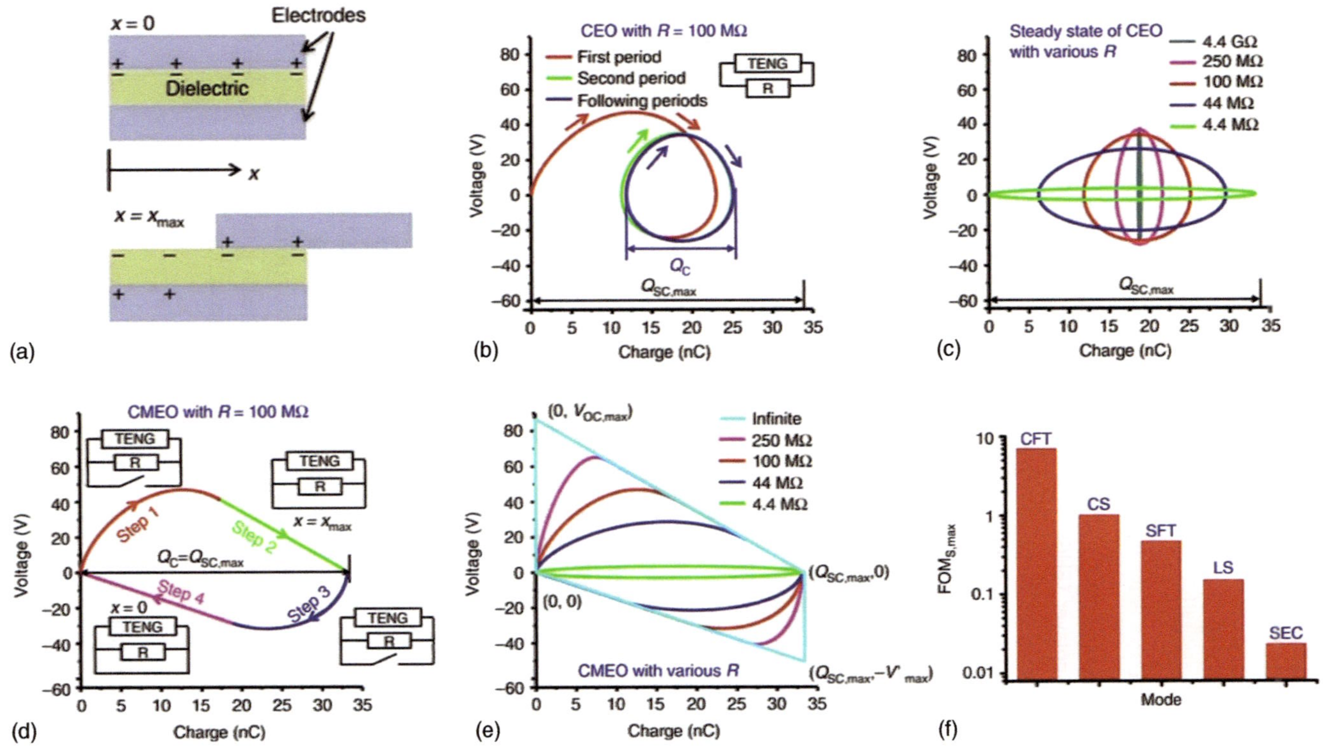

Figure 7.1 FOM of TENG. (a) Schematic diagram of LS-TENG with horizontal sliding distance of 0 and x_{max}. (b) Curves of V-Q with an external 100 MΩ resistor. (c) V-Q curves with different external resistances. (d) Maximum energy output cycle obtained by a switch connected in parallel with an external 100 MΩ resistor. (e) Maximum energy output cycles at different external resistances. (f) FOM_S of TENG with different modes/structures. (Reproduced with permission of Ref. [41], 2015 © Springer Nature).

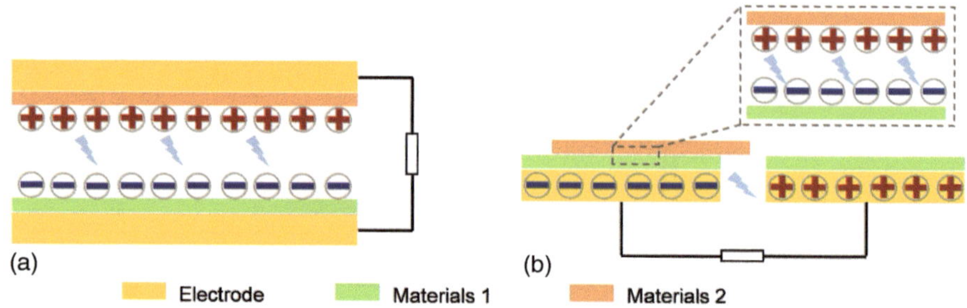

Figure 7.2 Air breakdown in TENG devices. Air breakdown in (a) contact-separated TENG and (b) free-standing tribolayer TENG.

Figure 7.3b. The authors advanced an elucidative air breakdown model for TENG, elegantly captured by Equation (7.5), thereby providing a theoretical framework that comprehends the attrition of surface charge during air breakdown, as visually expounded in Figure 7.3c.

$$\frac{A(Px)}{ln(Px)+B} - \frac{d\sigma_0 x}{\varepsilon_0(d+x\varepsilon_r)} > 0 \tag{7.5}$$

The determination of constants A and B is reliant on gas composition and pressure, with atmospheric pressure denoted by P. The convergence of air breakdown is notably correlated with the point of intersection between the TENG output voltage and Paschen's curve, as vividly illustrated in Figure 7.3d. Furthermore, Equation (7.5) provides a lucid depiction of the structural impact of the TENG, specifically the dielectric film thickness (d), on σ_{max}. Through numerical depiction of the σ_{max} versus d relationship as per this equation, it becomes evident that σ_{max} escalates with decreasing film thickness, as evidenced in Figure 7.3e. By curtailing d from the range of tens of microns to a few hundred nanometers, a remarkable elevation of σ_{max} by two orders of magnitude is achieved, transcending the 10 mC/m^2 threshold.

In conventional settings, a reduction in the equivalent thickness of the tribolayer has been noted to elevate the air breakdown threshold, thereby culminating in an elevated surface charge density limit. Nonetheless, the constrained frictional charge transfer characteristics of the material impede the attainment of the device's theoretical limit value. Consequently, this section will concentrate on strategies aimed at fostering charge accumulation and achieving a heightened maximum charge density via innovative electromechanical structural design approaches.

7.2 Charge Excitation Mechanism

7.2.1 External Charge Excitation

Xu et al. achieved an unprecedentedly high surface charge density in ambient conditions through a judicious structural design of the TENG [43]. The configuration of the self-charge-pumping TENG (SCP-TENG) device is depicted in Figure 7.4a,

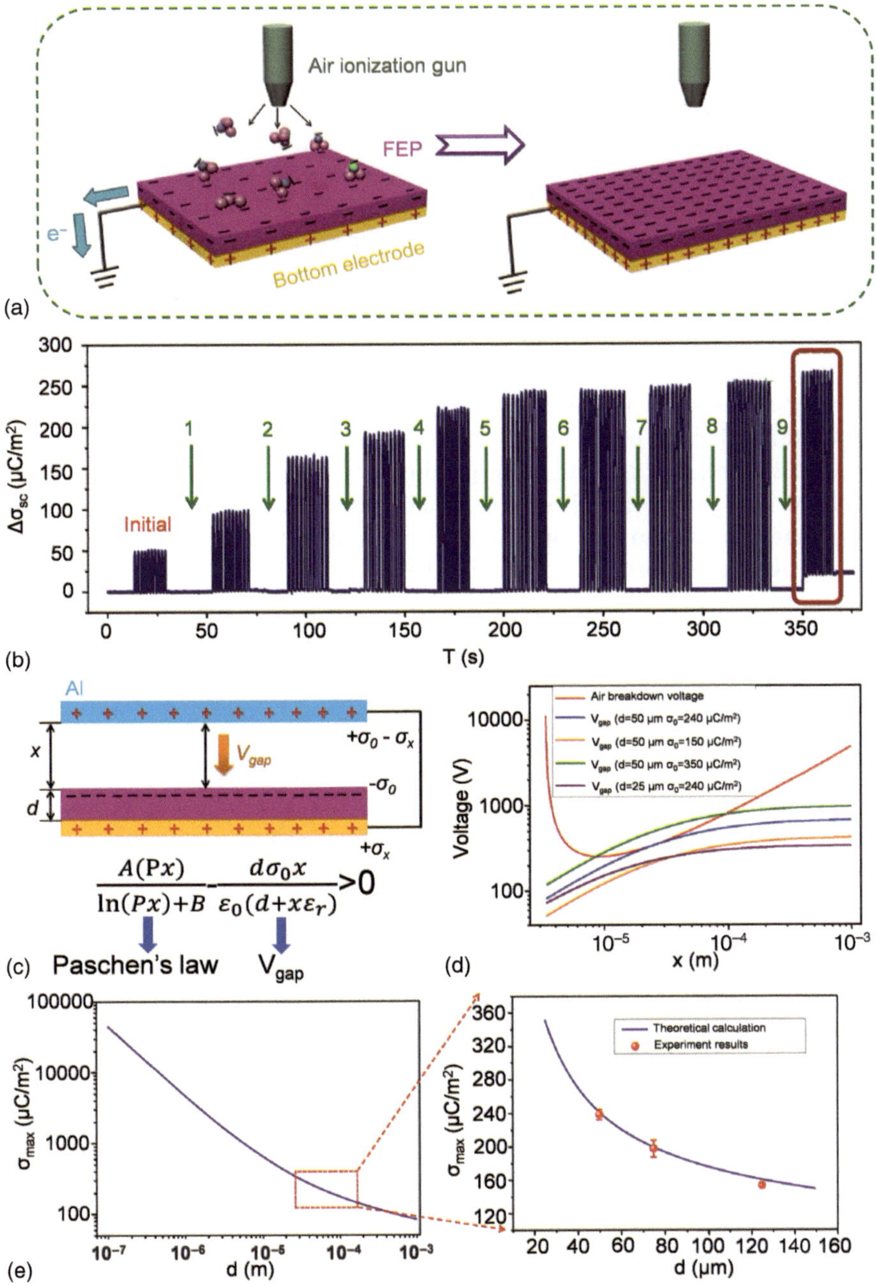

Figure 7.3 Maximum surface charge density of contact-separated TENG. (a) Ionized air injects an electrical charge. (b) The maximum surface charge density of PTFE (50 μm) was determined by charge injection. (c) Air breakdown model for TENG. (d) Voltage curve of TENG and the Paschen's curve. (e) Theoretical maximum surface charge density for different film thicknesses. Reproduced with permission of Ref. [42], 2014 © John Wiley & Sons.

featuring a primary TENG and a pump TENG. The charge generated by the operation of the pump TENG is rectified and introduced as an external excitation charge onto the two electrodes of the main TENG. As charge progressively accumulates on the main TENG electrode, the overall device output steadily amplifies (Figure 7.4b). Notably, the pump TENG harnesses the principles of contact electrification and electrostatic induction, whereas the main TENG solely relies on electrostatic induction. The magnitude of the charge output from the pump TENG during a single operational cycle is proportional to its surface area, as depicted in Figure 7.4c. Figure 7.4d illustrates the correlation between the output voltage of the main TENG and the operation time of the pump TENG for varying areas. A larger pump TENG area results in accelerated enhancement of the main TENG output voltage in a shorter duration, analogous to charging a capacitor. Similarly, the accumulation rate of output charge on the main TENG is expedited with the aid of a larger pump TENG (Figure 7.4e). Utilizing a pump TENG measuring $7 \times 7\,\mathrm{cm}^2$, the transferred charge of the main TENG reaches $10.2\,\mu\mathrm{C}$, corresponding to a charge density of $1020\,\mu\mathrm{C/m}^2$, as depicted in Figure 7.4f. This external charge excitation effectively elevates the surface charge density to four times the air breakdown threshold [42], offering a promising avenue to substantially enhance the power output of TENGs for practical applications.

7.2.2 Self-Charge Excitation

To enhance both the integrability and the efficiency of charge accumulation within the framework of external charge excitation, Liu et al. introduced and materialized a self-charge-excited (SCE) TENG system characterized by a robust and consistent high output [44]. Drawing inspiration from the principles underlying conventional magnetically excited generators, the architecture of the SCE-TENG is illustrated in Figure 7.5a, and its operational concept centers on charge exchange between two distinct sets of capacitors. The integration of an external capacitor bank facilitates an automatic transition between parallel and serial connections during the phases of contact and separation. This dynamic configuration permits the double charge generated by the parallel capacitors to be effectively injected into the TENG component, thereby engendering a self-excitation phenomenon. Within this study, the self-voltage-multiplying circuit (SVMC) was harnessed to realize the SCE-TENG system, as depicted in Figure 7.5b. For a more comprehensive comprehension of the self-charge excitation mechanism, the circuit is streamlined into an SVMC unit encompassing three rectifier diodes and two ceramic capacitors. In the initial configuration (Figure 7.5c), the capacitance and charge of both the TENG and the ceramic capacitor are designated as 2C, 2Q and C, Q, respectively. Upon the separation of the two electrodes, the TENG's capacitance experiences a pronounced reduction to C_L, causing a subsequent elevation in voltage V. This sequence leads to the establishment of an equilibrium state characterized by charge transfer 2Q-Q′ from the TENG to the ceramic capacitor, where Q′ denotes the real-time charge of the primary TENG (Figure 7.5d). Throughout this progression, the two ceramic capacitors are connected in series. Given larger separation distances, the TENG's

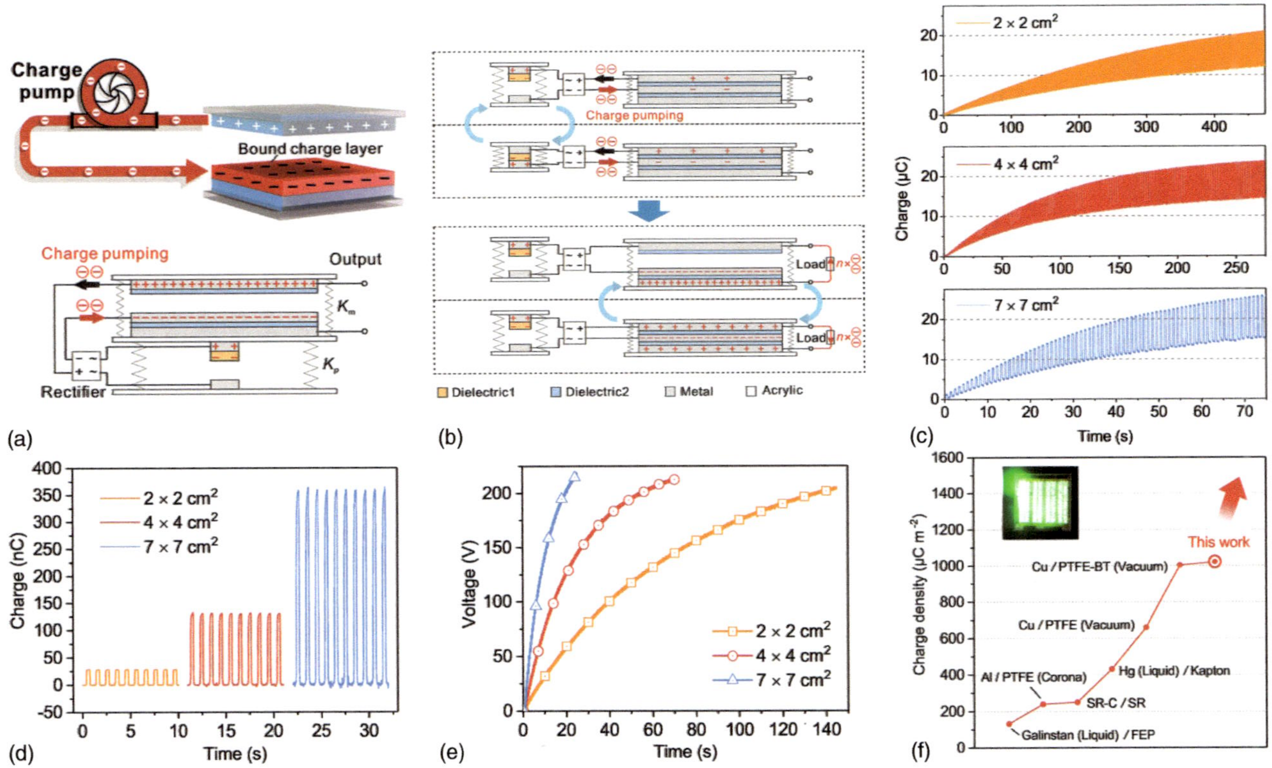

Figure 7.4 Charge pumping for ultrahigh charge density under ambient conditions. (a) Design of external charge excitation device, including pump TENG and main TENG. (b) The working mechanism of SCP-TENG. (c) The output charge of the pump TENG with different areas. (d) The efficiency of the pump TENG with different areas to boost the main TENG's voltage. (e) Relationship between the output charge of the main TENG versus time with different areas of the pump TENG. Performance comparison of SCP-TENG. (Reproduced with permission of Ref. [43], 2018 © Elsevier).

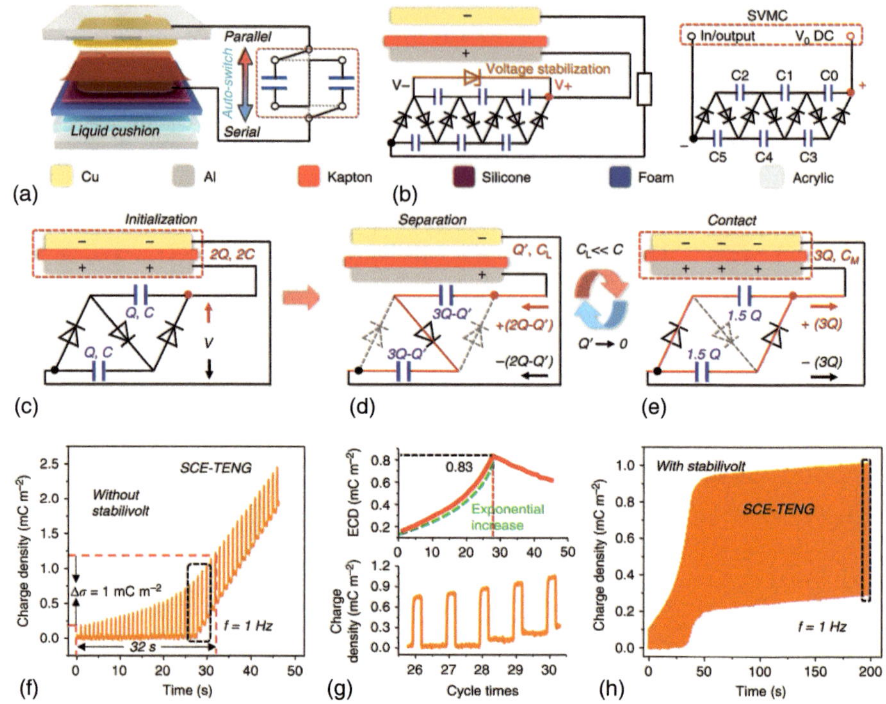

Figure 7.5 Structural design and performance characterization of SCE-TENG. (a) Structural design of SCE-TENG. (b) Self-excited voltage multiplication circuit. (c,d) Self-excited voltage multiplication circuit. (f,g) Charge density of SCE-TENG output under unstabilized voltage. (h) The charge density of the SCE-TENG output under stabilized voltage. (Reproduced with permission of Ref. [44], 2019 © Springer Nature).

capacitance C_L further diminishes to C, allowing for the assumption $Q' = 0$. In such cases, the entirety of the 2Q charge from the TENG is conveyed to the ceramic capacitors (equating to 3Q charge per capacitor). Upon the restoration of contact between the two electrodes (Figure 7.5e), the TENG's capacitance reverts to 2C, leading to a corresponding decline in voltage V. Consequently, a charge transfer from the ceramic capacitor to the TENG ensues. During this phase, the unidirectional nature of the diode causes the automatic parallel connection of the two ceramic capacitors. This orchestration facilitates the return of 3Q charges to the TENG, thereby effectuating self-induced excitation. Leveraging the inherent self-charge excitation facilitated by the SVMC, the SCE-TENG demonstrates a remarkable achievement, achieving a peak charge density of $1.0\,\mathrm{mC/m^2}$ at 1 Hz within a mere 32 seconds, as depicted in Figure 7.5f. This noteworthy outcome stands in stark contrast to the external charge excitation strategy, delivering a sixfold acceleration in charge accumulation. The self-excitation approach, distinct from its linear external-charge excitation, showcases an exponential charge accumulation profile, ensuring a more efficient saturation of charge density, as evidenced in Figure 7.5g. However, upon reaching a critical threshold, a rapid decline in output occurs due to the onset of air breakdown. To address this, the implementation of a suitable voltage regulator diode becomes essential to stabilize the charge output. Evidently, the integration of a voltage regulator diode significantly enhances the charge density stability of the SCE-TENG, as demonstrated in Figure 7.5h. Remarkably, the charge density of the SCE-TENG, when coupled with a voltage regulator diode, attains a steady state within a mere 42 cycles. This study introduces a novel avenue, establishing a robust framework for achieving ultrastable power generation through charge-excited TENG systems, thereby paving the way for extensive large-scale applications.

7.2.3 Charge Excitation for Sliding Mode TENG

In the context of sliding TENGs, the feasibility of both external charge excitation and self-charge excitation strategies remains viable. Bai et al. accomplished an exceptional surge in charge density output within a rotating sliding TENG through a well-considered design of a synchronous rotating structure, accompanied by a charge pumping approach [45]. The intricate configuration of the Rotary Charge Pumping Triboelectric Nanogenerator (RC-TENG) is vividly depicted in Figure 7.6a, constituting a pump TENG and a main TENG. The RC-TENG incorporates three functional electrodes, namely the pumping TENG electrode (E_p) responsible for furnishing bound charge to the main TENG, the electrode (E_s) designated for storing bound charge on the main TENG rotator, and the electrode (E_o) intended for electrical energy output on the main TENG stator (Figure 7.6b). The operational mechanism of the RC-TENG is elucidated in Figure 7.6c, wherein the rectified operation of the pump TENG introduces charge to the rotor electrode of the main TENG, subsequently releasing it through the stator electrode. As indicated in Figure 7.6d,e, the pump TENG attains peak output charge and current values of 1.08 μC and 20 μA, respectively. Notably, the RC-TENG exhibits the capacity to generate an output charge of 4.5 μC and a current of 470 μA (Figure 7.6f,g). With equivalent rotational frequency and contact area, a normal S-TENG facilitates a charge transfer of 0.5 μC. In stark contrast, the RC-TENG showcases an impressive ninefold surge in charge density, underscoring its potential to address the power output bottleneck in mechanical energy harvesting. This groundbreaking achievement in adapting the strategy to both rotational and sliding TENGs, particularly in scenarios involving low-frequency agitations, holds immense promise for catalyzing advancements in high-power TENG design across diverse application domains.

Long et al. introduced a pioneering concept, the floating self-excited sliding TENG (FSS-TENG), which engenders autonomous enhancement of charge density through self-excited amplification interactions between the rotor and stator [46]. The intricate composition of the FSS-TENG is vividly depicted in Figure 7.6h, comprising a circular rotor and stator. To obviate direct air breakdown between metal electrodes, the stator surface is enveloped by a 25-μm-thick nylon film. Notably, the rotor surface accommodates staggered attachment of a PTFE film and a nylon film. Figure 7.6i delineates the simplified operational principle of the FSS-TENG, involving a high-voltage source and a noncontact mode sliding TENG with a charge-supplementing electrode. Introducing a unidirectional conduction Voltage

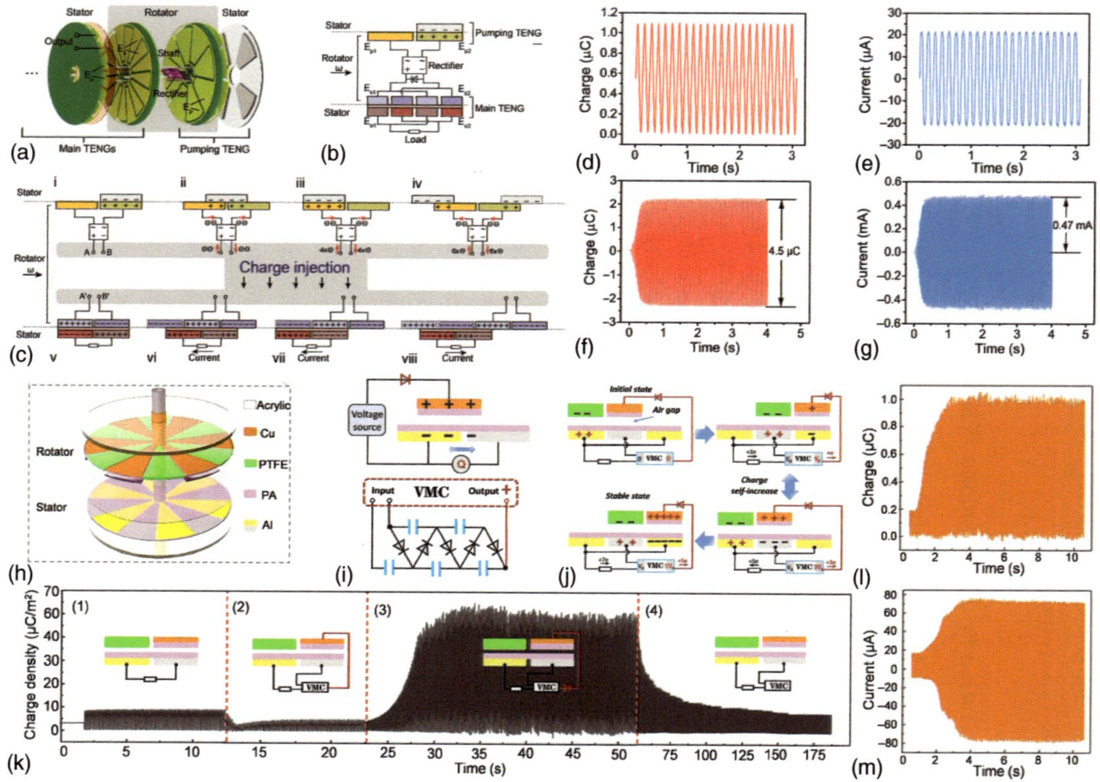

Figure 7.6 External charge excitation and self-charge excitation strategies for sliding mode TENG. (a,b) Structural design and simplified circuitry of RC-TENG. (c) Mechanism of external charge excitation for RC-TENG. (d,e) The output charge and current of the pump TENG. (f,g) The output charge and current of the main TENG. (h,i) Structural design and simplified circuitry of FSS-TENG. (j) Self-charge excitation mechanism of FSS-TENG. (k) Comparison of the output charge density of several different floating TENGs. (l,m) The output charge and current of the FSS-TENG. (Reproduced with permission of Ref. [46], 2021 © Springer Nature).

Multiplication Circuit (VMC) injects charge into the supplementary electrode. The FSS-TENG's self-excitation cycle is graphically depicted in Figure 7.6j. In the initial phase, a minor negative charge is preloaded onto the PTFE, with the VMC's terminal voltage at equilibrium. Subsequent rotation of the rotor initiates AC output from the stator electrodes, through the VMC to charge the rotor electrodes. A collaborative effect between the negatively charged PTFE and positively charged rotor electrodes compels electron movement within the stator electrodes, consequently augmenting the induction output from the FSS-TENG. Following several cycles, equilibrium is established at both VMC terminals, resulting in saturation of injected charge within the rotor electrode, thereby culminating in maximal output. This innovation harbors profound implications, particularly in harnessing self-amplification mechanisms to elevate charge density in sliding TENG systems. Figure 7.6k graphically presents the dynamic trajectories of output charge density under four distinct operational scenarios: (i) Standard noncontact TENG output, (ii) Partial VMC-directed output, lacking diode within excitation path (depicted as the red line), (iii) Employing a unidirectional diode, and (iv) Absence of an excitation path. This delineation unequivocally underscores the robust augmentation potential of the meticulously devised FSS-TENG. Figure 7.6l,m illustrates the dynamic temporal evolution of charge and current output (at 10 MΩ), respectively, during the process of self-charge excitation at 300 rpm. Remarkably, a charge of 1 μC and a current of 76 μA are expeditiously attained within a mere five seconds. Subsequently, the surface charge density of the copper (Cu) electrode attains an impressive 71.53 $\mu C/m^2$ in its floating state. This research presents an effective and ingenious strategy that not only surmounts the challenge of sliding TENG self-charge excitation but also effectively addresses the issue of low floating output.

7.3 Other Promotion Strategies

7.3.1 Charge-Space Accumulation Mechanism

In the realm of conventional S-TENG systems, the inevitable emergence of air voids sandwiched between the two tribolayers serves possibility for air breakdown occurrences, thus imposing significant limitations upon surface charge density and overall output performance. Addressing this limitation, a pioneering design termed the charge space accumulation S-TENG (CSA-S-TENG) was ingeniously introduced by He et al. with the primary aim of elevating the surface charge density [47]. The structural blueprint of the CSA-S-TENG is unveiled in Figure 7.7a, featuring PTFE and nylon as the designated friction materials for the sliding layers, with their respective electrodes being grounded. To further curtail the formation of air voids, a strategically incorporated sponge foam acts to optimize the interfacial contact dynamics between the tribolayers. The charge transfer potential of a conventional S-TENG, displayed in Figure 7.7b, is inherently curtailed by air breakdown within the interstitial gap, an issue promptly mitigated by the introduction of a grounded electrode upon the sliding layer. The consequential alterations in potential distribution within the air voids, as evinced by the simulated results in Figure 7.7c, attest to

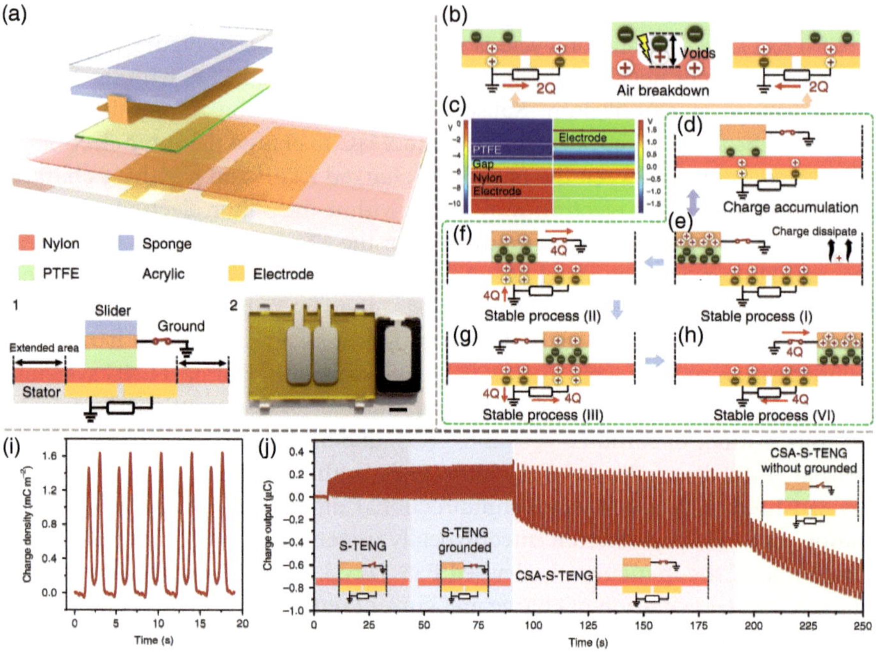

Figure 7.7 Structure and mechanism of CSA-S-TENG. (a) The 3D schematic of CSA-S-TENG. (b) Working mechanism and air breakdown of traditional S-TENG. (c) Potential simulation of air gap existence in CSA-S-TENG with/without shielding electrode. (d) Schematic of CSA-S-TENG with the same charge distribution as traditional S-TENG in the initial stage. (e–h) Working mechanism and relative charge transfer quantity of CSA-S-TENG in four states during half working period. (i) The output charge density curve of CSA-S-TENG. (j) Dynamic output charge curve under four working circumstances. [47] / Springer Nature / CC BY-4.0.

a threefold amplification in potential difference between the upper PTFE layer and lower PA layer, with and without the intervening shielding layer. This augmentation considerably thwarts air breakdown events and, concurrently, enables greater charge density accumulation across the tribolayers. In addition, the presence of the shielding weakens the electrostatic induction effect of the bottom electrodes, thus affecting the output charge. Apparent from the comparative cases depicted in Figure 7.7j, whether the shielding electrodes are grounded or ungrounded, there exists limited enhancement in the effective charge output between the two lowermost electrodes of the S-TENG. Figure 7.7d delineates the schematic charge distribution of the CSA-S-TENG during its initial phase of evolution from the conventional S-TENG configuration. The CSA-S-TENG benefits markedly from the integrated shielding electrode and the judicious incorporation of an additional blank-triboarea structure. Following several sliding cycles, a transformative charge distribution scenario unfolds on the surface of the upper tribolayer (PTFE), whereby charges are meticulously accumulated from the tribosurface situated on the lower electrode and the supplementary triboarea. This intricate process culminates in the establishment of a resolute and stable charge distribution state, elegantly displayed in Figure 7.7e. Figure 7.7e–h shows the working mechanism and four charge distribution states in half a cycle of CSA-S-TENG when surface charge density reaches stable, which indicates that 4Q transfer charge quantity can be generated from the bottom electrodes as well as the shielding electrode. The third case (red background area) in Figure 7.7j shows the dynamic charge output curve of CSA-S-TENG with the grounded shielding electrode, and the charge space-accumulation process can be observed and the charge output quickly reaches a stable state, which is consistent with theoretical predictions. Experimental refinement bore fruit in the achievement of a commendable stable output charge density of 1.63 mC/m^2, as evinced by the fundamental charge output curve showcased in Figure 7.7i. This work provides a deep understanding of the working mechanism of S-TENG and an effective strategy for promoting its output performance.

7.3.2 Ternary Tribolayer Architecture

Deng et al. introduced a novel ternary electrification layered (TEL) configuration to enhance the output performance of the S-TENG systematically [48]. Distinguished from preceding binary electrification layered (BEL) TENGs, the TEL-TENG ingeniously employs three distinct friction materials, as visually portrayed in Figure 7.8a. With the introduction of the third material, two more electron transfer processes will occur among these three different materials in TEL-TENG system, as indicated by the arrows in Figure 7.8b. An incisive dissection of the operative principles of the TEL-TENG is undertaken, casting it against the backdrop of its traditional BEL-TENG counterpart (concretely laid out in Figure 7.8c–e). In the conventional BEL-TENG framework, the juxtaposition of electrification layer 1 (EL1), represented by copper, with electrification layer 2 (EL2), embodied by FEP, prompts charge transfer at the point of contact due to the disparate charge affinities within the triboelectric series. Upon the rotator's departure from electrode A (EA) to electrode B (EB), an exchange of electrons transpires between EA and EB driven by the inherent electric potential disparity between these two electrodes – an elucidation graphically presented in Figure 7.8f. A comprehensive exploration is conducted to scrutinize the implications of varying air gaps on the resulting voltage dynamics, as underscored by the simulation findings elucidated in Figure 7.8g. As the interelectrode air gap contracts to a mere 0.03 mm, a corresponding voltage of approximately 130.97 V emerges – an equivalency akin to 4.37 mV/m^1. This calculated result serves to enrich the understanding of the system's performance attributes and their relevance to the S-TENG's overall functionality. In response to the potential air breakdown issue precipitated by the formidable electric field, the third electrification layer (EL3), Kapton, is introduced to prevent direct contact between EL1 and EL2. Since the charge affinity of EL3 is between EL1 and EL2, this can effectively reduce the electric field caused by the existence of the air gap. When the system achieves a stable rotational state, the voltage differential between the two electrodes notably experiences a 26.5% diminution (as illustrated in Figure 7.8i), effectively countering the potential air breakdown phenomenon. The intricate dynamics of charge accumulation within the TEL-TENG framework

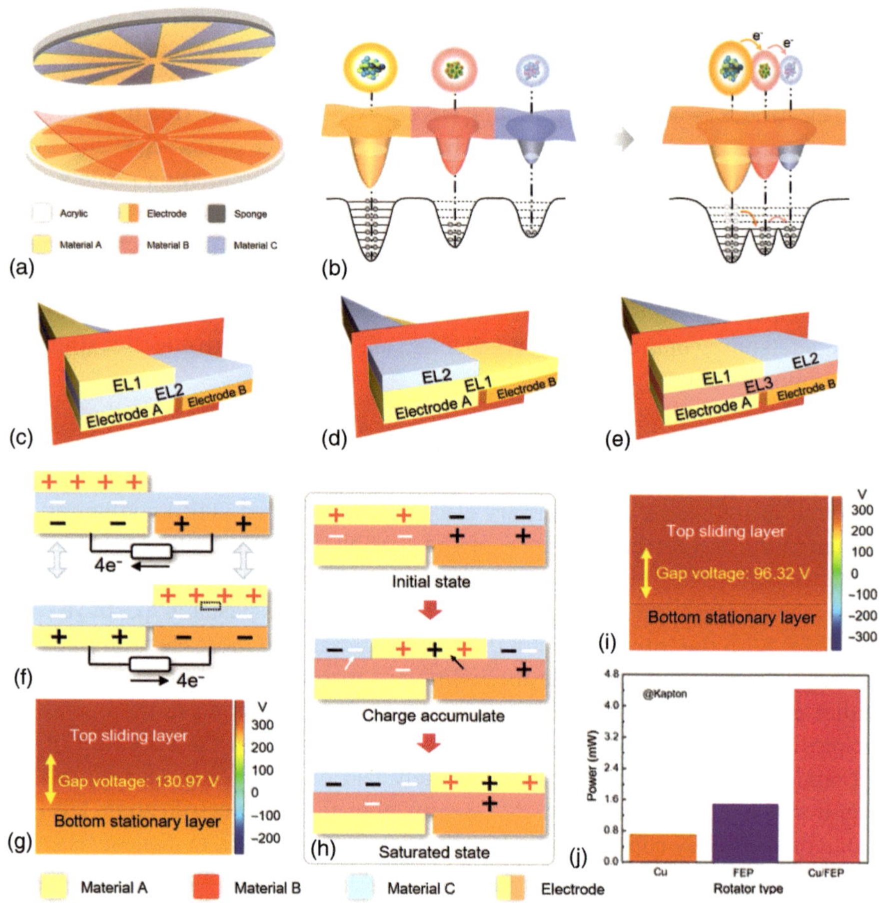

Figure 7.8 Structural design and performance characterization of TEL-TENG. (a) Schematic illustration of a TEL-TENG, which consists of an upper rotator and a lower stator. (b) The electron-cloud-potential-well model depicts the charge transfer at the surface state between different materials before and during contact. One repeating unit of BEL-TENG with different materials of either (c) electrification layer 1 (EL1) or (d) electrification layer 2 (EL2) on the rotator. (e) Electrification material (EL3) is sandwiched between EL1/EL2 and electrode A/B to form one repeating unit of TEL-TENG. (f) The operating principle of BEL-TENG. (g) The potential distribution between top sliding layer and bottom stationary layer of BEL-TENG by COMSOL. (h) The electron transfer processes of TEL-TENG. (i) The potential distribution between top sliding layer and bottom stationary layer of TEL-TENG by COMSOL. (j) Comparison of the output performance of TEL-TENG and BEL-TENG. (Reproduced with permission of Ref. [48], 2020 © American Chemical Society).

are masterfully delineated in Figure 7.8h. In its nascent configuration, EL1 and EL2 establish contact with EL3 ensconced on the stator. This initiates an electron migration from EL1 to EL3, followed by a sequential transference from EL3 to EL2 within the corresponding contact region, persisting until material-specific charge affinity equilibria are attained. Upon EL1's rotational transit from EA to

EB, the electron flux endures, culminating in saturation achieved through the dynamic interplay of distinct electrical balance zones manifesting during rotation. Evident in Figure 7.8j, a comparative appraisal of the output prowess between Cu/FEP@Kapton and Cu@Kapton, TEL-TENG arrangements reveals a 6.29-times enhancement in output power. The basic concept and design demonstrated in this work can be widely extended beyond current binary TENG systems with improved performance for mechanical energy harvesting. Its versatility and conceptual underpinnings not only forge a pioneering path but also offer an alternative design principle for the creation of high-performance TENG systems.

7.3.3 Interfacial Insulating Liquid

Addressing the dual challenges of constrained power output attributed to air/dielectric breakdown and diminished durability arising from material wear, He et al. innovatively introduced the concept of a liquid lubrication-promoted TENG (LP-TENG) [49]. The structural blueprint of the LP-TENG, showcased in Figure 7.9a, underscores a comprehensive submergence of the CSA-TENG within silicone oil. Notably departing from the antecedent CSA-TENG framework, a voltage balancing (VB) bar is adroitly integrated to counteract the dielectric breakdown of the frictional stratum (Figure 7.9b). Demonstrated in Figure 7.9c, the introduction of an equipotential VB, endowed with an equivalent surface charge density, effectively ameliorates the potential variance between the two electrodes. Ergo, the VB bar operates as an agent of potential dispersion, mitigating localized electric field intensity centered around the lower electrode. The operational trajectory of the LP-TENG unfolds sequentially: initial charge generation materializes upon the PTFE and PA surfaces, harnessed through a process of solid–solid direct contact charging (Figure 7.9d-I). Subsequently, a double electric layer is formed on the noncontacting surfaces through a solid–liquid charging process (Figure 7.9d-II), and then the flowing oil convects and recombines the opposing charges. Consequently, a net charge reservoir is thereby engendered upon the noncontacting surfaces, conclusively depicted in Figure 7.9d-III and 9d-IV. Under the influence of liquid lubrication, the mechanical robustness of the LP-TENG undergoes a notable enhancement. This augmentation is succinctly depicted in Figure 7.9e, where the electrical yield within the milieu of silicone oil demonstrates a remarkable preservation of 90% with sustained stability even after 500 000 operational cycles – a marked contrast to the mere 40% retention observed after a relatively scant 40 000 cycles under ambient air conditions. Furthermore, a discernible amplification in the electric output of the LP-TENG emerges. Theoretically, output power is proportional to the square of surface charge density. Consequently, the charge density of the CSA-TENG within the silicone oil milieu attains a value of 2.0 mC/m^2. This achievement translates to a 20% augmentation in comparison to its CSA-TENG counterpart operating in an air environment, and a striking 216% amplification in relation to the conventional sliding mode TENG as empirically established (Figure 7.9f). This work has a significance for the commercialization of TENG-based self-powered systems.

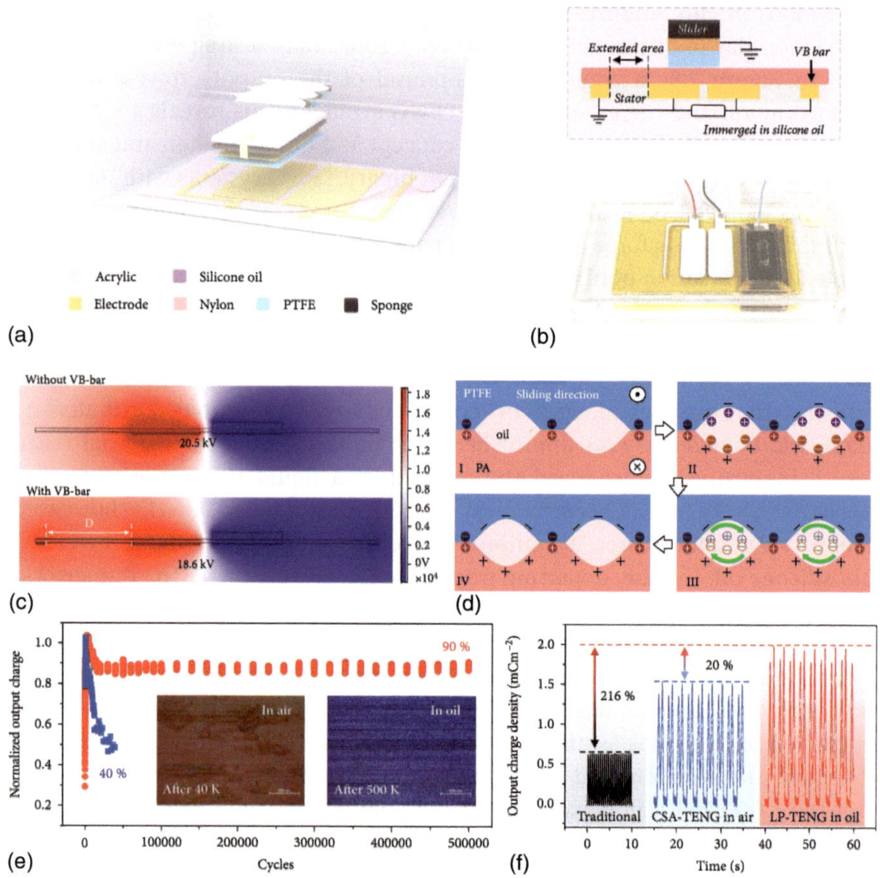

Figure 7.9 Interfacial insulating liquid used to enhance TENG's output performance. (a) Structural design of LP-TENG. (b) Simplified schematic and physical drawing of LP-TENG. (c) Simulation results of VB effect on voltage. (d) The working mechanism of LP-TENG. (e) Cyclic stability test of LP-TENG. (f) Output charge density of the LP-TENG compared with traditional device. [49] / American Association for the Advancement of Science / CC BY-4.0.

7.4 Summary

Air breakdown often occurs in TENG devices, leading to a reduction in their output efficiency. Wang et al. calculated the limiting theoretical value of the surface charge density of TENGs by introducing Paschen's law describing air breakdown in parallel plates. However, due to the limited frictional charge transfer of the material, it is not possible to operate the device at the theoretical limit value. Therefore, this chapter presents methods to achieve further surface charge density accumulation through electromechanical structural design, including external charge excitation, self-charge excitation, space charge accumulation, ternary tribolayers, and interfacial insulating liquids. These methods are combined with the operating mode of TENG to further approximate the theoretical limit value of surface charge density through mechanical structure and circuit design, which provides the possibility of large-scale commercialization of TENG.

References

1 Fan, F.-R., Tian, Z.-Q., and Wang, Z.L. (2012). Flexible triboelectric generator! *Nano Energy* 1 (2): 328–334.

2 Wang, Z.L. (2017). On Maxwell's displacement current for energy and sensors: the origin of nanogenerators. *Mater. Today* 20 (2): 74–82.

3 Niu, S., Wang, S., Lin, L. et al. (2013). Theoretical study of contact-mode triboelectric nanogenerators as an effective power source. *Energy Environ. Sci.* 6 (12): 3576–3583.

4 Niu, S., Liu, Y., Wang, S. et al. (2013). Theory of sliding-mode triboelectric nanogenerators. *Adv. Mater.* 25 (43): 6184–6193.

5 Niu, S., Liu, Y., Wang, S. et al. (2014). Theoretical investigation and structural optimization of single-electrode triboelectric nanogenerators. *Adv. Funct. Mater.* 24 (22): 3332–3340.

6 Niu, S., Liu, Y., Chen, X. et al. (2015). Theory of freestanding triboelectric-layer-based nanogenerators. *Nano Energy* 12: 760–774.

7 Chen, J. and Wang, Z.L. (2017). Reviving vibration energy harvesting and self-powered sensing by a triboelectric nanogenerator. *Joule* 1 (3): 480–521.

8 Chandrashekar, B.N., Deng, B., Smitha, A.S. et al. (2015). Roll-to-roll green transfer of CVD graphene onto plastic for a transparent and flexible triboelectric nanogenerator. *Adv. Mater.* 27 (35): 5210–5216.

9 Wang, Z.L., Chen, J., and Lin, L. (2015). Progress in triboelectric nanogenerators as a new energy technology and self-powered sensors. *Energy Environ. Sci.* 8 (8): 2250–2282.

10 Hinchet, R., Yoon, H.J., Ryu, H. et al. (2019). Transcutaneous ultrasound energy harvesting using capacitive triboelectric technology. *Science* 365 (6452): 491–494.

11 Parida, K., Thangavel, G., Cai, G. et al. (2019). Extremely stretchable and self-healing conductor based on thermoplastic elastomer for all-three-dimensional printed triboelectric nanogenerator. *Nat. Commun.* 10: 2158.

12 Xu, W., Zheng, H., Liu, Y. et al. (2020). A droplet-based electricity generator with high instantaneous power density. *Nature* 578 (7795): 392–396.

13 Seol, M.-L., Jeon, S.-B., Han, J.-W., and Choi, Y.-K. (2017). Ferrofluid-based triboelectric-electromagnetic hybrid generator for sensitive and sustainable vibration energy harvesting. *Nano Energy* 31: 233–238.

14 Sano, C., Mitsuya, H., Ono, S. et al. (2018). Triboelectric energy harvesting with surface-charge-fixed polymer based on ionic liquid. *Sci. Technol. Adv. Mater.* 19 (1): 317–323.

15 Shankaregowda, S.A., Ahmed, R.F.S.M., Liu, Y. et al. (2019). Dry-coated graphite onto sandpaper for triboelectric nanogenerator as an active power source for portable electronics. *Nanomaterials* 9 (11): 1585.

16 Taghavi, M., Sadeghi, A., Mazzolai, B. et al. (2014). Triboelectric-based harvesting of gas flow energy and powerless sensing applications. *Appl. Surf. Sci.* 323: 82–87.

17 Ouyang, H., Liu, Z., Li, N. et al. (2019). Symbiotic cardiac pacemaker. *Nat. Commun.* 10: 1821.

18 Proto, A., Penhaker, M., Conforto, S., and Schmid, M. (2017). Nanogenerators for human body energy harvesting. *Trends Biotechnol.* 35 (7): 610–624.

19 Zhou, L., Liu, D., Ren, L. et al. (2022). Reconfigurable fiber triboelectric nanogenerator for self-powered defect detection. *ACS Nano* 16 (5): 7721–7731.

20 Yin, W., Xie, Y., Long, J. et al. (2018). A self-power-transmission and non-contact-reception keyboard based on a novel resonant triboelectric nanogenerator (R-TENG). *Nano Energy* 50: 16–24.

21 Li, S., Zhao, Z., Liu, D. et al. (2022). A self-powered dual-type signal vector sensor for smart robotics and automatic vehicles. *Adv. Mater.* 34 (14): 2110363.

22 Luo, J., Han, K., Wu, X. et al. (2021). Self-powered mobile sterilization and infection control system. *Nano Energy* 88: 106313.

23 Pang, Y., Zhu, X., Lee, C., and Liu, S. (2022). Triboelectric nanogenerator as next-generation self-powered sensor for cooperative vehicle-infrastructure system. *Nano Energy* 97: 107219.

24 Sun, N., Wang, G.-G., Zhao, H.-X. et al. (2021). Waterproof, breathable and washable triboelectric nanogenerator based on electrospun nanofiber films for wearable electronics. *Nano Energy* 90: 106639.

25 Liao, J., Zou, Y., Jiang, D. et al. (2020). Nestable arched triboelectric nanogenerator for large deflection biomechanical sensing and energy harvesting. *Nano Energy* 69: 104417.

26 Chen, C., Chen, L., Wu, Z. et al. (2020). 3D double-faced interlock fabric triboelectric nanogenerator for bio-motion energy harvesting and as self-powered stretching and 3D tactile sensors. *Mater. Today* 32: 94–93.

27 Zheng, N., Xue, J., Jie, Y. et al. (2022). Wearable and humidity-resistant biomaterials-based triboelectric nanogenerator for high entropy energy harvesting and self-powered sensing. *Nano Res.* 15 (7): 6213–6219.

28 Guo, H., Chen, J., Wang, L. et al. (2021). A highly efficient triboelectric negative air ion generator. *Nat. Sustain.* 4 (2): 147–153.

29 Cheng, J., Ding, W., Zi, Y. et al. (2018). Triboelectric microplasma powered by mechanical stimuli. *Nat. Commun.* 9: 3733.

30 Gu, G.Q., Han, C.B., Lu, C.X. et al. (2017). Triboelectric nanogenerator enhanced nanofiber air filters for efficient particulate matter removal. *ACS Nano* 11 (6): 6211–6217.

31 Zhou, L., Liu, L., Qiao, W. et al. (2021). Improving degradation efficiency of organic pollutants through a self-powered alternating current electrocoagulation system. *ACS Nano* 15 (12): 19684–19691.

32 Yao, S., Zhao, X., Wang, X. et al. (2022). Bioinspired electron polarization of nanozymes with a human self-generated electric field for cancer catalytic therapy. *Adv. Mater.* 34 (15): 2109568.

33 Wei, X., Wang, Y., Tan, B. et al. (2022). Triboelectric nanogenerators stimulated electroacupuncture (EA) treatment for promoting the functional recovery after spinal cord injury. *Mater. Today* 60: 41–51.
34 Liu, Z., Nie, J., Miao, B. et al. (2019). Self-powered intracellular drug delivery by a biomechanical energy-driven triboelectric nanogenerator. *Adv. Mater.* 31 (12): 1807795.
35 Yang, Y., Luo, R., Chao, S. et al. (2022). Improved pharmacodynamics of epidermal growth factor via microneedles-based self-powered transcutaneous electrical stimulation. *Nat. Commun.* 13 (1): 6908.
36 Wang, S., Lin, L., and Wang, Z.L. (2015). Triboelectric nanogenerators as self-powered active sensors. *Nano Energy* 11: 436–462.
37 Parida, K., Xiong, J., Zhou, X., and Lee, P.S. (2019). Progress on triboelectric nanogenerator with stretchability, self-healability and bio-compatibility. *Nano Energy* 59: 237–257.
38 Li, X., Xu, G., Xia, X. et al. (2019). Standardization of triboelectric nanogenerators: progress and perspectives. *Nano Energy* 56: 40–55.
39 Cheng, X., Tang, W., Song, Y. et al. (2019). Power management and effective energy storage of pulsed output from triboelectric nanogenerator. *Nano Energy* 61: 517–532.
40 Chen, G., Li, Y., Bick, M., and Chen, J. (2020). Smart textiles for electricity generation. *Chem. Rev.* 120 (8): 3668–3720.
41 Zi, Y., Niu, S., Wang, J. et al. (2015). Standards and figure-of-merits for quantifying the performance of triboelectric nanogenerators. *Nat. Commun.* 6: 8376.
42 Wang, S., Xie, Y., Niu, S. et al. (2014). Maximum surface charge density for triboelectric nanogenerators achieved by ionized-air injection: methodology and theoretical understanding. *Adv. Mater.* 26 (39): 6720–6728.
43 Xu, L., Bu, T., Yang, X. et al. (2018). Ultrahigh charge density realized by charge pumping at ambient conditions for triboelectric nanogenerators. *Nano Energy* 49: 625–633.
44 Liu, W., Wang, Z., Wang, G. et al. (2019). Integrated charge excitation triboelectric nanogenerator. *Nat. Commun.* 10: 1426.
45 Bai, Y., Xu, L., Lin, S. et al. (2020). Charge pumping strategy for rotation and sliding type triboelectric nanogenerators. *Adv. Mater.* 10 (21): 2000605.
46 Long, L., Liu, W., Wang, Z. et al. (2021). High performance floating self-excited sliding triboelectric nanogenerator for micro mechanical energy harvesting. *Nat. Commun.* 12 (1): 4689.
47 He, W., Liu, W., Chen, J. et al. (2020). Boosting output performance of sliding mode triboelectric nanogenerator by charge space-accumulation effect. *Nat. Commun.* 11 (1): 4277.
48 Deng, W., Zhou, Y., Zhao, X. et al. (2020). Ternary electrification layered architecture for high-performance triboelectric nanogenerators. *ACS Nano* 14 (7): 9050–9058.
49 He, W., Liu, W., Fu, S. et al. (2022). Ultrahigh performance triboelectric nanogenerator enabled by charge transmission in interfacial lubrication and potential decentralization design. *Research* 2022: 9812865.

8

Power Management and Effective Energy Storage

8.1 Introduction

Due to the rapid advancements in the Internet of Things (IoT) and 5G technologies, the need for sustainable energy supply in widely distributed sensor networks and transmission electronic devices has become increasingly urgent [1–3]. The traditional centralized power generation, according to the distributed energy entropy theory, may not be sufficient to meet the power supply demands of these numerous and widely dispersed electronic devices [4, 5]. Moreover, relying on chemical batteries for power supply is not sustainable and can cause environmental pollution [6, 7]. In order to ensure a continuous power supply for these distributed sensor networks and communication electronic devices, researchers have developed various methods to harvest energy from sunlight, wind, and vibrational mechanical energy by utilizing the electromagnetic induction effect, photovoltaic effect, thermoelectric effect, and piezoelectric effect.

Although the triboelectrification effect was discovered as far back as 600 BC, it was not until 2012 that Zhonglin Wang and his team pioneered the application of this principle in the invention of the triboelectric nanogenerator (TENG) [8–10]. This energy harvesting method operates when two friction materials of the TENG come into contact under an external force, leading to the generation of equal positive and negative electrostatic charges on the contact surfaces due to triboelectrification [2, 11]. The separation of these materials by a mechanical force results in a spatial separation of the positive and negative charges, thereby creating a potential difference between the two electrodes [12, 13]. This induced potential difference, when the electrodes are either short-circuited or connected to a load, drives the flow of electrons between the electrodes, culminating in an energy output. The TENG operates in four main modes: vertical contact-separation mode, sliding mode, single-electrode mode, and freestanding-layer-based mode [14, 15].

Rooted in the second term of Maxwell's displacement current, the TENG has demonstrated its capability to harvest a broad spectrum of environmental mechanical energies, encompassing human motion, raindrop, and wave energy [16, 17]. TENG offers several advantages, including low cost, a wide array of material choices,

Triboelectric Nanogenerators: Technology, Applications, and Commercialization, First Edition.
Yunlong Zi, Hengyu Guo, Jie Wang, Chi Zhang, Xiangyu Chen, and Qing Zhao.

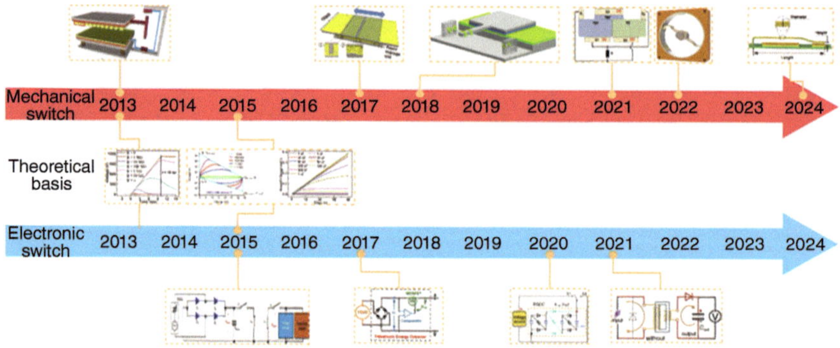

Figure 8.1 The advancement history and theoretical basis of power management and effective energy storage methods in TENG. From 2013 to 2024, TENG's energy management strategy follows two main lines of evolution, one is a mechanical switch-based energy management strategy, and the other is an electronic switch-based management strategy. Reproduced with permission of Ref. [26], 2013 © American Chemical Society. Reproduced with permission of Ref. [27], 2017 © Elsevier. Reproduced with permission of Ref. [28], 2018 © John Wiley & Sons. Reproduced with permission of Ref. [24], 2021 © Springer Nature. Reproduced with permission of Ref. [29], 2022 © Springer Nature. Reproduced with permission of Ref. [30], 2013 © Royal Society of Chemistry. [31] /Springer Nature/ CC BY-4.0. Reproduced with permission of Ref. [32], 2015 © Simiao Niu. [33] / Springer Nature / CC BY-4.0. Reproduced with permission of Ref. [34], 2017 © Elsevier. [35] / Springer Nature / CC BY-4.0. Reproduced with permission of Ref. [36], 2021 © Elsevier. Reproduced with permission of Ref. [46], 2024 © Elsevier.

and flexibility. Notably, under conditions of low frequency, microamplitude, and weak-force mechanical excitation, TENG can yield a higher electromechanical conversion efficiency and a more substantial power energy output than traditional electromagnetic generators [18–20]. However, due to its inherent capacitive characteristics, TENG exhibits a higher open-circuit voltage (~kV), larger impedance (~MΩ/GΩ), lower short-circuit (SC) current (~μA), and average output power (~mW). Furthermore, the amplitude and frequency of the electrical signals output from the TENG are unstable, owing to the high randomness of the ambient mechanical energy [21, 22]. These electrical characteristics do not meet the input requirements for a stable DC power supply for conventional electronic devices [23]. To mitigate the voltage and impedance and enhance the power supply efficiency, researchers have proposed a variety of switching power management methods for TENGs. Figure 8.1 illustrates the evolution of power management methods from 2013 to 2023 [24, 25].

In this chapter, we will delve into the latest advancements in TENG switching power management. Initially, we will discuss the research progress made in switching management circuits and their theoretical foundations. Following that, we will categorize the different switching power management methods into mechanical and electronic switches, based on the type of switch used. We will then proceed to categorize and describe the design concepts and power management outcomes of the reported work. Toward the end of our chapter, we will summarize the limitations and research challenges faced in switching power management and provide a brief insight into the future development of TENG power management.

8.2 Theoretical Basis of Energy Management for TENG

The V-Q-x relationship is an important theoretical equation for describing the TENG state, which is as follows [30, 37, 38]:

$$V = -\frac{1}{C(x)}Q + V_{OC}(x) \tag{8.1}$$

In this equation, V represents the output voltage, C is the capacitance between the two electrodes, Q is the amount of charge transferred between the electrodes, V_{OC} is the open-circuit voltage of the TENG, and x is the separation distance between the two friction layers. Equation (8.1) allows us to derive the equivalent circuit model of the TENG. The two terms on the right side of the equation can be represented by two circuit elements in this model. The first term originates from the inherent capacitance between the two electrodes, which can be represented by a capacitor (C). The second term represents the separation of polarized tribocharges, which can be expressed as an ideal voltage source. Therefore, the equivalent circuit model of the TENG can be represented by a series connection of a capacitor and an ideal voltage source, as shown in Figure 8.2a.

The equivalent circuit of the system composed of a load resistance (R) connected with a TENG is shown in Figure 8.2a. The governing equation of the whole system can be obtained by Kirchhoff's law [39]:

$$R\frac{dQ}{dt} = V = -\frac{1}{C(x)}Q + V_{OC}(x) \tag{8.2}$$

The peak values of current, voltage and power under different load resistances are shown in Figure 8.2b,c. The TENG has three working regions as shown in Figure 8.2b. In region I with low resistance (0.1–1000 Ω), the peak current is not significantly decay compared to the SC condition because the charge transfer process is similar to the SC condition. Under the condition that only the peak value of the output current is considered, the TENG can be regarded as an ideal current source. In region III where the resistor is greater than 1 GΩ, the output characteristics are close to the OC condition, and the TENG can be used as a constant voltage source in terms of load property. In region II, as the resistance increases, the current decreases and the voltage increases. The instantaneous output power of the TENG reaches the maximum value at approximately 100 MΩ in this region (Figure 8.2c). The three working-region characteristics of the TENG under different load resistances can be explained by the impedance matching between the inherent capacitance of the TENG and the load resistance. When the load resistance is much less than the impedance of the TENG inherent capacitance, the total impedance is determined by the impedance of the TENG inherent capacitance. As a result, the output current remains almost constant. When the load resistance

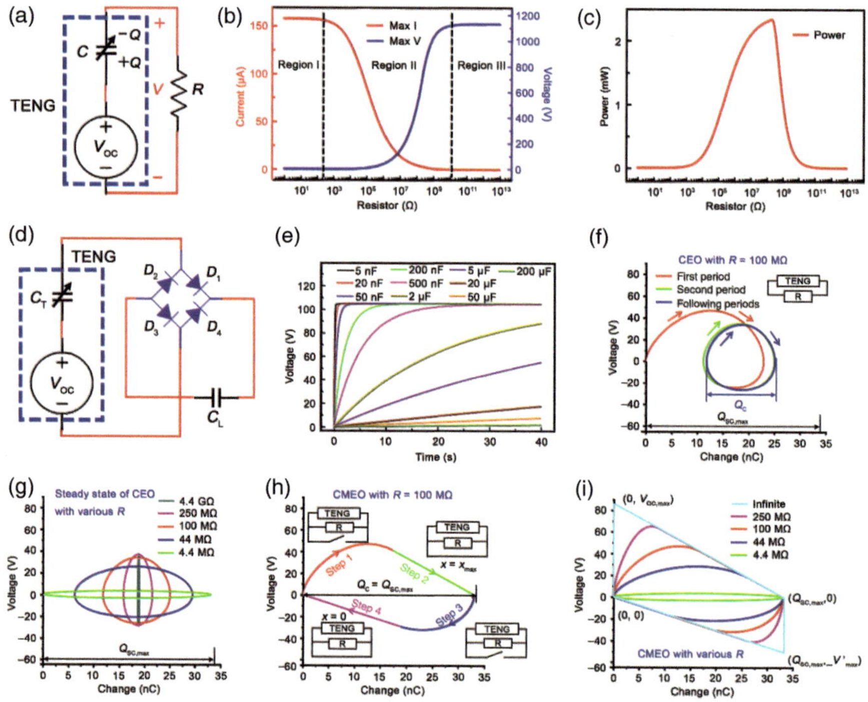

Figure 8.2 Theoretical basis of energy management for TENG. (a) Equivalent circuit model of the TENG connected a load resistance. [39] / Elsevier / CC BY-4.0. (b) The three working regions of the output current and voltage under the different load resistances. (c) The influence of the load resistance on the instantaneous power output. Reproduced with permission of Ref. [30], 2013 © Royal Society of Chemistry. (d) Full-wave rectifier circuit for periodic charging calculation. [39] / Elsevier / CC BY-4.0. (e) The influence of the load capacitance on the final stored energy. Reproduced with permission of Ref. [32], 2015 © Simiao Niu. (f) The CEO under load resistance. (g) The steady state of CEO under different load resistances. (h) Four steps of cycles for maximized CMEO under a load resistance R = 100 MΩ. (i) The CMEO under different load resistances. [31] / Springer Nature / CC BY-4.0.

is much greater than the impedance of the TENG inherent capacitance, almost all the voltage is applied to the load resistance, and the TENG works under quasi-OC conditions. When the load resistance matches the impedance of the TENG inherent capacitance, the instantaneous output power reaches the maximum. Therefore, optimizing the load resistance is the key to maximizing the effective efficiency of the TENG. The optimum resistance (R_{opt}) of the contact-separation mode TENG is calculated by Equation (8.3):

$$R_{opt} = \frac{d_0^2}{F_{opt}^2 S v \varepsilon_0} \approx \frac{(d_0 + x_{max})^2}{S v \varepsilon_0} \tag{8.3}$$

where d_0 is the sum of the thickness to relative permittivity ratios of all the dielectric materials, x_{max} is the maximum separation distance, S is the contact area, v is the average separation speed, and ε_0 is the permittivity of vacuum. Equation (8.3) shows that the optimal impedance is independent of the surface charge density.

For the TENG with AC output, a rectifier is usually required to improve the charging efficiency. The equivalent circuit model for a load capacitor (C_L) charging by the TENG after rectification is shown in Figure 8.2d. Under the condition of external periodic motion, charging the capacitor with TENG connected to the rectifier is equivalent to charging the capacitor with a DC voltage source connected in series with a resistor, and both of them follow the same exponential saturation trend, as shown in Figure 8.2e. The saturation voltage (V_{sat}) value can be calculated by Equation (8.4) [32]:

$$\mathrm{V}_{sat} = \lim_{k\to\infty} \left| V_{k,2\mathrm{end}}^{C} \right| = \frac{Q_{\mathrm{SC,max}}}{C_{\min} + C_{\max}} \tag{8.4}$$

where V_{sat} is a function of TENG parameters and has nothing to do with C_L. In addition, there is also an optimal load capacitor that maximizes the stored energy. The optimal load capacitance can be obtained by the following equation:

$$C_{\mathrm{L,opt}} = 1.592k\left(C_{\min} + C_{\max}\right) \tag{8.5}$$

where k is the charging cycle and $k > 10$. The $C_{\mathrm{L,opt}}$ and maximum energy stored increase with the number of charging cycles, and there is a linear relationship between the $C_{\mathrm{L,opt}}$ and the number of charging cycles.

When the TENG performs a continuous periodic motion, its electrical output signal is also periodic. For a certain period of time (T), the output energy per cycle (E) can be calculated by Equation (8.6):

$$E = \overline{P}\mathrm{T} = \int_0^T VIdt = \int_{t=0}^{t=T} VdQ = \oint VdQ \tag{8.6}$$

where $\overline{P}$ represents the average output power of the TENG. The relationship between the voltage and the transferred charges (V-Q curve) in Equation (8.6) can be used to represent the energy output per cycle of the TENG. As can be seen from the V-Q curve in Figure 8.2f, the electrical output signal of the TENG reaches a steady state after a few cycles of operation. The V-Q curve in the steady state should be a closed loop, and its area is the output energy per cycle of the TENG according to Equation (8.6) [31]. The energy output per cycle can be optimized by matching a load resistance (Figure 8.2g), and the cycles are called "cycles for energy output" (CEO). The total cycling charge (Q_C) is the difference between the maximum and minimum transferred charges for each CEO under steady-state conditions.

The Q_C of each CEO is always less than the maximum transferred charges $Q_{\mathrm{SC,max}}$ (Figure 8.2f,g). The output energy per cycle can be further improved by $Q = Q_{\mathrm{SC,max}}$ under SC condition. As shown in Figure 8.2h, the instantaneous SC condition is achieved by a switch in parallel with the external resistor, and the $Q = Q_{\mathrm{SC,max}}$ occurs in steps 2 and 4. Figure 8.2i shows the V-Q curve of different external resistors and switches in parallel. These cycles are called "cycles for maximized energy output" (CMEO). The energy output per cycle of the CMEO is always greater than that of the CEO with the same external resistance. The output energy per cycle of the CMEO increases with the increase of the external resistance. When the external resistance is infinite, the area enclosed by the curve is close to the trapezoid, which is greater than the area under any external resistance. The output energy per cycle of the CMEO can be calculated by Equation (8.7):

$$E_{\text{m}} = \frac{1}{2}Q_{\text{SC,max}}\left(V_{\text{OC,max}} + V'_{\text{max}}\right) \tag{8.7}$$

where $V_{\text{OC,max}}$ is the maximum open-circuit voltage, V'_{max} is the absolute value of maximum voltage in the condition that $Q = Q_{\text{SC,max}}$. The above theoretical basis provides guidance for energy management of TENG.

8.3 Mechanical Switched Converter

8.3.1 Travel Switch

8.3.1.1 Series Switch

In order to improve the output performance of the TENG while reducing the matching impedance, Cheng et al. proposed a travel switch scheme in 2013, the principle of which is shown in Figure 8.3a [26]. Based on the original TENG, a switch is introduced in the external circuit to close the circuit at full contact and at separation to the farthest position, respectively, so that the transient current is maximized. This travel switch can be used in both contact separation TENGs and sliding TENGs. The generator structure designed by the team is shown in Figure 8.3b. An aluminum pin is introduced at the upper electrode of the generator, and the travel switch is closed or opened by the contact or separation of the aluminum pin with the upper and lower electrodes. The generator with the travel switch introduced has a significant impedance reduction effect, as shown in Figure 8.3c. The power increase is especially obvious at low loads. By introducing the travel switch, the maximum peak current of this generator is about five orders of magnitude higher than that of the generator without the travel switch, and the maximum power density is increased by more than 1100 times.

Using the principle of travel switch, Qin et al. designed a TENG scheme capable of rectification effect, as shown in Figure 8.3d [28]. Figure 8.3e shows the image and enlarged illustration of the output current under a 20 MΩ load using a series travel switch. As shown in Figure 8.3g, the peak output voltage of this generator is essentially constant in the case of the load circuit, while the peak voltage width increases with the load. By design of the energy management circuit shown in Figure 8.3f, the generator energy is stored in capacitor C2, and the energy storage efficiency of this scheme is up to 48% by experimental analysis. With the introduction of the travel switch, the design has a significant impedance reduction effect and the output energy has almost no dependence on the matching resistance, as shown in Figure 8.3h,i.

Wu et al. designed a contact-separated TENG with a three-layer structure and introduced travel switches as shown in Figure 8.3j [40]. By means of a dual-channel

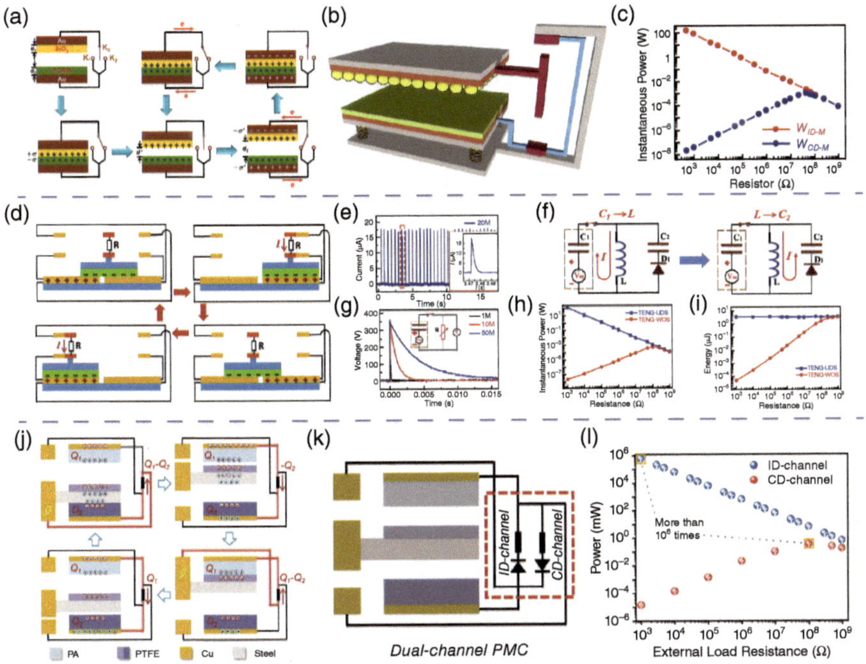

Figure 8.3 The power management strategy utilizes series travel switches. (a) Operating principle of the travel switch in an on–off–on contact process. (b) Three-dimensional structure display of the travel switch. (c) Dependence of instantaneous power on resistance for with/without series travel switch. Reproduced with permission of Ref. [26], 2013 © American Chemical Society. (d) Operating principle of series travel switch with rectification function. (e) Image and enlarged illustration of the output current under a 20 MΩ load using a series travel switch. (f) Operating mode of the energy management module. (g) Output voltage curves for loads of 1, 10, and 50 MΩ, respectively. The inset shows the output circuit diagram. (h) Dependence of peak power (h) and output energy (i) on resistance. Reproduced with permission of Ref. [28], 2018 © John Wiley & Sons. (j) Operating principle of a triple structure TENG with trip switch. (k) Structure of the dual-channel energy management scheme. (l) Variation of the peak output power with respect to the external load. Reproduced with permission of Ref. [40], 2023 © John Wiley & Sons.

PMC with rectifier diodes, as shown in Figure 8.3k, the ID current and CD current generated by the TENG can be finely divided into two DC outputs, enabling the accumulation of dual charges and the coupling of mechanical switches. This generator shows a large peak current and transfer charge in phases (i), (iii) of Figure 8.3j, but small in phases (ii), (iv) of Figure 8.3j So, the total transfer charge of this generator is almost three times that of a traditional TENG in one cycle. Through the resistance matching experiment, the output power of ID channel of this TENG increases sharply with the decrease of load resistance, and the output power of CD channel is consistent with the characteristics of the conventional TENG. The output power of ID channel can reach 616 W, which is 106 times the CD discharge state under the corresponding load.

8.3.1.2 Parallel Switch

Zi et al. designed a scheme to achieve TENG output enhancement by connecting switches in parallel. As shown in Figure 8.4a I–IV, the switch is closed to balance the charge before the start of a motion cycle, and then the switch is opened to start the motion of the friction layer, and the switch is closed again when the motion reaches the maximum displacement position, and so on. Zi et al. [41]. proposed a complete charge transfer by generating a transient SC current through the parallel switch through theoretical analysis and calculation, so that in the next half cycle, more charges can be used to charge the energy storage unit. In Figure 8.4b,c, the energy cycle curve without the parallel switch is marked as I, while the energy in each storage cycle after the addition of the switch includes both I and II so that the energy in each cycle has a larger boost. By calculation, the energy storage efficiency is increased from 25% to 50% in a single cycle, and the saturation voltage is increased by at least two times. This theory was also verified from an experimental point of view, as shown in Figure 8.4d.

8.3.1.3 Switch Capacitor Converter

Tang et al. proposed a TENG based on the capacitive conversion principle in 2014, and the schematic is shown in Figure 8.5a [42]. When the TENG starts to output electrical energy, the capacitors collect electrical energy in series; when the output of the

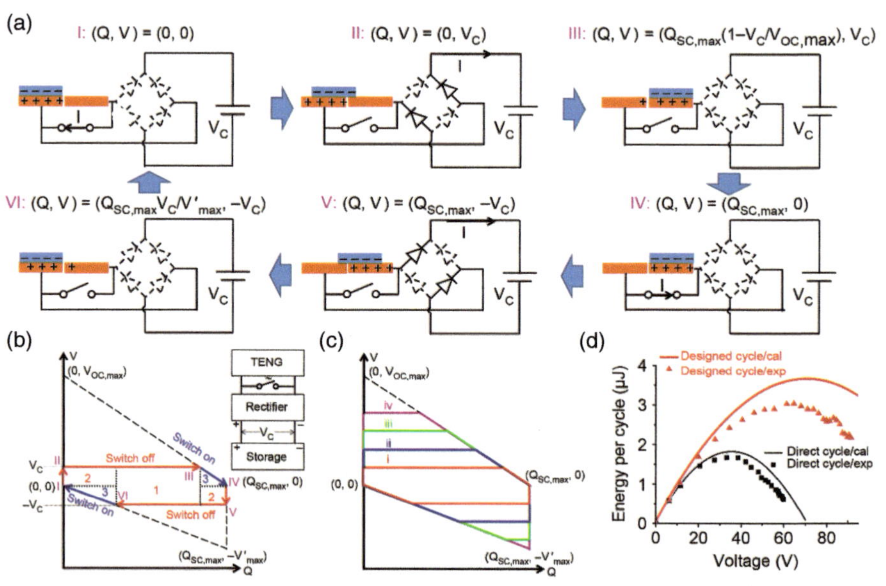

Figure 8.4 Energy management for parallel switching schemes. (a) The working principle of a TENG with a parallel switch. (b) *V-Q* diagram. The illustration shows the circuit for designing the charge cycle. The dashed line in b represents a cycle with maximum energy output at infinite impedance. (c) Design charging cycles with different charging voltages, charging voltage intensity of iv > iii > ii > i. (d) Experimental (point) and calculation (line) results for storing energy per cycle energy. Reproduced with permission of Ref. [41], 2016 © Springer Nature.

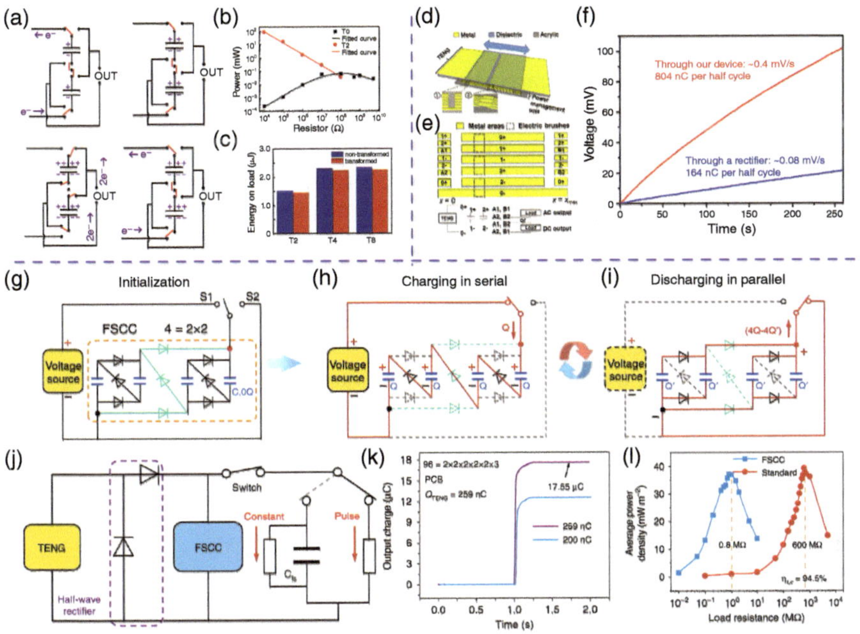

Figure 8.5 Energy management scheme based on capacitive conversion. (a) Operating principle of the power conversion principle TENG. (b) Image of output power variation with load. (c) Comparison of energy after conversion and without conversion. Reproduced with permission of Ref. [42], 2023 © IOP publishing. (d) Schematic (d) and structural diagram (e) of automatic power management TENG. (f) Comparison of this scheme with the charging rate of a rectifier bridge to a 5 mF capacitor. Reproduced with permission of Ref. [27], 2023 © Elsevier. (g) The operating process of the FSCC system in different phases, (h) charging in serial and (i) discharging in parallel. (j) Power management scheme structure. (k) The maximum output charges of 96-unit FSCC per cycle, indicating 67.8 times charge boosting. (l) The variation of output power density with load resistance for FSCC scheme with standard TENG used. [35] / Springer Nature / CC BY-4.0.

TENG ends, the electrical energy is output in parallel. It can achieve T-fold enhancement of the output charge through the conversion of capacitors in series and parallel (T is number of capacitors). At the same time, the scheme also has a good ability to reduce impedance, as shown in Figure 8.5b. Under low load resistance, the output power of this scheme is much higher than that of ordinary TENG, and the output power decreases as the load rises. Figure 8.5c compares the energy lost in the power conversion of this scheme, the converted energy is reduced, indicating that there is energy loss in the switching conversion, but the loss is weak, and the energy conversion efficiency can reach more than 95%. Zi et al. proposed a power management scheme based on series–parallel conversion of capacitors and changed the switch to a motion switch integrated into the TENG to achieve automatic series–parallel conversion of capacitors without additional operations or devices. The structure and schematic of this TENG are shown in Figure 8.5d,e [27]. The capacitor switching performed in this way can reduce the remaining charge to increase the charge output per cycle. Charging the same capacitor, the efficiency of this power management scheme is greatly improved compared to charging using a rectifier bridge, as shown in Figure 8.5f.

Series–parallel conversion by capacitors can effectively reduce the voltage and boost the output charge, but the number of capacitors is limited by the presence of diode on-state voltage drop. To address this challenge, Liu et al. proposed a fractal switched capacitor converter (FSCC), the basic principle of which is shown in Figure 8.5g–i [35]. Through fractal design, this scheme can effectively reduce the on-state voltage drop of the diode and can expand the number of capacitors through fractal structure evolution for efficient energy management. As shown in Figure 8.5k, the output charge has been boosted 67.8 times from 259 nC to 17.55 μC under the effect of 96 units of FSCC. The power management module is formed by accessing the rectifier circuit and switches, as shown in Figure 8.5j. The power management module can reduce the matching resistance from 600 MΩ to 0.8 MΩ, as shown in Figure 8.5l, which strongly promotes the application of TENG.

8.3.2 Electrostatic Switch

8.3.2.1 Spark Switch

Cheng et al. designed a self-powered air discharge switch in which the switching state of the circuit is controlled by the voltage of the TENG itself. The switching part consists of a tip tungsten electrode and a stainless-steel plate. When the distance between the two is relatively short, the electric field around the tip and the plate ionizes the air and the generated plasma bridges the two electrodes Thus, the switch closes, and this discharge method is called arc discharge (Figure 8.6a) [43], when the distance between the two is far, the ionized air cannot bridge the two electrodes, but the current is formed by the drift diffusion of ions, resulting in the switch closing, and this discharge method becomes corona discharge (Figure 8.6b). This type of discharge becomes corona discharge. The relationship between the discharge energy per cycle and the electrode spacing d is shown in Figure 8.6c. Due to the change in the discharge method, the image has two inflection points at the 0.4 mm and 0.72 mm spacing. As shown in Figure 8.6d, the matching impedance reduction effect of corona discharge is not obvious, while the arc discharge has a significant impedance reduction effect.

Zhang et al. fabricated a self-sustained and automatic hysteresis plasma switch made from silicon micromachining to improve the energy management efficiency of the TENG. The circuit structure of the scheme is shown in Figure 8.6e [44]. First, the energy from the TENG is stored in the buffer capacitor (C_{buf}) through the Bennet doubler; second, when the voltage in C_{buf} reaches a certain value, the hysteresis switch can be closed by contact or spark discharge, and the energy from C_{buf} is stored in the energy storage capacitor C_{store} after the buck circuit; finally, the energy in C_{store} is finally passed through a commercial voltage regulator chip to drive the load. Figure 8.6f,g shows the schematic diagrams and SEM images of the fixed electrode switch and the movable electrode switch, respectively, in which the fixed electrode switch is conducted by spark discharge and the movable electrode switch

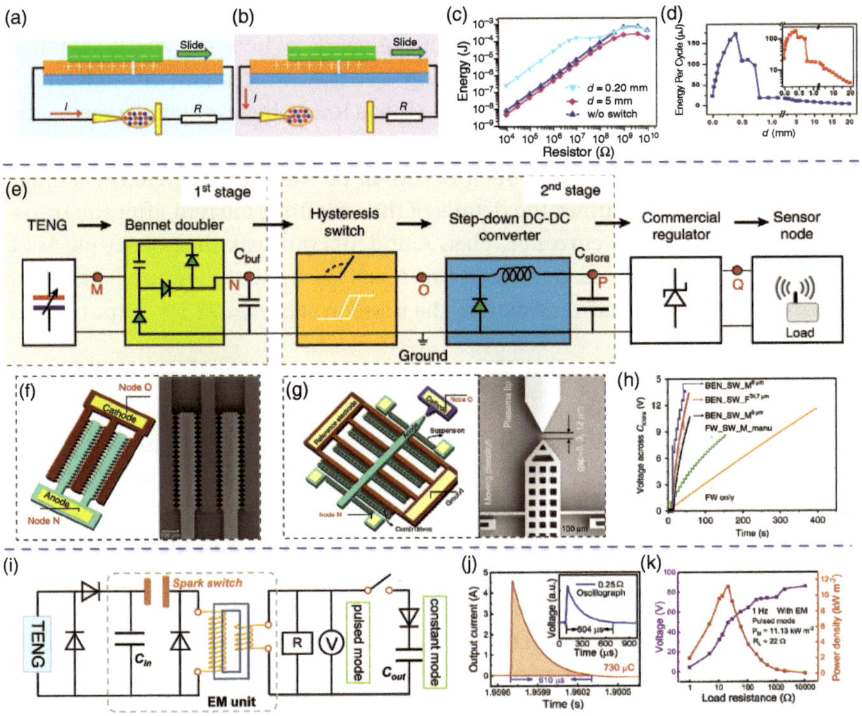

Figure 8.6 Energy management scheme based on electronic discharge switch. (a) Electronic switching structure for arc discharge mode and (b) corona discharge mode. (c) Arc discharge mode ($d = 0.2$ mm), corona discharge mode ($d = 5$ mm), and TENG output energy without electrostatic switch changes with load resistance increasing. (d) The dependence of the output energy per cycle on d. Reproduced with permission of Ref. [43], 2018 © Elsevier. (e) Circuit block diagram for energy harvesting and regulation. (f) Structural diagram of the MEMS of the fixed plasma switch and SEM image. (g) Structural diagram of the active MEMS switch and SEM image. (h) Voltage curves of charging a capacitor of 22 μF with different conditioning methods, including the methods of: a two-stage system using a movable switch with gap 6 μm (BEN_SW_M6 μm), using a movable switch with a gap of 9 μm (BEN_SW_M9 μm), using a fixed flat switch (BEN_SW_FTri, 7 μm), using manually fabricated full-hysteresis switch and FW rectifier (FW_SW_M_manu), and a classical one-stage system using only one full-wave rectifier (FW_only). [44] / Springer Nature / CC BY-4.0. (i) Power management circuit diagram with spark switch. (j) Detailed plot of output current using a 3000 : 40 transformer and peak time image measured by an oscilloscope. (k) The relationship between voltage and energy density after passing through the management module and load resistance. Reproduced with permission of Ref. [36], 2018 © Elsevier.

is conducted by contact. As shown in Figure 8.6h, the movable switch with a tip gap of 6 μm has higher charging efficiency and on/off voltage for the capacitor. The efficiency of the buck converter with MEMS plasma switch was experimentally measured to be 35%.

Wang et al. proposed a spark discharge switch scheme with a regulator, and the structure diagram is shown in Figure 8.6i [36]. The scheme deposits the energy from the TENG into the capacitor C_{in} through the half-wave rectifier, and when the voltage C_{in} reaches the discharge voltage of the spark switch, the spark switch is excited to conduct, thus transferring the energy into the transformer, through which the high voltage from TENG can be converted into a low voltage while increasing the current and reducing the impedance. Finally, the energy can be directly connected to the load at the back end of the buck or stored or stored in the energy storage capacitor C_{out}. Figure 8.6j shows the details of the amplified current after the transformer, which amplifies the current to class A and hits the peak time of 610 μs. Also, the scheme has excellent performance at low loads, as shown in Figure 8.6k. After calculation and experimental verification, the scheme achieves 78.5% output power under load resistance.

8.3.2.2 Electrostatic Switch

Yang et al. proposed an electrostatic-based vibration switch loaded onto the TENG to enhance the output performance of the TENG, the schematic diagram of the scheme is shown in Figure 8.7a [45]. The two ends of the switch consist of a copper wire and a flat plate, which are connected to the two electrodes of the TENG. When the TENG moves periodically, a periodic potential difference is generated between the copper wire and the flat plate to drive the contact separation between the copper wire and the flat plate periodically, thus realizing the periodic closure of the switch and increasing the output performance of the TENG. This scheme can greatly reduce the matching impedance while increasing the output, as shown in Figure 8.7b, and shows excellent performance at low loads. At the same time, the just solution can also be applied to rotary TENGs. Figure 8.7c shows the output of an electrostatic vibration switch equipped with a rotary TENG, which exhibits superb performance enhancement. The corresponding physical design of electrostatic switch is shown in Figure 8.7d. The strategy for the TENG power management is to add an Autonomous Cantilever Buck Switch (ACBS) between TENG and LC filter circuit, which was reported by Zhou et al. in 2024 (Figure 8.7e). A freestanding-layer-based sliding TENG as a demonstration was used, in which polymethyl methacrylate (PMMA) and copper are employed as stator base and electrode layer, fluorinated ethylene propylene (FEP) film attached to the sponge for sufficient contact is employed as a triboelectric layer (slider). The core circuit consists of a print circuit board (PCB) with three electric ports and a copper cantilever beam. This device is similar to the field effect transistor or relay. The hinged end of the cantilever is named as drain which is soldered on the pad and connected to the right electrode of the TENG. The electrode that is about 100 μm below the free end of the cantilever is named as source. In order to establish an electrostatic field that surrounds the cantilever, we set a gate electrode below the middle section of the cantilever and connected it to the left electrode of TENG. The input of the LC circuit is connected with the source and gate of the ACBS to build a complete electrical circuit. Furthermore, e integrated the ACBS and the LC filter circuit into a PCB with only 5.3 cm in length and 2.8 cm in width (Figure 8.1d), which is significantly more versatile

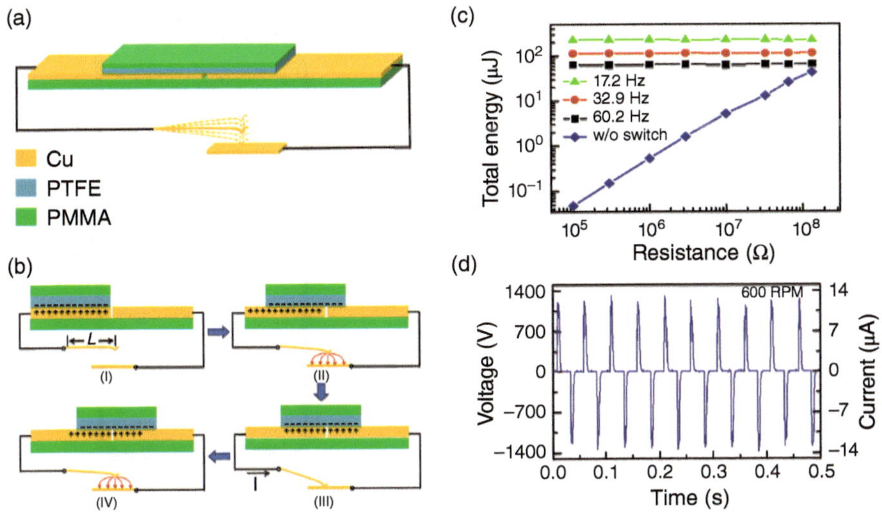

Figure 8.7 Power management scheme of electrostatic switch principle. (a) The structure of electrostatic vibration switch. (b) The change of output energy of TENG with adding load resistance at different output frequencies of the electrostatic vibration switch, the relationship between the output energy of TENG without energy management and the load resistance. (c) Output voltage and current curve of rotary TENG with electrostatic vibration switch (40.3 Hz switch, load resistance 100 MΩ, speed 650 rpm). (d) The structure of electrostatic vibration switch for rotary TENG. Reproduced with permission of Ref. [45], 2018 © Elsevier. (e) Schematic diagram of TENG, cantilever buck switch and LC-filter circuit. (f) Ultrahigh energy ratio realized in this work. Reproduced with permission of Ref. [46], 2024 © Elsevier.

compared with the previous management methods in the form of a travel switch (Figure 8.7f) [46].

8.3.3 Application Demonstration

For the design of the travel switch, Wu et al. gave an all-in-one vibration sensor, which achieves stable vibration monitoring as well as signal transmission and visualization on cell phones through wireless network. Figure 8.8a shows the physical image of this sensing system and the workflow diagram is shown in Figure 8.8b [40]. Once the vibration sensor is triggered, the generated status information is transmitted from the infrared LED to the repeater, and the repeater MCU performs analog-to-digital conversion and sends the data to the cell phone through the built-in WiFi module, thus realizing wireless vibration monitoring as well as data visualization (Figure 8.8c).

For the scheme of energy management through capacitive conversion, Liu et al. built a power management solution based on the 96-cell FSCC as shown in Figure 8.8d [35]. In the experiment, after this power management scheme, the energy from TENG successfully drove the buzzer to emit 96 dB (Figure 8.8e), made the thermohygrometer with three sensors work properly (Figure 8.8f),

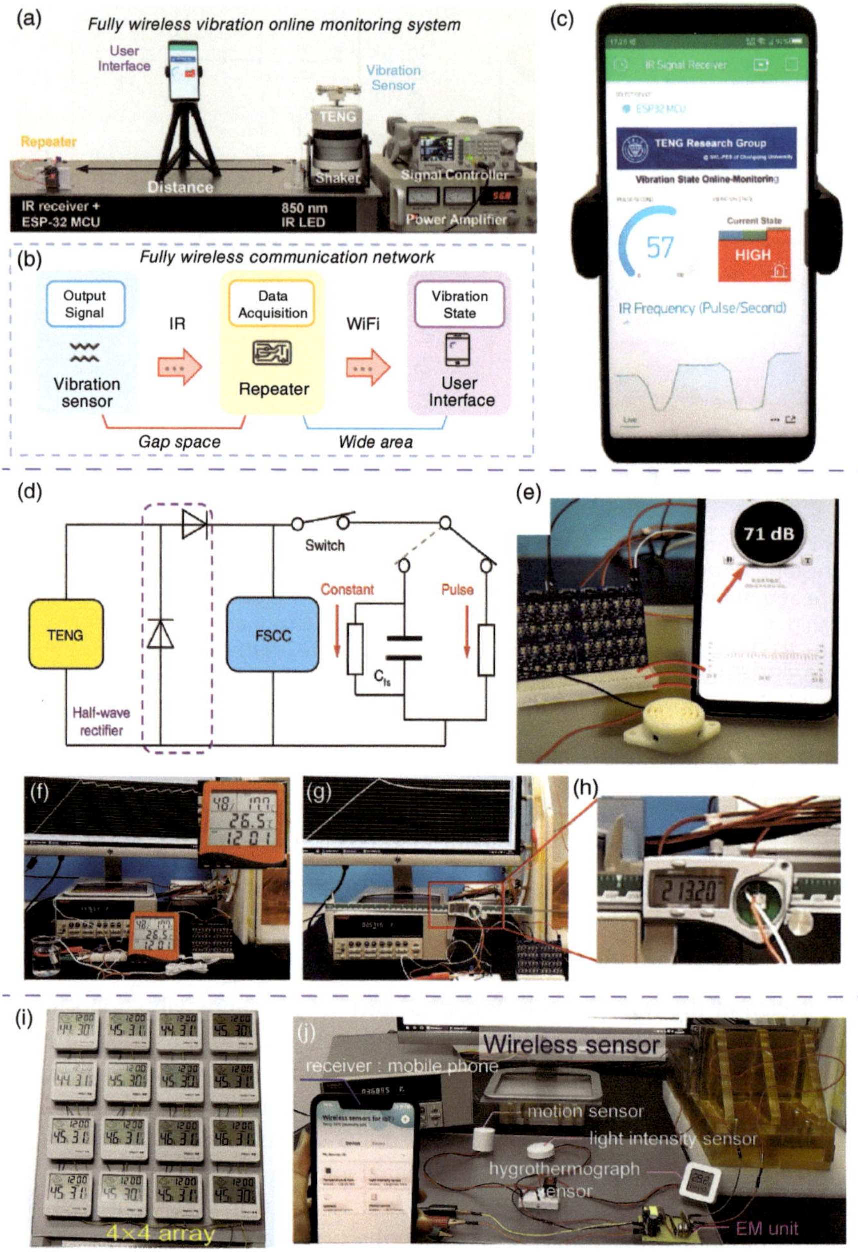

Figure 8.8 Application demonstration of the mechanical switch energy management scheme. Physical photos of the integrated wireless vibration monitoring sensing system, (a) Wireless network communication flow (b) and vibration monitoring information on the cell phone side (c). Circuit schematic diagram of FSCC-based energy management scheme. Reproduced with permission of Ref. [40], 2023 © John Wiley & Sons. (d) Driving buzzer with 1 Hz pulse output mode (e) Actual image of driving temperature hygrometer (f–h) Continuous driving of medium-sized digital vernier caliper along with detail enlargement. [35] / Springer Nature / CC BY-4.0. (i) Simultaneously drive 16 digital temperature and humidity meters. (j) Powering wireless sensors to continuously transmit sensed data. Reproduced with permission of Ref. [36], 2021 © Elsevier.

moreover powered the digital vernier caliper to measure the length continuously (Figure 8.8g,h).

For the electronic switch scheme, Wang et al. greatly increased the output performance and reduced the voltage by using a power management module with spark switches. In the application part, with the on scheme, the energy from the TENG enable to power a 4×4 array of thermohygrometers work normally connected in parallel (Figure 8.8i) [36]. Based on this power management scheme, the application was further extended to drive three wireless sensors (including motion sensor, light intensity sensor and humidity sensor) to operate properly at a TENG motion frequency of 1 Hz, as shown in Figure 8.8j, which strongly demonstrates the superior working performance of the scheme.

8.4 Electronic Switch Converter

8.4.1 Integrated Circuit

Boisseau et al. reported a self-starting power management integrated circuit (IC) for harvesting piezoelectric and triboelectric energy with the circuit structure, as shown in Figure 8.9a [47]. The startup circuit is composed of an energy harvester, a rectifier bridge, a depletion MOSFET (dMOS) Kbp, and a capacitor C_s. When the energy in C_s is not enough to power the control circuit, the Kbp is turned on so that the current bypasses the flyback circuit and flows directly into C_s. When the energy in C_s can power the control circuit, the two switches K_p and K_s in the flyback circuit are closed at the voltage near the generator voltage peak to extract the generator voltage with maximum efficiency. The IC is manufactured in an AMS 350 nm complementary metal–oxide–semiconductor (CMOS) process. The off-chip components include a flyback circuit, a rectifier bridge, two buffer capacitors, derivative capacitors, and dMOS.

Inho Park et al. designed a high-voltage dual-input IC converter for power management of TENG, as shown in Figure 8.9b [48]. The positive half-wave and negative half-wave energy from TENG are stored in the capacitors $C_{IN,P}$ and $C_{IN,M}$, respectively, by the dual-output rectifier. Through reasonable control circuit design, the above two capacitors are accurately controlled to release energy to the buck circuit when the maximum output power is reached. This method could extract TENG energy with the highest efficiency. The above-mentioned IC adopts the 180 nm Bipolar-CMOS-DMOS (BCD) process with an effective area of 2.482 mm^2. The off-chip components include an inductor, five input capacitors, one output capacitor, and four resistors. After connecting the TENG and the load, the author measured the accuracy and efficiency of the MPPT. The accuracy is higher than 96.39%, and the MPPT efficiency reaches the highest of 94.86% when the input is 17.13 μW. When the input power is 20.9 μW, the overall end-to-end efficiency is 52.9%. This work proves the feasibility and effectiveness of the maximum power point tracking method for triboelectric energy harvesting.

Kara et al. designed a triboelectric energy harvesting circuit based on a synchronous inductor parallel switch and DC step-down conversion (Figure 8.9c) [49],

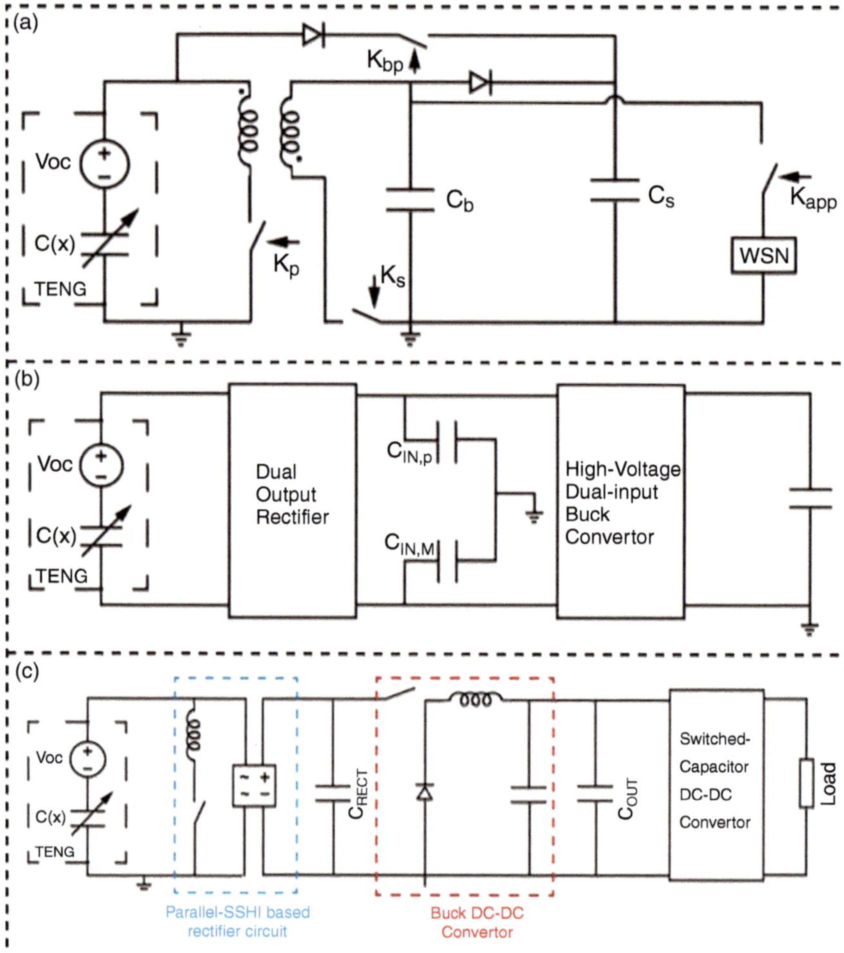

Figure 8.9 Power management with integrated circuit. (a) A self-starting power management integrated circuit for triboelectric energy harvesting. Adapted from [47]. (b) A high-voltage dual-input integrated circuit converter for power management. Adapted from [48]. (c) A triboelectric energy harvesting circuit based on synchronous inductor parallel switch and DC step-down conversion. Adapted from [49].

which can convert 70 V input into 2 V DC output. The circuit is composed of the following three step-down modules, synchronous inductance parallel switching rectifier circuit, DC step-down conversion circuit, and switched capacitor conversion circuit. When the voltage of capacitor C_{RECT} reaches 70 V, the step-down conversion is started, the voltage of C_{OUT} is stabilized at 10 V, and the load voltage is stabilized at 2 V. Similar to the idea of piezoelectric management, when the switch is closed at the time of zero-current the voltage can be reversed instantaneously, so as to avoid energy loss.

8.4.2 MOSFET

Xi et al. invented a power management circuit combining TENG and buck converter maximized energy output, as shown in Figure 8.10a [34]. The series switch in the loop, the inductor, and capacitor at the back end form a standard buck circuit. Figure 8.10b shows the structure of the switch control loop. The comparator is used to detect the TENG output voltage, and when the TENG voltage is greater than the reference voltage, the control signal turns on the MOSFET. Significantly, the power for the logic circuit comes from the TENG, without any other external resource. Figure 8.10c shows the average power curve with resistances in comparison with the direct output characteristic of the TENG. At a low frequency of 1 Hz, the maximal average power directly from the TENG is 11.2 μW at the matched resistance of 35 MΩ while the maximum is 9.0 μW with the PMM at the matched resistance of 1 MΩ, which indicates that the output impedance has been greatly reduced and the output power is retained at 80.4% efficiency. The charging curve for a 1 mF capacitance at 1 Hz is characterized in comparison with the direct charging by the TENG. As shown in Figure 8.10d, by direct charging for five minutes, the voltage only increases from 4.632 V to 4.636 V with an increment of 18.5 μJ while the voltage increases from 4.632 V to 5.118 V with an increment of 2.37 mJ by the PMM with 128 times improvement in stored energy. The excellent increase in energy extraction efficiency comes from the energy extractor. Figure 8.10e shows the voltage waveform UGS by the comparator for controlling the MOSFET switch, which is a narrow pulse voltage within 1 ms. When the voltage exceeds the threshold voltage Uth of the MOSFET (5 V), the switch is turned on and the opening time t_1 is 38 μs for this controlling signal. It is worth noting that t_1 is much less than t_2 in the switching sequence, which is very suitable for the TENG in continuous periodic motion at low frequency. The peak voltage is about ±250 V and the peak charge is about 100 nC. The U-Q curve is plotted in Figure 8.10f for comparing the measurement result and the theoretical maximum. The comparison indicates that 84.6% of energy can be autonomously transferred from the TENG by the tribotronic energy extractor (TEE).

Song et al. proposed a self-charging smart bracelet based on the integration of freestanding sliding TENG and a flexible PCB-based power management module. The series switch in the circuit is closed only when the open-circuit voltage of the TENG reaches the peak value, and the structure of the control circuit is shown in Figure 8.10g [50]. The rectified signal is differentiated first and then input to the noninverting terminal of the comparator. The differential signal is compared with the zero potential of the inverting input terminal. The switch control signal output by the comparator is connected to the gate of MOSFET switch after being processed by the delay module composed of RC delay circuit and AND gate circuit. The function of the delay module is to accurately control the closing time of the switch to achieve the maximum energy transfer efficiency of the LC oscillation circuit. The comparison between standard charging and charging with PMM is shown in Figure 8.10h, with the stored capacitor increasing, the charging through PMM demonstrated the high-efficiency energy capacity. The stored energy becomes

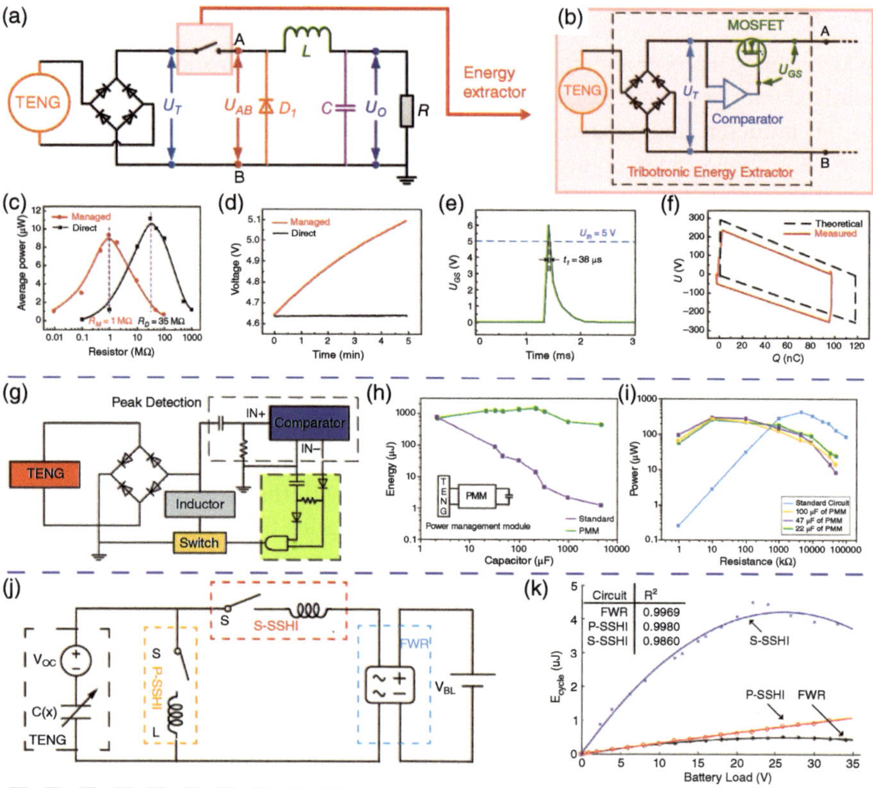

Figure 8.10 Power management with MOSFET switch. (a) The schematic circuit diagram of AC–DC buck conversion by coupling TENG, rectifier and classical DC–DC buck converter. (b) TEE for autonomous energy transfer from the TENG, which is the enlarged switch feature of Figure a. (c) Comparison of direct and managed charging for a 1 mF capacitance. (d) Comparison of direct and managed charging for a 1 mF capacitance. (e) The voltage waveform U_{GS} by the comparator for controlling the MOSFET switch. (f) The U-Q plot of the TENG by the TEE with 84.6% efficiency. Reproduced with permission of Ref. [34], 2017 © Elsevier. (g) The schematic circuit diagram of switch signal generator composed of peak detection, switch signal generator, switch and other elements. (h) Charging voltage comparison between the standard charging and charging with a power management module (PMM). (i) Power curves for standard circuits and PMMs connected to different capacitors when connected to different impedances. Reproduced with permission of Ref. [50], 2017 © Elsevier. (j) Circuit topology of P-SSHI, S-SSHI and full wave rectifier (FWR). (k) The dependence of energy per cycle on voltage of battery load. Reproduced with permission of Ref. [51], 2019 © Madhav Pathak.

gradually lower with the increase of the storage capacitance, while the energy charged by the PMM is maintained at about 1000 μJ. The output power of TENG for charging different capacitors through the standard circuit and PMM at different load resistances is shown in Figure 8.10i. The trend of output power through PMM with increasing resistance is almost the same, while the trend for standard circuit archived the maximum power at 10 MΩ, for the output power through PMM, the maximum power was obtained at 8 kΩ, which has been reduced significantly.

Pathak et al. deduced the characteristics of charging battery using full wave rectifier (FWR) circuit, parallel synchronous switched harvesting on inductor (P-SSHI) circuit and series synchronous switched harvesting on inductor (S-SSHI) circuit. (Figure 8.10j) [51]. The parameters studied include the following three: the energy output per cycle, the optimal battery voltage, and the maximum voltage that the battery can reach. Comparative experiment shows that whether the inductor is connected in parallel or in series in the loop, the energy storage effect of the battery is better than the FWR (100 times) gain. The optimal load voltage of the S-SSHI circuit is not affected by the quality factor of the inductor and is smaller than that of the P-SSHI circuit. The author built a power management circuit to verify the theoretical derivation and simulation (Figure 8.10k). The generation principle of MOSFET control signal is the same as that in Song's work. The difference is that the author designed two independent contact-separation TENGs in order to avoid the mutual interference between energy harvesting and signal generation.

The original output current is enhanced from 50 μA to 12 mA, which is improved 240 times. Only 200 LEDs can be lit up, with SCAC plugging into the system, 1280 LEDs can be lit up simultaneously. The energy extraction efficiency of SCAC is about 88.89%. This is a new type of TENG power management circuit, which creatively applies mechanical switches and electronic switches together in the power management of TENG. When TENG is contacted by the switch, the collected mechanical energy is directly converted into DC, and then the management of buck-boosting is carried out, which further improves the utilization rate of energy.

8.4.3 SCR and Triode

Harmon et al. designed a power management circuit based on a silicon-controlled rectifier (SCR), as shown in Figure 8.11a [52]. The energy flow in the circuit can be divided into four stages: First, the electrical energy is stored in the capacitor C_{in} through the rectifier bridge. Then, when the voltage in C_{in} reaches the reverse bias voltage threshold of the regulator D5, the current passing through can turn on the SCR, and the energy flows from C_{in} into the inductor L, the output capacitor C_{out}, and the resistor R. In the third stage, after all the energy in C_{in} is transmitted to the back end, the SCR cuts off, and the inductive energy continues to transfer to C_{out} and R. Finally, the energy in Cout is continuously consumed by the resistor R. Simulation results show that the power loss of SCR decreases significantly with increasing inductance, but the increase in inductance will also increase the equivalent series resistance in the circuit. Due to the appropriate turn-on time and very low energy loss of SCR, this switching circuit can reduce the output matching impedance

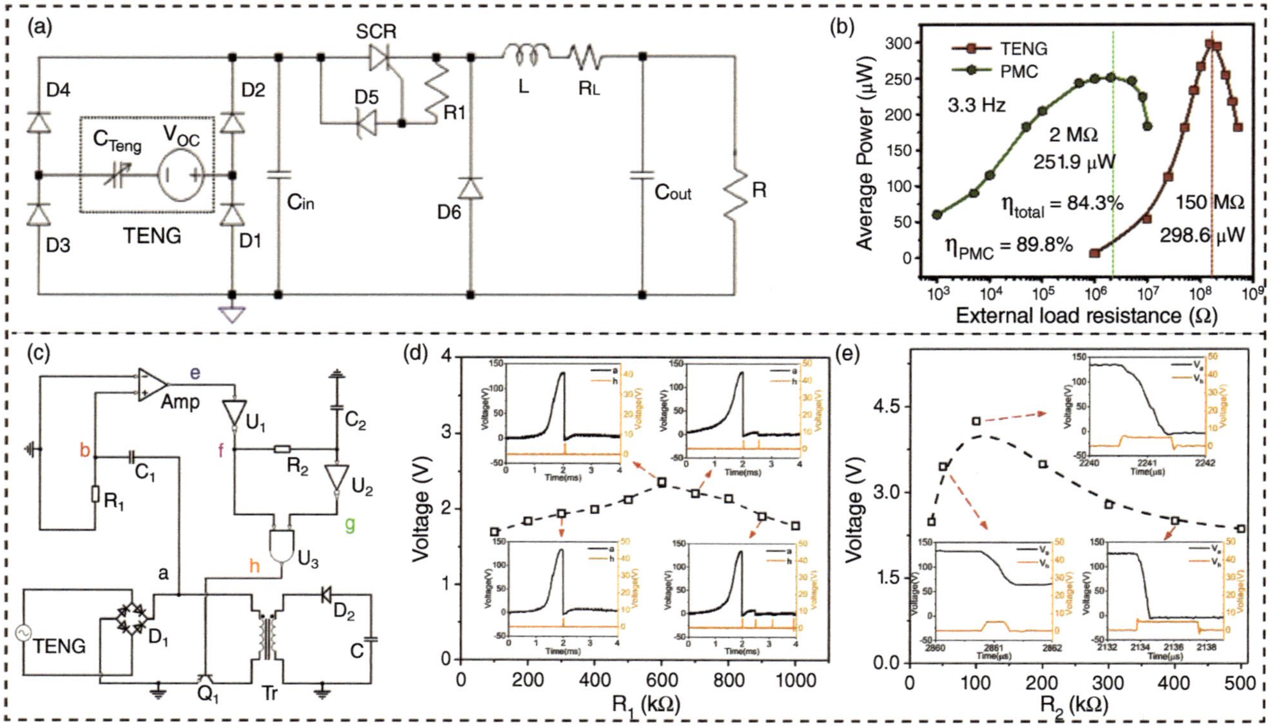

Figure 8.11 Power management with silicon-controlled rectifier (SCR) and triode. (a) Power management circuit with SCR. (b) Average output power under different loads with/without power management. Reproduced with permission of Ref. [52], 2020 © Elsevier. Institute of Electrical and Electronics Engineers. (c) Power management circuit with triode. (d) The optimum value of R_1. (e) The optimum value of R_2. Reproduced with permission of Ref. [53], 2014 © IOP publishing.

of contact separation TENG from 150 MΩ to 2 MΩ, while maintaining 84.3% of AC peak power under 150 MΩ load, as shown in Figure 8.11b.

Wu et al. proposed a method of power management for TENG using triode as a switch, and the circuit is shown in Figure 8.11c [53]. The peak detection circuit that is composed of R1, C1, and voltage comparator Amp is used to detect the voltage peak of TENG. The delay circuit that is composed of R2, C2, AND gate circuit and inverter circuit is used to accurately adjust the pulse duration of the switch control signal. The practical tests show that when the resistance increases to more than 600 kΩ, multiple pulses are generated (Figure 8.11d). Therefore, R1 is set at 600 kΩ as the optimal value of the differential circuit. R2 has an important influence on the ON time of the switch in the time-delay circuit. By comparing the energy optimization in the end energy storage capacitor C after charging for 10 seconds, the optimal value of R2 could be obtained (Figure 8.11e). Through the above optimization, the power management efficiency can reach 37.8%.

8.5 Transformer Converter

8.5.1 Transformer Converter for EMS

TENG based on Maxwell displacement current has the characteristics of high voltage and low current, which does not match the needs of many electronic devices. Similar to the traditional electromagnetic generator, the purpose of voltage reduction and current increase for TENG is achieved through energy management. Zhu et al. designed a power management circuit, including a 40 : 1 electromagnetic induction coil, capacitors (0.001, 1000, and 0.01 μF) and a rectifier bridge, which can significantly boost the output current at the cost of lower voltage and reduce the internal impedance (Figure 8.12a) [54]. Only after 0.5 charging, the TEG combined with power management circuit can output a stable 5 V DC signal, which has been depicted in Figure 8.12b. As shown in Figure 8.12c, a two-stage TENG power management method was first introduced by Niu et al. [33]. By theoretical derivation, the average power can be peaked when the charged capacitor voltage reaches 0.7153 times the saturation voltage. Therefore, two switches controlled was designed by a logic circuit to periodically extract the energy from the buffer capacitor. Whenever the voltage in the buffer capacitor reaches the optimal value mentioned above, an electronic switch closes and transfers the energy to the rear storage capacitor through electromagnetic conversion, this can effectively improve energy efficiency. The voltage of storied energy capacitors is dependent on the power consumption of R_L, in Figure 8.12d, which means the V_{store} has a positive slope with the charging time increasing as the R_L is high enough. As an obvious comparison, the stored voltage between the directed measured and board measured is depicted in Figure 8.12e, about 3.7 times enhancement is recorded. For another energy storage device, lithium-ion batteries, Pu et al. designed a charging circuit as shown in Figure 8.12f, which consisted of $n : 1$ electromagnetic induction coil and rectifier. The main purpose of designing this circuit is to achieve resistance matching

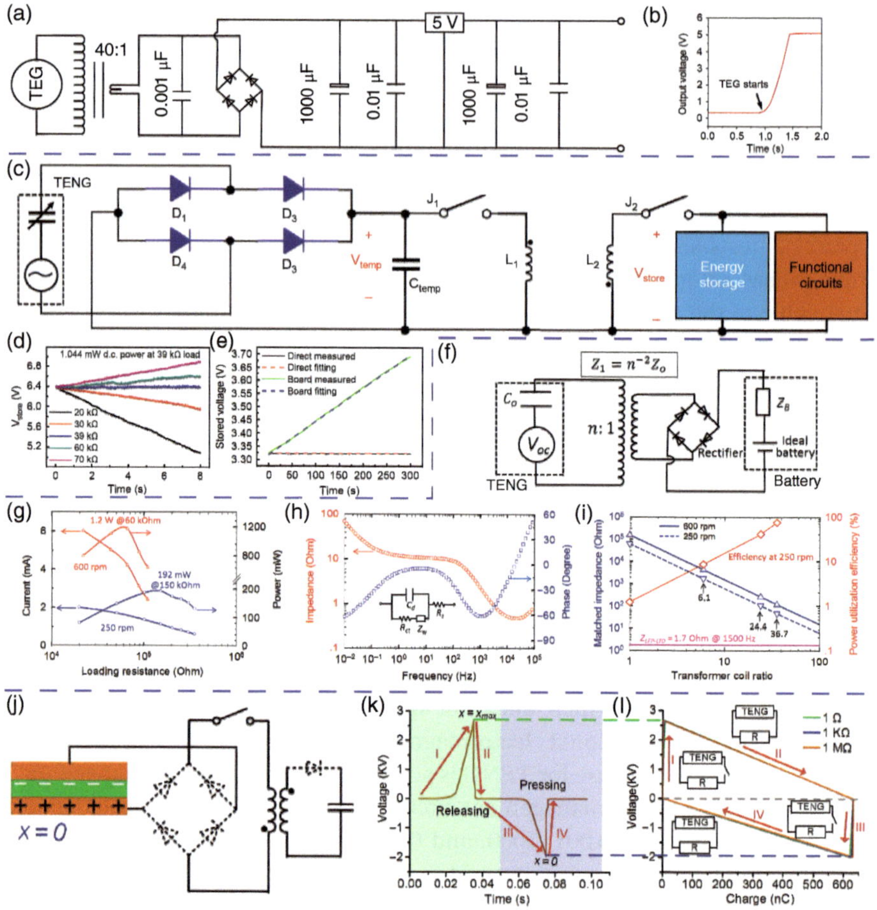

Figure 8.12 The power management module with using of a transformer converter. (a) The circuit diagram of transformer converter-based power management module consisting of TENG and power management circuit. (b) As the rotating TENG starts working, the system outputs a steady 5 V DC signal. Reproduced with permission of Ref. [54], 2014 © Springer Nature. (c) The two-stage energy-releasing power management circuit. (d) The output voltage curve on different load when collecting human biomechanical energy. (e) The charging curve compassion between directing charging and board charging. [33] / Springer Nature / CC BY-4.0. (f) The circuit diagram shown the TENG charging for the ideal battery managed by a transformer. (g) Variation of output current and output power with the load resistance increase. (h) Bode plot of impedance and phase variation of an LFP–LTO cell (battery pouch cell with $LiFePO_4$ cathode and $Li_4Ti_5O_{12}$ anode). (i) Variation of matched impedance and power utilization efficiency with the aid of transformer coil. Reproduced with permission of Ref. [55], 2016 © John Wiley & Sons. (j) TENG scavenging maximum mechanical energy with the aid of Inductor-Capacitor (LC) oscillating circuit. (k) The simulated result of output voltage for TENG by SPICE software. (l) The cycles for maximized energy output (CMEO) using switches to connect different load resistors. Reproduced with permission of Ref. [56], 2017 © Elsevier.

and increase the efficiency of energy utilization. Without employing the charging circuit, the output power is only 192 mW under the load resistance of 150 kΩ, after utilizing a charging circuit, the output power increased to 1.2 W and the internal resistance decreased to 60 kΩ (Figure 8.12g) [55]. The AC impedance of lithium-ion battery tested in this research demonstrated the impedances of 200 mAh is less than 1 Ω (Figure 8.12h). Therefore, a charging circuit could be a bridge between battery and the TENG to compensate for the difference in impedance between the two. With a charging circuit (the turns coil is 36.7), the matching impedance decreases from 10^5 Ω to 115 Ω, and the charging power increases from 2.4 mW to 139 mW, which is equivalent to the energy utilization efficiency increased to 72.4% (Figure 8.12i). For the characteristics of TENG to collect low-frequency mechanical energy, Cheng et al. designed an efficient energy management strategy. As shown in Figure 8.12j [56], using this energy management strategy, the energy stored in the capacitor increases and the optimal load resistance decreases from 30 MΩ to 100 kΩ, with a corresponding η_{total} of 71.9%. The basic circuit design strategy is a transformer converter combined with a switching circuit, the simulated result of output voltage for TENG by SPICE software is shown in Figure 8.12k. The V-Q plot at different load resistance ranging from 1 Ω to 1 MΩ shows that the CMEO could be reached (Figure 8.12l).

8.5.2 Application Demonstration

The TENG can power a wide range of portable electronics with effective power management. As shown in Figure 8.13a, Zhu et al. have demonstrated a practical and effective integrated power-supplying system for realizing diverse applications. TEG rotated at the speed of 3000 r/min, the power-supplying system can simultaneously power 10 LEDs in parallel to full brightness (Figure 8.13a) [54]. An alarm fully powered by a power-supplying system can emit a panic signal (Figure 8.13b). A multi-functional digital clock and a cellphone can work normally and charge, respectively (Figure 8.13c,d). Niu et al. designed TENG based self-charging unit first harvesting mechanical energy such as vibration of vehicles, wave energy, and wind energy. Then the collected AC electrical energy managed by a power management circuit can be used as a universal power source for portable electronics, including digital watches, cell phones, and weighing electronics, which is a great addition in addressing future IoT sensor network nodes and portable electronics power supply (Figure 8.13e) [33]. Meanwhile, flexible TENG savage biomechanical energy and managed by a power management module to supply electrical energy for battery or capacitor. Then the functional tools and wearable electronics can be solo powered by energy derived from the body itself (Figure 8.13f) [54].

8.6 Conclusion and Perspective

Power management strategies build a bridge between mechanical energy harvesters based on the principles of contact initiation and electrostatic induction and widely

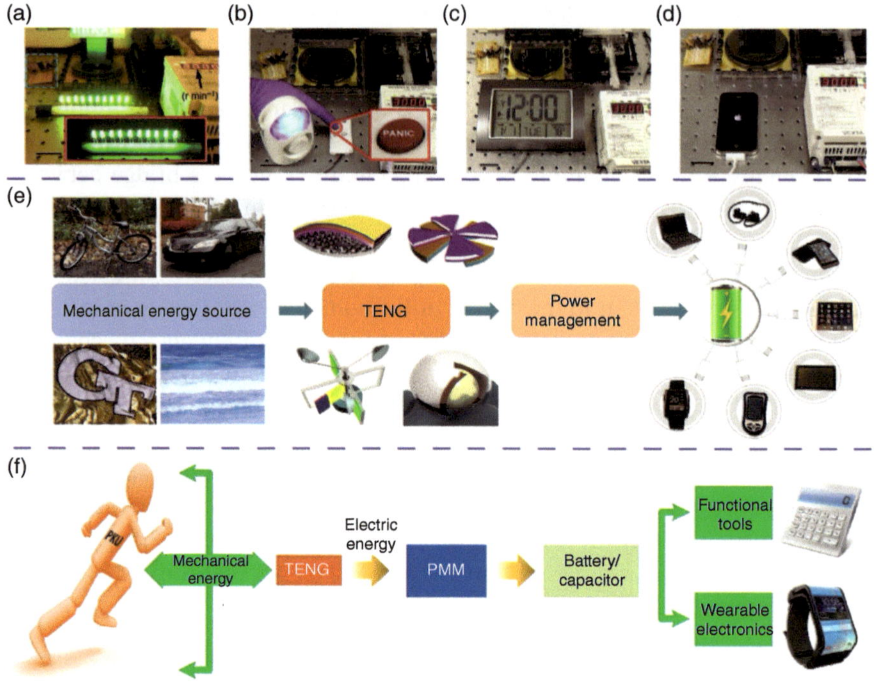

Figure 8.13 Application demonstration utilizing transformer converter. (a) Radial-arrayed rotary TENG scavenging mechanical energy with power management circuit as a power-supplying system to simultaneously power 10 LEDs. The power supply system directly power (b) an alarm emitter, (c) digital clock and (d) a cellphone. Reproduced with permission from Ref. [54] © 2014, Springer Nature. (e) Multiple mechanical energy sources can be scavenged by TENG and transferred to electrical energy with the aid of power management, which can be supplied to various portable electronic devices. Reproduced from Ref. [33], 2015 © Springer Nature. (f) Biomechanical energy collected by TENG and transferred to the electrical energy, it can continue to replenish electrical energy to battery or capacitor after being managed with management module. Reproduced with permission of Ref. [56], 2017 © Elsevier.

distributed IoT sensors and portable electronics (Figure 8.14). This review focuses on the latest switching power management for TENG. First, the research progress of switching management circuits and their theoretical foundations is introduced, respectively. Then, based on the types of switches, the management methods of switching power supply are divided into mechanical switches and electronic switches. The design ideas and power management results of the reported work are described in categories. At the end of the review, we summarize the shortcomings and research challenges of switching power management and briefly look into the future development of TENG power management., Next, I will describe the future direction of TENG power management and the urgent problems to be solved in four aspects.

A) Standard energy extraction characterization strategies
   The development of TENG's power management strategies has reached a critical point, and its ultrahigh energy extraction efficiency has been reported in some

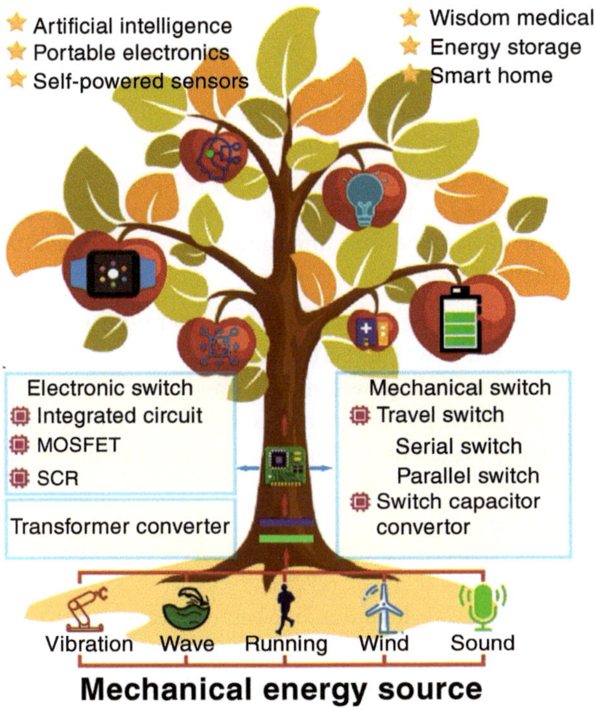

Figure 8.14 Conclusion and perspectives for power management strategy and effective energy storage strategy.

works. However, the characterization process of efficiency is still ambiguous. Resulting in the inability to form an effective comparison and provide guidance for the next research. The next step is to put forward a unified characterization method, characterization equipment and characterization conditions for energy extraction efficiency, which will provide reliable theoretical guidance for the continuous promotion of TENG power management.

B) Multiple energy extraction mechanisms
A single energy extraction mechanism cannot adapt to the energy management needs of multiple forms of TENG. The latest reported work of multimode hybrid TENG is used to harvest multiple forms of mechanical energy, while a solo energy extraction strategy miss some of the energy, leading to energy dissipation. In order to adapt to the current development trend of TENGs, we should design multifunctional management strategies, including gradient extraction, multisource extraction, and combined extraction, to deal with energy collectors with different outputs, energy collectors with different acquisition modes and various energy collectors.

C) High degree of integration
The highly integrated management strategy is suitable for the next step of packaging and is also easy to be embedded in the energy harvester. In addition, the highly integrated management module is easy to commercialize, which lays the foundation for the next step of industrialization of the management strategy.

D) Reliable encapsulation strategy
   TENG collects multiple forms of mechanical energy and faces multiple complex environments. For example, when collecting blue energy, the humid and corrosive offshore environment can easily damage the structure of the circuit, affecting the stability and durability of the device. Another example is the collection of wind energy, larger winds are often accompanied by high temperature, high humidity, and other repeated changes in the environment. So, a good package can maintain the stability of the device's performance and avoid the secondary trouble brought about by frequent maintenance.

As a significant technology in the development of TENG, the power management circuit can effectively improve the output power and efficiency of TENG. We believe that in the near future, efficient and versatile power management circuits can be developed for TENGs and used in most energy harvesting occasions, which become a key technology to promote the development of IoT and 5G communication.

References

1. Zhang, C., Zhou, T., Tang, W. et al. (2014). Rotating-disk-based direct-current triboelectric nanogenerator. *Adv. Energy Mater.* 4 (9): 1301798.
2. Wang, Z.L., Zhu, G., Yang, Y. et al. (2012). Progress in nanogenerators for portable electronics. *Mater. Today* 15 (12): 532–543.
3. Chen, B., Tang, W., and Wang, Z.L. (2021). Advanced 3D printing-based triboelectric nanogenerator for mechanical energy harvesting and self-powered sensing. *Mater. Today* 50: 1369–7021.
4. Meng, Y., Zhao, J., Yang, X. et al. (2018). Mechanosensation-active matrix based on direct-contact tribotronic planar graphene transistor array. *ACS Nano* 12 (9): 9381–9389.
5. Zhou, T., Zhang, L., Xue, F. et al. (2016). Multilayered electret films based triboelectric nanogenerator. *Nano Res.* 9 (5): 1442–1451.
6. Zhu, G., Lin, Z.H., Jing, Q. et al. (2013). Toward large-scale energy harvesting by a nanoparticle-enhanced triboelectric nanogenerator. *Nano Lett.* 13 (2): 847–853.
7. Zhou, H., Liu, G., Gao, Y. et al. (2021). Dual mode rotary triboelectric nanogenerator for collecting kinetic energy from bicycle brake. *Adv. Energy Sustain. Res.* 2: 2000113.
8. Wang, Z.L. (2021). On the expanded Maxwell's equations for moving charged media system – general theory, mathematical solutions and applications in TENG. *Mater. Today* 52: 348–363.
9. Wang, Z.L. (2020). Triboelectric nanogenerator (TENG)—sparking an energy and sensor revolution. *Adv. Energy Mater.* 10 (17): 2000137.
10. Wang, Z.L. (2018). Nanogenerators, self-powered systems, blue energy, piezotronics and piezo-phototronics: a recall on the original thoughts for coining these fields. *Nano Energy* 54: 477–483.
11. Wang, L., Fei, Z., Qi, Y. et al. (2022). Overview of human kinetic energy harvesting and application. *ACS Appl. Energy Mater.* 5 (6): 7091–7114.

12 Wang, H., Xu, L., and Wang, Z. (2021). Advances of high-performance triboelectric nanogenerators for blue energy harvesting. *Nanoenergy Adv.* 1 (1): 32–57.

13 Tian, J., Chen, X., and Wang, Z.L. (2020). Environmental energy harvesting based on triboelectric nanogenerators. *Nanotechnology* 31 (24): 242001.

14 Liu, J., Gu, L., Cui, N. et al. (2019). Fabric-based triboelectric nanogenerators. *Research (Wash D C)* 2019: 1091632.

15 Khan, A.A., Mahmud, A., and Ban, D. (2019). Evolution from single to hybrid nanogenerator: a contemporary review on multimode energy harvesting for self-powered electronics. *IEEE Trans. Nanotechnol.* 18: 21–36.

16 Dong, K., Peng, X., Cheng, R. et al. (2022). Advances in high-performance autonomous energy and self-powered sensing textiles with novel 3D fabric structures. *Adv. Mater.* 34: e2109355.

17 Dong, K., Hu, Y., Yang, J. et al. (2021). Smart textile triboelectric nanogenerators: current status and perspectives. *MRS Bull.* 46 (6): 512–521.

18 Cheng, T., Gao, Q., and Wang, Z.L. (2019). The current development and future outlook of triboelectric nanogenerators: a survey of literature. *Adv. Mater. Technol.* 4 (3): 1800588.

19 Chen, J. and Wang, Z.L. (2017). Reviving vibration energy harvesting and self-powered sensing by a triboelectric nanogenerator. *Joule* 1 (3): 480–521.

20 Chen, G., Li, Y., Bick, M., and Chen, J. (2020). Smart textiles for electricity generation. *Chem. Rev.* 120 (8): 3668–3720.

21 Zhang, C., Zhao, J., Zhang, Z. et al. (2023). Tribotronics: an emerging field by coupling triboelectricity and semiconductors. *Int. J. Extreme Manuf.* 5 (4): 042002.

22 Pu, X., Zhang, C., and Wang, Z.L. (2023). Triboelectric nanogenerators as wearable power sources and self-powered sensors. *Natl Sci. Rev.* 10 (1): nwac170.

23 Choi, D., Lee, Y., Lin, Z.H. et al. (2023). Recent advances in triboelectric nanogenerators: from technological progress to commercial applications. *ACS Nano* 17 (12): 11087–11219.

24 Wu, H., Wang, S., Wang, Z., and Zi, Y. (2021). Achieving ultrahigh instantaneous power density of 10 MW/m^2 by leveraging the opposite-charge-enhanced transistor-like triboelectric nanogenerator (OCT-TENG). *Nat. Commun.* 12 (1): 5470.

25 Yu, X., Ge, J., Wang, Z. et al. (2022). High-performance triboelectric nanogenerator with synchronization mechanism by charge handling. *Energy Convers. Manag.* 263: 115655.

26 Cheng, G., Lin, Z.-H., Lin, L. et al. (2013). Pulsed nanogenerator with huge instantaneous output power density. *ACS Nano* 7 (8): 7282–7291.

27 Zi, Y., Guo, H., Wang, J. et al. (2017). An inductor-free auto-power-management design built-in triboelectric nanogenerators. *Nano Energy* 31: 302–310.

28 Qin, H., Cheng, G., Zi, Y. et al. (2018). High energy storage efficiency triboelectric nanogenerators with unidirectional switches and passive power management circuits. *Adv. Funct. Mater.* 28 (51): 1805216.

29 Yu, X., Wang, Z., Zhao, D. et al. (2022). Triboelectric nanogenerator with mechanical switch and clamp circuit for low ripple output. *Nano Res.* 15: 2077–2082.

30 Niu, S., Wang, S., Lin, L. et al. (2013). Theoretical study of contact-mode triboelectric nanogenerators as an effective power source. *Energy Environ. Sci.* 6 (12): 3576.

31 Zi, Y., Niu, S., Wang, J. et al. (2015). Standards and figure-of-merits for quantifying the performance of triboelectric nanogenerators. *Nat. Commun.* 6: 8376.

32 Niu, S., Liu, Y., Zhou, Y.S. et al. (2015). Optimization of triboelectric nanogenerator charging systems for efficient energy harvesting and storage. *IEEE Trans. Electron Devices* 62 (2): 641–647.

33 Niu, S., Wang, X., Yi, F. et al. (2015). A universal self-charging system driven by random biomechanical energy for sustainable operation of mobile electronics. *Nat. Commun.* 6: 8975.

34 Xi, F., Pang, Y., Li, W. et al. (2017). Universal power management strategy for triboelectric nanogenerator. *Nano Energy* 37: 168–176.

35 Liu, W., Wang, Z., Wang, G. et al. (2020). Switched-capacitor-convertors based on fractal design for output power management of triboelectric nanogenerator. *Nat. Commun.* 11 (1): 1883.

36 Wang, Z., Liu, W., He, W. et al. (2021). Ultrahigh electricity generation from low-frequency mechanical energy by efficient energy management. *Joule* 5 (2): 441–455.

37 Niu, S., Liu, Y., Wang, S. et al. (2013). Theory of sliding-mode triboelectric nanogenerators. *Adv. Mater.* 25 (43): 6184–6193.

38 Niu, S., Zhou, Y.S., Wang, S. et al. (2014). Simulation method for optimizing the performance of an integrated triboelectric nanogenerator energy harvesting system. *Nano Energy* 8: 150–156.

39 Niu, S. and Wang, Z.L. (2015). Theoretical systems of triboelectric nanogenerators. *Nano Energy* 14: 161–192.

40 Wu, H., Wang, Z., Zhu, B. et al. (2023). All-in-one sensing system for online vibration monitoring via IR wireless communication as driven by high-power TENG. *Adv. Energy Mater.* 13 (16): 2300051.

41 Zi, Y., Wang, J., Wang, S. et al. (2016). Effective energy storage from a triboelectric nanogenerator. *Nat. Commun.* 7: 10987.

42 Tang, W., Zhou, T., Zhang, C. et al. (2014). A power-transformed-and-managed triboelectric nanogenerator and its applications in a self-powered wireless sensing node. *Nanotechnology* 25 (22): 225402.

43 Cheng, G., Zheng, H., Yang, F. et al. (2018). Managing and maximizing the output power of a triboelectric nanogenerator by controlled tip–electrode air-discharging and application for UV sensing. *Nano Energy* 44: 208–216.

44 Zhang, H., Marty, F., Xia, X. et al. (2020). Employing a MEMS plasma switch for conditioning high-voltage kinetic energy harvesters. *Nat. Commun.* 11 (1): 3221.

45 Yang, J., Yang, F., Zhao, L. et al. (2018). Managing and optimizing the output performances of a triboelectric nanogenerator by a self-powered electrostatic vibrator switch. *Nano Energy* 46: 220–228.

46 Zhou, H., Liu, G., Bu, T. et al. (2024). Autonomous cantilever buck switch for ultra-efficient power management of triboelectric nanogenerator. *Appl. Energy* 357: 122475.

47 Boisseau, S., Gallardo, P.G.M., and Despesse, G. (2013). Self-starting power management circuits for piezoelectric and electret-based electrostatic mechanical energy harvesters. *J. Phys. Conf. Ser.* 476: 2080.

48 Park, I., Maeng, J., Shim, M. et al. (2020). A high-voltage dual-input buck converter achieving 52.9% maximum end-to-end efficiency for triboelectric energy-harvesting applications. *IEEE J. Solid State Circuits* 55 (5): 1324–1336.

49 Kara, I., Becermis, M., Kamar, M.A.-A. et al. (2021). A 70-to-2 V triboelectric energy harvesting system utilizing parallel-SSHI rectifier and DC–DC converters. *IEEE Trans. Circuits Syst. I Regul. Pap.* 68 (1): 210–223.

50 Song, Y., Wang, H., Cheng, X. et al. (2019). High-efficiency self-charging smart bracelet for portable electronics. *Nano Energy* 55: 29–36.

51 Pathak, M. and Kumar, R. (2021). Synchronous inductor switched energy extraction circuits for triboelectric nanogenerator. *IEEE Access* 9: 76938–76954.

52 Harmon, W., Bamgboje, D., Guo, H. et al. (2020). Self-driven power management system for triboelectric nanogenerators. *Nano Energy* 71: 104642.

53 Wu, H., Li, H., and Wang, X. (2020). A high-stability triboelectric nanogenerator with mechanical transmission module and efficient power management system. *J. Micromech. Microeng.* 30 (11): 115017.

54 Zhu, G., Chen, J., Zhang, T. et al. (2014). Radial-arrayed rotary electrification for high performance triboelectric generator. *Nat. Commun.* 5: 3426.

55 Pu, X., Liu, M., Li, L. et al. (2016). Efficient charging of Li-ion batteries with pulsed output current of triboelectric nanogenerators. *Adv. Sci.* 3 (1): 1500255.

56 Cheng, X., Miao, L., Song, Y. et al. (2017). High efficiency power management and charge boosting strategy for a triboelectric nanogenerator. *Nano Energy* 38: 438–446.

9

Tribotronics

9.1 Introduction

With the rapid development of electronic information technology, the research of functional, integrated, and intelligent electronic devices has attracted wide attention [1, 2]. The core part of the intelligent function of electronic devices is the interaction between electronic devices and external environment [3–7]. Therefore, it is important to study signal processing and the interaction between signal and device generated by external environment changes. However, the traditional approach is to use passive sensors to sense and detect signal changes in the external environment, which lacks the direct interaction mechanism between electronic devices and the external environment, limiting the information acquisition and interaction ability of electronic devices [8–11]. The realization of active mechanical perception of the external environment is an urgent need to promote the development of the Internet of Things (Figure 9.1) [12–14].

In 2012, Wang et al. invented the triboelectric nanogenerator (TENG) by the coupling of triboelectrification and electrostatic induction [15]. Based on four basic modes, the TENG can convert various forms of energy such as mechanical vibration energy, human motion energy, wind energy, and water energy into electrical energy for self-powered electronic devices [16–20]. In 2014, Zhang and Wang first proposed the emerging research field of tribotronics by coupling triboelectric effects with semiconductors [21]. Tribotronics uses the tribo-potential generated by the TENG instead of the traditional gate voltage as a signal to regulate and control the carrier transport in the semiconductor. Various kinds of tribotronic devices have been developed and applied to human-machine interfaces, establishing a direct interaction mechanism between the external environment and semiconductor devices [22–29]. Tribotronics realizes the active mechanical perception of the external environment, providing a new idea and pathway in sensors and self-powered systems.

Here in this chapter, we review the recent progress of tribotronics. First, we analyze the working mechanism of tribotronics, in which the tribotronic potential can be used to control electrical transport in semiconductors. Second, we introduced the contact electrification field-effect transistor (CE-FET), which is the fundamental unit of tribotronics. On this basis, several tribotronic functional devices

Triboelectric Nanogenerators: Technology, Applications, and Commercialization, First Edition.
Yunlong Zi, Hengyu Guo, Jie Wang, Chi Zhang, Xiangyu Chen, and Qing Zhao.

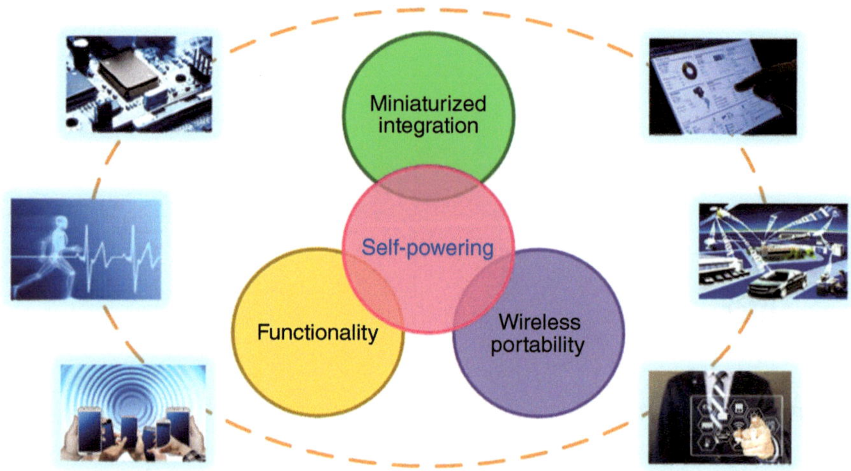

Figure 9.1 A summary of four major technological drives for electronics: miniaturized integration, functionality, wireless portability, and self-powering. The self-powerbility serves as the base of the other three fields. Adapted from [12].

have been generalized for active control and information sensing, demonstrating triboelectricity-controlled electronics and establishing active mechanosensation for the external environment. Based on the research prospects for interactions between triboelectricity and semiconductors, tribotronics is expected for a significant impact and potential applications in micro-electromechanical systems/nanoelectromechanical systems (MEMS/NEMS), flexible electronics, robotics, wireless sensor network, and Internet of Things.

9.2 Tribo-Potential

The invention of the TENG provides a new electro-mechanical transformation method. When the TENG acts on a capacitive load, it will generate a holding voltage on the load, which is called the tribo-potential. The tribo-potential generated by the TENG can be used to control tribotronic devices. Figure 9.2 shows the tribo-potential generation principle of vertical-contact-separation mode TENG [30]. Kapton and polymethyl methacrylate (PMMA) are used as examples to illustrate the working principle, which can be applied to any other material. The two materials are separated in the initial state and no charge is generated. There is no electric potential difference (EPD) between the two electrodes (Figure 9.2a). The two material surfaces gradually approach and contact each other under external stress. Compared with PMMA, Kapton has more ability to gain electrons. Thus, the charge transfer occurs on the contact surface of Kapton and PMMA under the action of the contact electric effect. Electrons on the surface of PMMA are transferred to the Kapton, resulting in a negative charge on the surface of Kapton and a positive charge on the surface of PMMA. The charge on the surface can be maintained for a long

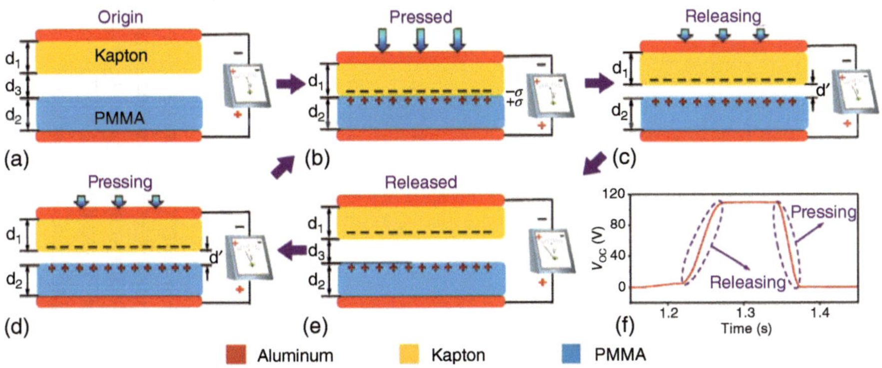

Figure 9.2 Tribo-potential generation principle of vertical contact separation mode TENG. Reproduced with permission of Ref. [30], 2012 © American Chemical Society. (a–e) Operation principle of the tribopotential on a capacitive load. (f) Open circuit voltage of TENG.

time, and these charges are only limited to the surface of the material, due to the inherent insulation characteristics of the insulator. Two equal amounts of opposite charges are at the same interface, so there is no EPD between the two electrodes (Figure 9.2b). When the force applied to the generator is released, Kapton returns to its original position due to the elasticity of the material itself, so the two charged surfaces separate, creating EPD between the two electrodes due to the electrostatic induction effect (Figure 9.2c). As the force applied to the generator is gradually released, the distance between the two films increases. The open circuit voltage of the generator gradually increases until the Kapton film completely returns to its initial position, at which point the EPD reaches its maximum value (Figure 9.2e). If the force is reapplied, the Kapton and PMMA layers are gradually brought closer together. The EPD between the two electrodes begins to gradually decrease from the maximum (Figure 9.2d). When the two polymers are fully in contact again, the EPD between the two electrodes is zero. During this process, the friction potential is a holding voltage due to the capacitive impedance characteristics of the TENG, which can remain stable when the two layers of polymer are separated. The friction potential depends on the separation distance between the two layers and can be approximately calculated as:

$$U_{\mathrm{T}} = -\frac{Q}{(C_{\mathrm{load}} + C_{\mathrm{d}})} \tag{9.1}$$

where Q is the triboelectric charge, C_{d} is the total capacitance of TENG when released, and C_{load} is the load capacitance. The open circuit voltage of TENG is shown in Figure 9.2f.

9.3 Triboelectricity Modulate Field Effect

According to the above analysis of the tribo-potential, it can be seen that the TENG is particularly suitable for regulating capacitive electronic devices. Metal-oxide

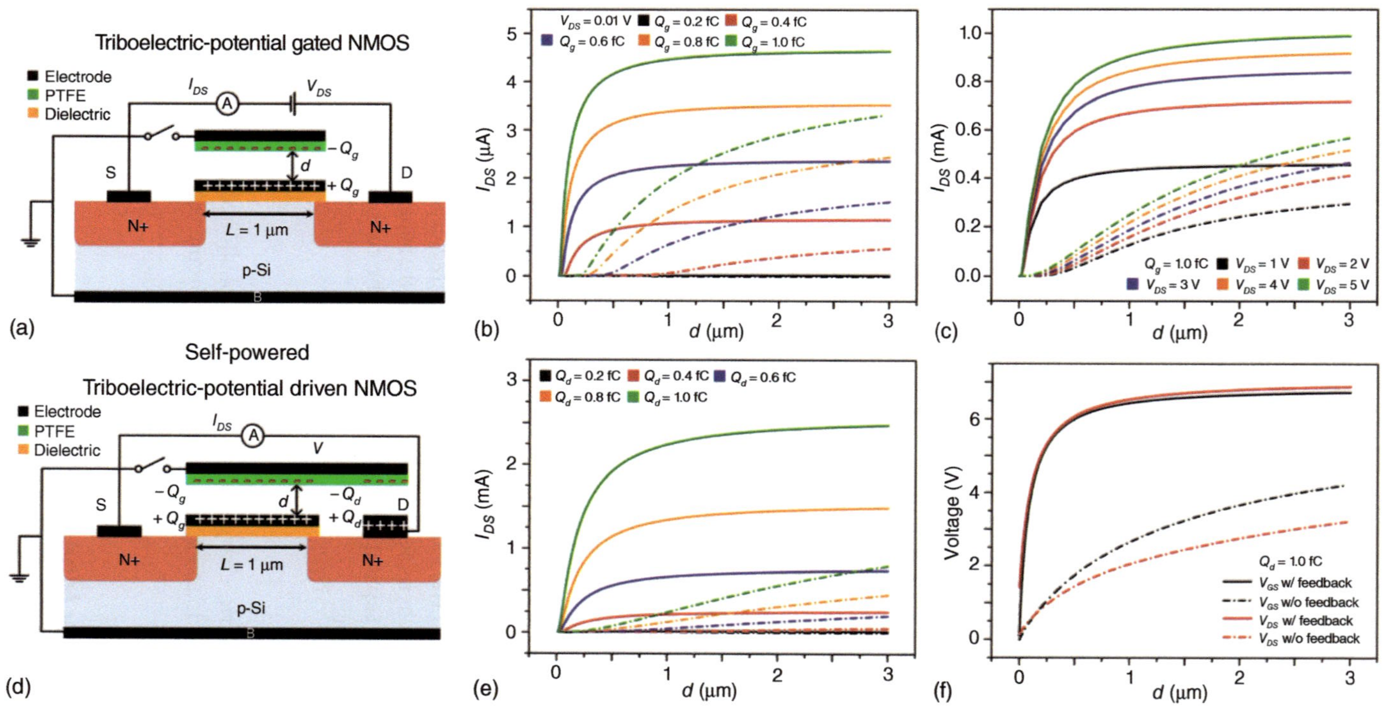

Figure 9.3 Theoretical analysis of the performance of tribotronics MOSFET. Reproduced with permission of Ref. [31], 2014 © American Chemical Society. (a) Schematic illustration of TPG NMOS, which is composed of a traditional silicon-based NMOS and a PTFE layer with aluminum electrode coated at its back as a mobile electrode. (b) I_{DS}-d characteristic of TPG NMOS with a small V_{DS} bias of 0.01 V under different gate triboelectric charges (Q_g). (c) I_{DS}-V_{DS} output characteristic of TPG NMOS with Q_g of 1.0 fC under different d values from 0.5 to 3.0 μm. (d) Schematic illustration of TPD NMOS, where PTFE layer contacts both gate and drain electrodes for contact electrification. (e) I_{DS}-d characteristic under different Q_d values from 0.2 to 1.0 fC. (f) triboelectric charges induced V_{GS} and V_{DS} with Q_d of 1.0 fC and d ranging from 1 nm to 3 μm.

semiconductor field-effect transistor (MOSFET), as a typical capacitive electronic device, is coupled with TENG to simulate the triboelectric modulating field effect. A MOSFET model has been established in finite element analysis, which consists of a conventional N-type MOSFET (NMOS) and a 25 nm thick polytetrafluoroethylene (PTFE) layer (Figure 9.3a) [31]. Based on the contact electrification effect, an equal number of triboelectric charges with opposite polarity is generated when the PTFE layer is in contact with the gate electrode. When the distance d increases to the threshold distance d_{th}, the gate voltage generated by the triboelectrification also reaches the threshold voltage V_{th} and the NMOS turns off. The first mode of operation is to connect the mobile electrode to the grounded source and the block electrode, and the second mode of operation is to isolate the mobile electrode from all other electrodes as a "floating" electrode, which are defined as "w/feedback" and "w/o feedback", respectively. The dependence between the threshold distance d_{th} and the triboelectric potential generated on the gate is simulated by varying the separation distance d (1 nm to 3 μm) between the PTFE layer and the gate at a bias voltage V_{DS} of 0.01 V in Figure 9.3b. In the "w/o feedback" mode (dashed line), the initial value of I_{DS} is 0 ("off" state), and as d increases gradually to d_{th}, the I_{DS} increases gradually ("on" state). However, in the "w/feedback" mode (solid line), triboelectric-potential gated (TPG) NMOS turns on at very small d values and the I_{DS} increases rapidly to saturation, unlike that in the "w/o feedback" mode. In addition, as shown in Figure 9.3c, d_{th} decreases with the increase of d for both modes, because of the larger Q_{g} and the stronger electrostatic induction effect. By fixing Q_{g} to 1.0 fC, the I_{DS}-V_{GS} relationship is simulated at different d values. For each fixed d value (0.5 to 3.0 μm), the I_{DS} exhibits a two-stage growth behavior in the "w/o feedback" mode.

On this basis, the diagram of the operation of a self-powered triboelectric-potential driven (TPD) NMOS is shown in Figure 9.3d. All the parameters of silicon-based NMOS-including conduction channel width and all other device geometric parameters should be the same, except that the length of the PTFE layer is increased to vertically align and contact with both gate and drain electrodes. Since the gate and drain electrodes are both Al, the contact electrification between the PTFE layer and these two electrodes should be the same. Therefore, assuming that the triboelectric surface charge densities generated at the gate and drain electrodes are the same, the total charge at the drain is defined as Q_{d}. The transfer characteristics of the TPD NMOS are simulated by varying Q_{d} from 0.2 fC to 1.0 fC, as shown in Figure 9.3e. In the "w/o feedback" mode, the TPD NMOS turns on after reaching a specific separation distance d_{th} and the I_{DS} increases linearly without appearing saturation. In the "w/feedback" mode, the TPD NMOS turns on instantaneously when the PTFE layer is separated from the gate and drain and the I_{DS} increases linearly in the initial stage and then gradually reaches saturation. Since the negative triboelectric charges on the PTFE layer are almost electrostatically shielded, as shown in Figure 9.3f, the V_{GS} and V_{DS} caused by the frictional charge increasing rapidly to saturation.

In Figure 9.4a, a finite element method model of a sliding-electrification-gated NMOS (SG NMOS) in the enhanced mode is developed, which consists of a conventional silicon-based top-gate NMOS and a PTFE layer located on the gate electrode?

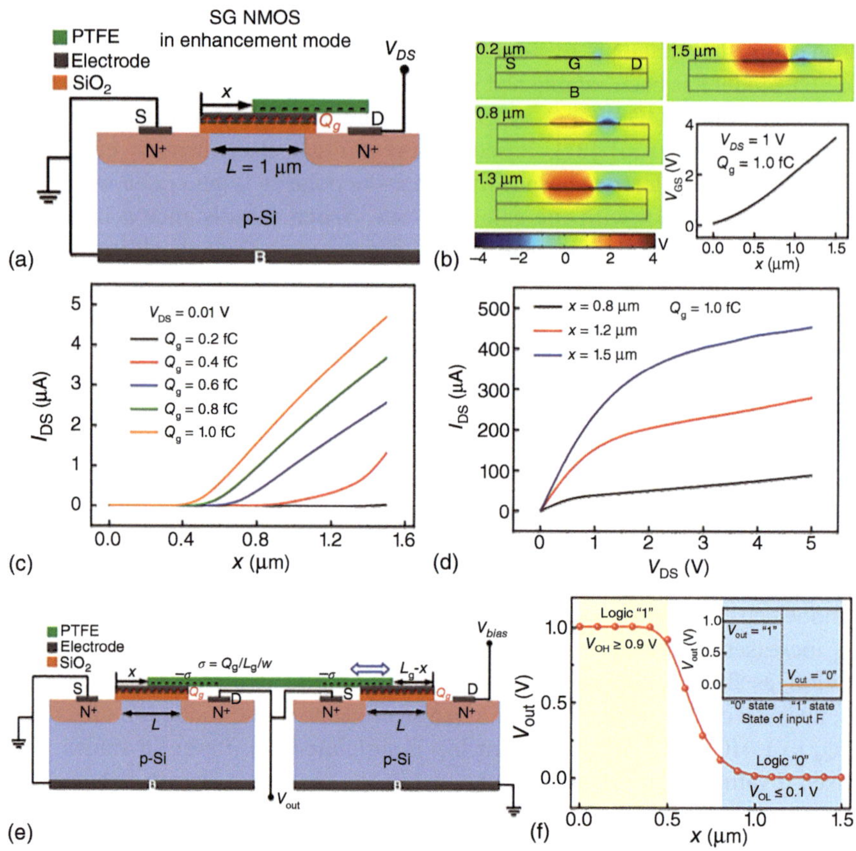

Figure 9.4 Theory of nanoscale tribotronics. Reproduced with permission of Ref. [32], 2018 © John Wiley & Sons. (a) Schematic diagram of sliding-electrification-gated NMOS in enhancement mode. (b) Electric potential distributions at different x and induced gate voltage V_{GS} with respect to x at $V_{DS} = 1$ V and $Q_g = 1.0$ fC. (c) I_{DS}-x relationship of SG NMOS at $V_{DS} = 0.01$ V under different gate triboelectric charges Q_g. (d) I_{DS}-V_{DS} relationship for various x. (e) Structure of a sliding-electrification-gated tribotronic logic device based on two SG NMOSs in enhancement mode. (f) V_{out}-x transfer characteristic of the logic device at a fixed bias of 1 V. The logic "1" output region at VOH ≥ 0.9 V and logic "0" output region at VOL ≤ 0.1 V are marked. The insert shows the output logic levels at "0" and "1" states of the input force.

[32]. The width of the NMOS is 1 μm and the conductive channel length is 1 μm. The width of both the gate and the PTFE film is 1.5 μm, which leads to the sliding motion of the PTFE film limited to 1.5 μm. The potential difference is generated between the gate and the PTFE film due to the different charge affinities, which can be used to regulate the carrier transport in the NMOS. When the drain-source voltage is set to 1 V and the charge amount on the gate electrode is given with 1 fC, the simulated electrostatic potential values are calculated in Figure 9.4b. The results show that the gate electrode voltage increases with the increasing sliding distance. The I_{DS}-x curves for different values of Q_g is shown in Figure 9.4c. Due to the existence of a threshold voltage V_{th} of the silicon-based NMOS, a threshold sliding distance x_{th} for modulating the drain-source current could be analogized. When x increases to

x_{th}, the triboelectric charges induced the gate voltage reaches V_{th} and the inverted conduction channel between drain and source, thus the SG NMOS is turned on. The x_{th} increases with the increase of Q_g, consequently, the induced V_{GS} can reach V_{th} faster at a larger Q_g. In addition, the I_{DS} is larger at a higher value of Q_g, which is attributed to the larger V_{GS}. At a very small Q_g (0.2 fC), the SG NMOS cannot be turned on during the sliding process, which means that a minimum Q_g is required to turn on the device. When the charge amout on the gate is set to 1 fC, the relationship between I_{DS} and V_{DS} for different sliding distance x is shown in Figure 9.4d. I_{DS} increases with the increasing of V_{DS}, showing a saturation trend at a larger x. The I_{DS} at different gate triboelectric charges and drain voltages has also been investigated. Subsequently, SG NMOS in depletion mode has been constructed to confirm that the triboelectric potential can be used as the external gate voltage to control carrier transport in semiconductors, which is successful in replacing the conventional gate voltage. In addition, the p-type channel sliding-electrification-gated MOS has been verified to validate the mechanism of tribotronics.

Considering the size effect in large-scale integrated manufacturing, the effectiveness of tribotronics must be discussed. By reducing the channel length of NMOS to 10 nm, the I_{DS} characteristics were investigated for a proportional reduction of the relevant parameters. As the doping concentration increases, the triboelectric surface charge density and thickness of the PTFE layer remain constant, and the I_{DS} keeps constant for shorter channel lengths, increasing with x gradually. The difference is that the threshold gate voltage drops sharply at shorter conduction channels and there is no longer a threshold sliding distance. Then simulations are also performed for the application of tribotronics in logic control. A sliding-electrification-gated tribotronics logic device (SGL) that converts external mechanical stimuli into logic-level signals is proposed based on two SG NMOS devices in enhancement mode. The schematic diagram of SGL is shown in Figure 9.4e, where a 6 μm-long PTFE film is laid on the two gate electrodes of both SGLs. The V_{out}-x is shown in Figure 9.4f. V_{out} decreases as x increases. When x increases from 0 to 1.5 μm, the logic “1” can be converted to a logic “0”. The simulation results show that the triboelectric potential can still regulate the carrier transport of the semiconductor at the nanoscale without the limitation of the size effect.

The performance of tribotronics MOSFETs based on the two basic modes of operation is simulated through finite element analysis, guiding an in-depth understanding of the physical mechanism and design applications of tribotronic electronics. Meanwhile, the simulation results show the great potential of tribotronics for applications in large-scale array integration and fingerprint recognition, wearable electronics, etc.

9.4 Tribotronic Transistor

9.4.1 Theory of Tribotronic Transistor

Based on the theoretical of triboelectricity modulating field effect, the contact electrification field effect transistor (CE-FET) coupled with TENG and MOSFET is prepared by semiconductor micro-nanoprocessing technology. The CE-FET is first developed by the coupling of TENG and a back-gate silicon-on-insulator (SOI)

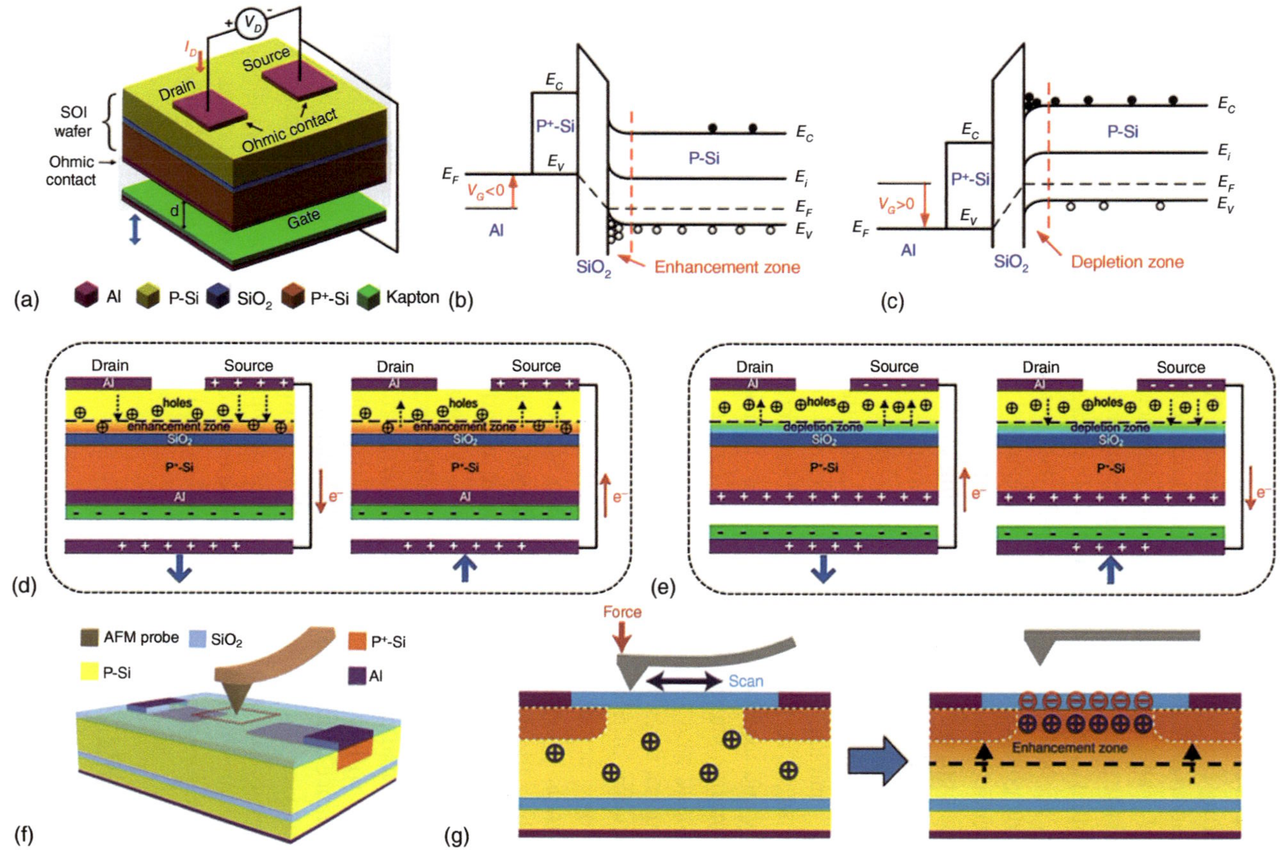

Figure 9.5 Theory of tribotronic transistor. (a) Structure of the CE-FET based on a SOI wafer with a heavy-doped silicon substrate and a mobile layer for vertical contact separation. (b) Energy diagram illustrating the MIS capacitor in the CE-FET at the negative gate voltage. (c) Energy diagram illustrating the MIS capacitor in the CE-FET at the positive gate voltage. (d) Schematic working principle of the CE-FET in enhancement mode. (e) Schematic working principle of the CE-FET in depletion mode. Reproduced with permission of Ref. [33], 2014 © American Chemical Society. (f) Schematic illustration of the NTT gated by nanoscale triboelectrification. (g) Schematic working principle of the NTT gated by nanoscale triboelectrification. Reproduced with permission of Ref. [34], 2020 © Springer Nature.

MOSFET [33]. It consists of a MOSFET and a mobile layer for contact electrification, as schematically illustrated in Figure 9.5a. By analyzing the process of contact and separation between the mobile electrode and gate area of the MOSFET, the working mechanism of the CE-FET has been demonstrated in enhancement mode and depletion mode. The equivalent circuit of the CE-FET in two different modes is shown in Figure 9.5b,c. In enhancement mode (Figure 9.5d), the mobile electrode is gradually separated from the Kapton film to the original state as the external force is released. The electrons flow from the source electrode to the mobile electrode and an inner negative gate voltage is generated. In this process, an inner charge polarization generated in the top silicon layer attracts the holes and repels the electrons in the bottom of the silicon layer. Therefore, an enhancement zone will be formed in this area, which will increase the conduction channel carrier concentration and the drain current. When the external force is vertically applied and the Al film is gradually approaching the Kapton film to the contact state, the electrons flow back from the mobile electrode to the source electrode with decreasing inner gate voltage. In this process, the inner charge polarization is depressed, which will decrease the conduction channel carrier concentration and drain current. In depletion mode (Figure 9.5e), when the Kapton film is gradually separated from the Al layer as the external force is released, the electric field formed by the triboelectric charges induces the electrons to flow from the mobile electrode to the source electrode, forming an inner gate voltage across the gate and source electrodes. In this process, the as-formed gate voltage attracts electrons and repels the holes in the bottom of the silicon layer. A depletion zone will be formed in this area, which will decrease the width of the conduction channel and the drain current. When the external force is vertically applied again and the Kapton film gradually approaches the Al layer, the electrons flow back from the source electrode to the mobile electrode with decreasing inner gate voltage. In this process, the inner charge polarization is depressed, which will increase the width of the conduction channel and the drain current.

Contact-mode atomic force microscopy (AFM) and scanning Kevin probe microscopy are used to study nanoscale triboelectric gated transistors (NTT) [34]. As shown in Figure 9.5f, the basic structure of NTT is similar to that of a conventional MOSFET without a top-gate electrode. The working mechanism of NTT can be divided into two stages in Figure 9.5g. During the process of nanoscale triboelectrification, the top SiO_2 surface above the channel region is scanned and rubbed by the AFM tip in contact mode. The AFM probe used in the experiment is made of Silicon Si, whose effective work function is smaller than that of SiO_2. According to the electron transfer mechanism, the electrons at the AFM tip will be transferred to the SiO_2 surface on the top during the triboelectric process. The transferred charges are bound to the top surface of SiO_2, resulting in an electric field inside the channel region, which attracts holes and repels electrons in the P-type silicon (P-Si) layer. Therefore, the concentration of the holes in the channel region is enhanced, and an enhancement region is formed, thereby increasing the drain current of the NTT.

9.4.2 Structure of Tribotronic Transistor

Based on the CE-FET, the MIS structure of the insulation layer can be simplified and directly used for contact electrification. A flexible organic tribotronic transistor (FOTT) without a top gate electrode has been proposed [35]. As shown in Figure 9.6a, the FOTT was fabricated on a flexible polyethylene terephthalate (PET)film substrate, consisting of a mobile triboelectric layer and an organic thin film transistor (OTFT) without a top gate electrode. The 300 nm Ta_2O_5 film is sputtered on the PET surface and a 45 nm thick pentacene film is deposited by thermal evaporation. Subsequently, two parallel Au drain/source electrodes are deposited on the pentacene film through shadow masks, and the thickness is controlled to about 45 nm. Thereafter, PMMA and Cytop are spun-coated as the dielectric layer, respectively. Based on this, pressure sensors have been developed by applying external pressure to generate relative movement inside the TENG, so as to realize the regulation of I_{DS}. By further adding magnetic composites in the triboelectric layer, the magnetic sensor based on FOTT has been developed. Figure 9.6b shows the equivalent circuit of the FOTT. The electrical potential difference, acting as the gate voltage, is generated from the movement of the TENG by the external pressure, which can modulate the I_{DS}. As shown in Figure 9.6c, the conductive channel of the FOTT is 60 μm in length and 1000 μm in width, respectively. The FOTT can be greatly modulated by contact electrification between the dielectric layer and a mobile triboelectric layer. Furthermore, the FOTT displays great sensing ability in pressure and magnetic detection with great sensitivity, fast response time, and excellent stability.

Besides, a sliding mode tribotronic transistor has been developed. The basic structure of the flexible transparent tribotronic transistor (FTT) is composed of an OTFT and a TENG in free-standing sliding mode, as schematically illustrated in Figure 9.6d [36]. The OTFT was fabricated on a flexible and transparent PET substrate with an indium tin oxide (ITO) layer as the bottom gate electrode. Then a Ta_2O_5 film was prepared by magnetron sputtering as the gate dielectric, which has a high dielectric constant and good mechanical stability. Subsequently, a pentacene film was deposited on the surface of the Ta_2O_5 film by thermal evaporation as the semiconductor layer. The Au drain/source electrodes were deposited on two sides of the pentacene film through shadow masks, which is of great importance to increase the transparency of the device. Two ITO electrodes are etched on the bottom of the PET along the y-axis, leaving an interval of less than 1 mm between them. The fluorinated ethylene propylene (FEP) film was purposely chosen as the freestanding sliding layer, which was cut into 1 cm × 1 cm and adhered on the surface of the bottom PET substrate. It can horizontally slide in the y-axis between the two ITO bottom electrodes by an external force. By the sliding-induced electrostatic potential of the TENG, an inner gate voltage is generated, which can be used to control the drain current in OTFT. As shown in Figure 9.6e, the whole FTT device has good optical transparency, which is about 71.6% in the visible wavelength. The inset is an optical image of the FTT, in which the logo on the paper behind the device is clearly visible. Figure 9.6f is an optical microscopy image of the FTT under a bending status in top view. To increase the effective contact area and

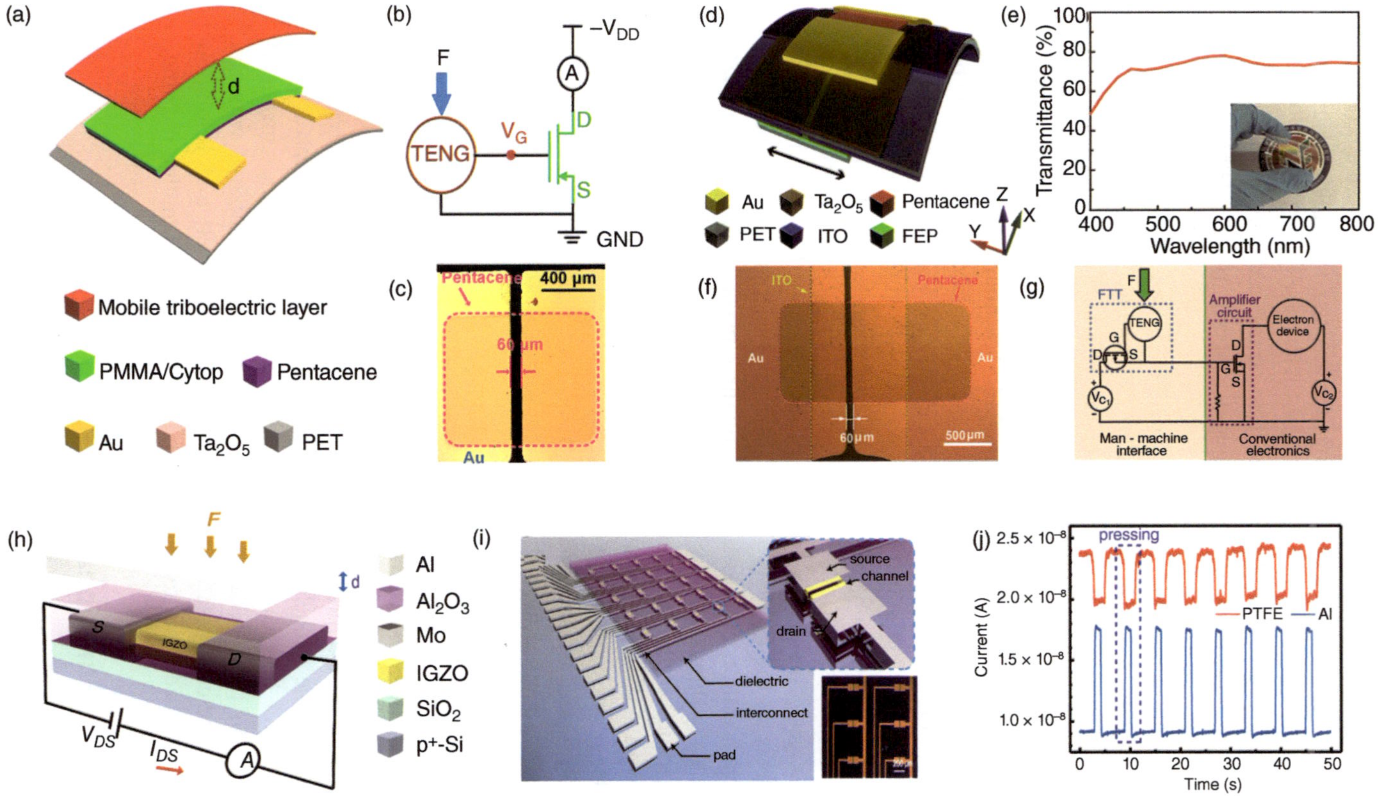

Figure 9.6 Structure of tribotronic transistor. (a) Structure of the FOTT without a top gate electrode. (b) Equivalent circuit of the FOTT. (c) Optical graph of the as-fabricated FOTT. Reproduced with permission of Ref. [35], 2017 © American Chemical Society. (d) Structure of the FTT based on an organic thin film transistor (OTFT) and a triboelectric nanogenerator (TENG) in free-standing sliding mode. (e) Photographic image and optical transmittance of the FTT. (f) Optical microscopy image of the FTT in top view. (g) Equivalent circuit of the portable FTT for active modulation of various conventional electronics. Reproduced with permission of Ref. [36], 2017 © Elsevier. (h) Structure diagram of the tribotronic TFT cell. (i) Schematic diagram of the sensor array and top-view photograph of a 5 × 5 monolithic integrated array. (j) Output current I_{DS} response of the cell with various friction materials (Al and PTFE) under the same separation distance of 100 μm. Reproduced with permission of Ref. [37], 2020 © John Wiley & Sons.

promote triboelectrification, a nanostructure was created on the FEP surface by the inductively coupled plasma (ICP). Figure 9.6g is the equivalent integrated circuit system, which consists of human–machine interface and conventional electronics. Compared with the contact-modulated tribotronic transistor, the sliding-modulated tribotronic transistor can be linearly modulated with a larger range.

Moreover, a miniature tribotronic thin film transistor (TFT) was developed by using MEMS processing technology, which is promising for an integrated tribitronic array. The structure of TFT is schematically shown in Figure 9.6h [37]. The tribotronic tactile sensing TFT is composed of a dual-gate indium-gallium-zinc-oxide (InGaZnO or IGZO) TFT and a TENG working in the contact separation mode. The Al_2O_3 layer in the TFT acts not only as the gate dielectric layer but also as one of the triboelectric layers for the TENG. The other triboelectric layer is a movable Al foil. The relative movement between the Al_2O_3 layer and the Al foil would induce an electrostatic potential applied on the TFT top gate and this potential can modulate the I_{DS} of TFT. In order to enhance the modulation effects of the tribotronic device, a positive bias V_B (≈4 V) is applied on the TFT bottom gate (the P-Si substrate) to help the device work in the subthreshold region. As shown in Figure 9.6h, the movable Al layer separates from the Al_2O_3 layer with a certain distance in the initial state. Since no contact electrification occurs, there are no charges generated on the Al_2O_3 and Al layers at first, thus having no modulation effects on the I_{DS}. The TFT is fabricated based on the state-of-the-art semiconductor fabrication processes and totally three masks are used, each IGZO TFT cell in the array shows uniform electrical performance. Figure 9.6i shows the schematic illustration of the mechanosensation-active array (5 × 5 IGZO TFT) on a Si substrate with a 100-nm thermal SiO_2 layer. In order to precisely control the separation distance between the two friction layers, this TFT array the array was assembled on a printed circuit board and then anchored on a fixed platform and a stepping motor was adopted to provide a precise movement of the movable Al foil relative to the TFT. Because PTFE has a stronger ability to attract electrons than the Al_2O_3 layer, the contact of PTFE with Al_2O_3 would result in negative charges accumulated at the PTFE surface and positive charges at the Al_2O_3 surface. In this case, a reverse inner electrical field across the TFT is established and leads to an opposite modulation effect on the I_{DS} compared to the Al foil. As demonstrated in Figure 9.6j, during the approaching process, an increase in the I_{DS} is observed with Al as the movable triboelectric layer, compared to the I_{DS} decreases when the triboelectric layer is replaced to the PTFE layer under the same movement operation. Based on the working principles of tribotronics combined with the monolithic integrated configuration, the array displays a relatively high resolution of 12 PPI and can realize a millimeter-level tactile perception and motion tracking.

9.5 Tribotronic Functional Devices

9.5.1 Tribotronics for Information Sensing

Human-machine interaction, which is used to translate natural information into computer language, is on the rise as big data grows. Moreover, a tactile sensor, as a crucial method in the field of information sensing, can detect external stimuli, which is the same as tactile perception. Many attempts have been made to achieve active tactile sensing until the emergence of tribotronic transistors, which realize direct interaction between humans and environment. This section reviews the typical work of tribotronic functional devices in the field of information sensing.

Recent studies on tribotronic tactile sensors focus on mimicking human skin, which is expected to be self-healable to satisfy the prolonged wearing demand and stretchable to cater to different mechanical conditions. As a result, the self-healable tribotronic transistor for bioinspired e-skin has been proposed, as schematically illustrated in Figure 9.7a–d. The structure of the self-healable tribotronic transistor is shown in Figure 9.7a, employing semi-conducting carbon nanotubes (s-CNTs) as robust electronic material for its high mobilities, excellent stretchability as well as exceptional durability, poly(urea-urethane) (PUU) as elastomers for its good stretchability and self-healable ability, and silver nanowires (Ag NWs) for source and drain electrodes due to its excellent conductivity. Further, the Ag NWs is sprayed through the shadow mask on the Si wafer, followed by spin-coating the PUU layer, which is then stripped from the Si wafer. Subsequently, the s-CNTs with high surface density is drop-casted onto the surface of the as-fabricated device. The top dielectric PUU is at last spun-coated on the s-CNTs to complete the basic structure of the tribotronic transistor. The whole fabrication process is simple and mass-industrialized produced. Moreover, the stretching ability has been carefully studied by conditioning the self-healable tribotronic transistor into different stress, indicating that the device only displays little current change at some specific separation distance. Note that the device under vertical strain direction also exhibits good stability. More importantly, the self-healable ability has been carefully tested and the mechanism is figured out to reveal the microcosmic nature, which involves two structures inside the PUU polymer, the aromatic disulfide and the urea groups, respectively. The aromatic disulfide keeps constant metathesis at room temperature and the urea groups can form a quadruple H-bonding, which is a kind of dynamic non-covalent crosslinking of polymer chains and can be reformed to restore the initial mechanical property. Additionally, the transfer curves of the tribotronic transistor after several cuts have been studied, demonstrating that the output performance of the tribotronic transistor can be restored after three complete cuts, with little performance decay being observed, exhibiting excellent self-healing ability. Consequently, the tribotronic transistor is engaged in robot skin for object grabbing. Figure 9.7b depicts the optical picture that the as-fabricated tribotronic transistor conformally adheres to a human knuckle. Moreover, the tribotronic transistor can be utilized in tactile sensing and robot control, as shown in Figure 9.7c, demonstrating that the robot posture can be controlled regardless of pristine and stretched state. In addition, the tribotronic transistor can also be engaged in tactile perception in object grabbing, as schematically shown in Figure 9.7d, noting that even after the cutting and healing process, the sponge ball can still be successfully sensed and grabbed by the robotic hand. In conclusion, this work shows broad prospects in future robust tribotronic electronics in human-machine interface,

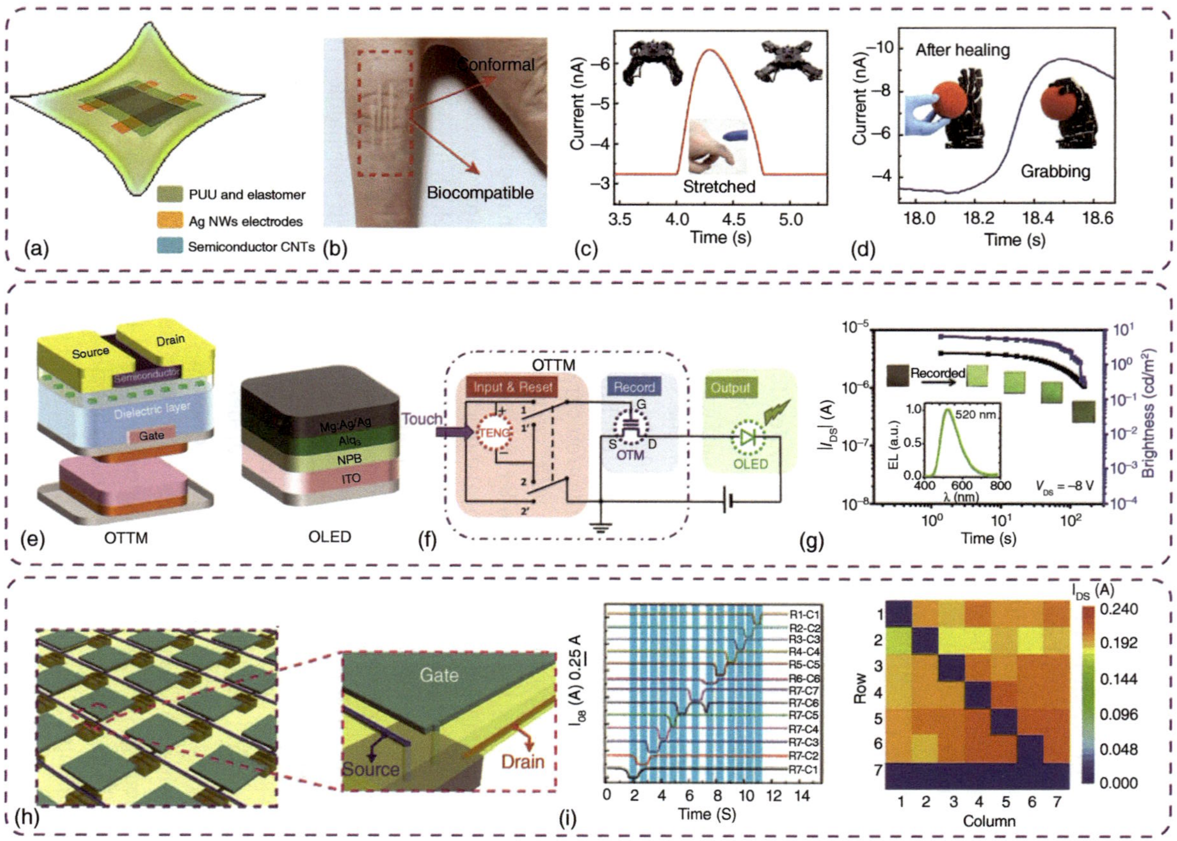

Figure 9.7 Tribotronics for information sensing. (a) Basic structure of the stretchable self-healable tribotronic transistor. (b) The transistor conformally adheres to a human knuckle for sensing the external touch. (c) Stretched transistor as bionic skin for robot control. (d) photograph and current changes of the transistor after healing for grabbing a sponge ball. Reproduced with permission of Ref. [38], 2022 © Elsevier. (e) Sketch of the OTTM and the OLED. (f) Schematics of the memory system. (g) The retention time of the input touch action recorded by the system, inset is the electroluminescent spectra of the OLED. Reproduced with permission of Ref. [39], 2016 © John Wiley & Sons. (h) Structure of the TTA. (i) Output currents of pixels in the moving trajectory of the ball. (j) Two-dimensional intensity map with the measured current signals during the ball moving. Reproduced with permission of Ref. [40], 2016 ©

wearable electronics, robotics, and bionics, and speed up the development of bionic human skin [38].

Besides, in order to multifunctionalize the tactile sensing process and realize the visualized information sensing and memory, a flexible organic tribotronic transistor memory (OTTM) as well as an organic light-emitting diode (OLED) have been proposed as shown in Figure 9.7e–g. The detailed structures of the OTTM and the OLED are demonstrated in Figure 9.7e, among which the OTTM consists of a contact separation mode TENG with a Cu/polyvinyl chloride (PVC)/Cu structure and an organic transistor memory (OTM) with an ITO/Ta_2O_5/PMMA/pentacene structure. The Ta floating gate is sandwiched into the PMMA layer which provides a storage function for OTM. In addition, the OLED possesses a structure of ITO/*N*, *N′*-diphenyl-*N*, *N′*-bis(1,1′-biphenyl)-4,4′-diamine (NPB)/Alq_3/Mg: Ag/Ag, which can be regarded as a signal output functional unit. During the whole fabrication process, sputtering, spin coating, and thermal evaporation methods are used. Moreover, a double pole double throw (DPDT) switch is employed to transfer the system between the monitoring and resetting modes. When the external stimulus is applied to the system, the PVC layer begins to periodically contact and separate with the Cu electrode, generating periodical gate voltages, which can be further employed to write and erase the signals in the OTTM. Consequently, the schematics of the multiple systems are illustrated in Figure 9.7f, employing the TENG as a triggering input and reset unit, the OTM as a storage unit to record the detect triggering from the TENG, and the OLED as a touch signal output functional unit. With the as-proposed system, the I_{DS} and the electroluminescence intensity of the OLED have been carefully studied in order to illustrate the retention time, as shown in Figure 9.7g. It shows that the emission is still visible after 102 seconds, which is still limited and remains to be developed in the future. By realizing direct interaction with the external environment, the OTTM and OLED system shows great perspectives in security monitoring, functional instruments, wearable functional electronics as well as human-machine interaction [39].

Large-scale tactile sensor array is another vital functional device in information sensing, which is of great importance in high-resolution sensing and shows broad potential in trajectory tracking. As a result, the tribotronic transistor array (TTA) has been developed by combining FET and TENG, which is shown in Figure 9.7h–j. The basic structure of TTA is schematically shown in Figure 9.7h, and each pixel has a size of $5 \times 5\ mm^2$, containing a square Cu pad and a transistor unit which are integrated on a flexible polyimide (PI) substrate. In addition, the gate electrodes of the pixels are connected to the Cu pad through the via hole, and the drain electrodes of the pixels in the same column are connected to an electrical wire. During the whole fabrication process, sputtering method is engaged to deposit Cu serving as electrical wires, and electroplating is used to form circuit configuration. Additionally, the chips of microfabricated n-type transistors and diodes using the standard silicon process are integrated onto the as-mentioned patterned PI substrate. With the as-proposed fabrication method, a 10×10 TTA has been developed, which is then used as a motion monitoring sensor, as shown in Figure 9.7i–f. For a certain motion path, first moving on the last row of the array, from the left to the right, followed by moving diagonally to the first pixel, the corresponding output currents of the pixels in the moving trajectory are shown in Figure 9.7i and the two-dimensional intensity map is shown in Figure 9.7j, validating that the TTA has good tracking ability and can real-time tracing with a relatively high resolution. In conclusion, the TTA has prospects in intelligent tactile sensing, wearable electronics as well as human-machine interface, which opens a door for the application of tribotronic devices [40].

9.5.2 Tribotronics for Active Control

Apart from the application in information sensing, intelligent electronic devices have emerged in large numbers with the aim of triggering/modulating electrical signals by using active and adaptive interactions between electronics and the environment. Among all, tribotronic functional devices play an important role in realizing direct interaction between electronics and the ambient environment, which simplifies the controlling process. In this section, tribotronic functional devices for active control have been reviewed.

In addition to a tribotronic transistor, tribotronic diode is another important part of the tribotronic functional devices with a simple structure and is easy to fabricate. By coupling a variable capacitance diode and a free-standing mode TENG, a tribotronic tuning diode (TTD) has been developed, as shown in Figure 9.8a–d. The basic structure of the TTD is schematically illustrated in Figure 9.8a, employing a PI layer as a flexible substrate, a free-standing mode TENG and a variable capacitance diode, among which the TENG contains two Cu pads as the two electrodes and a PTFE film as the triboelectric layer. When the external force is applied to the PTFE film, a relative movement between the PTFE and the Cu electrodes will generate an electrical signal, thus converting external mechanical signal into electrical signal. More importantly, different moving directions of the PTFE will generate a reverse voltage which is used to control the switch of the diode. Besides, the diode is made up of a Si-based junction which is constructed on the bottom layer, and the anode as well as the cathode of the diode is connected to the P-type and N-type sides of the diode, respectively. Moreover, one of the Cu pads is connected to the anode through the via hole. In order to improve the output performance of the TENG, ICP is engaged to pattern the surface of the PTFE film. Further, the output reverse voltage from the TENG and the junction capacitance under different sliding distances of the PTFE film have been studied, as shown in Figure 9.8b. With the increasing of the sliding distance, the reverse voltage increases almost linearly while the junction capacitance declines, noting that there is a sharp decline starting from 5 mm. In general, the diode can be effectively tuned by the TENG. With excellent output control, the as-proposed device is employed as an analog signal phase shift. The phase shift of the output signal with a varying distance from 0 to 20 mm, as shown in Figure 9.8c, indicates that the phase shift in analog circuits can be achieved by sliding with a finger. The schematic circuit diagram of the functional device is shown in Figure 9.8d, containing an operational amplifier, a TTD and three resistors R_P with the same value. Besides, the TTD can be utilized for frequency modulation and filter as well. In conclusion, the TTD has

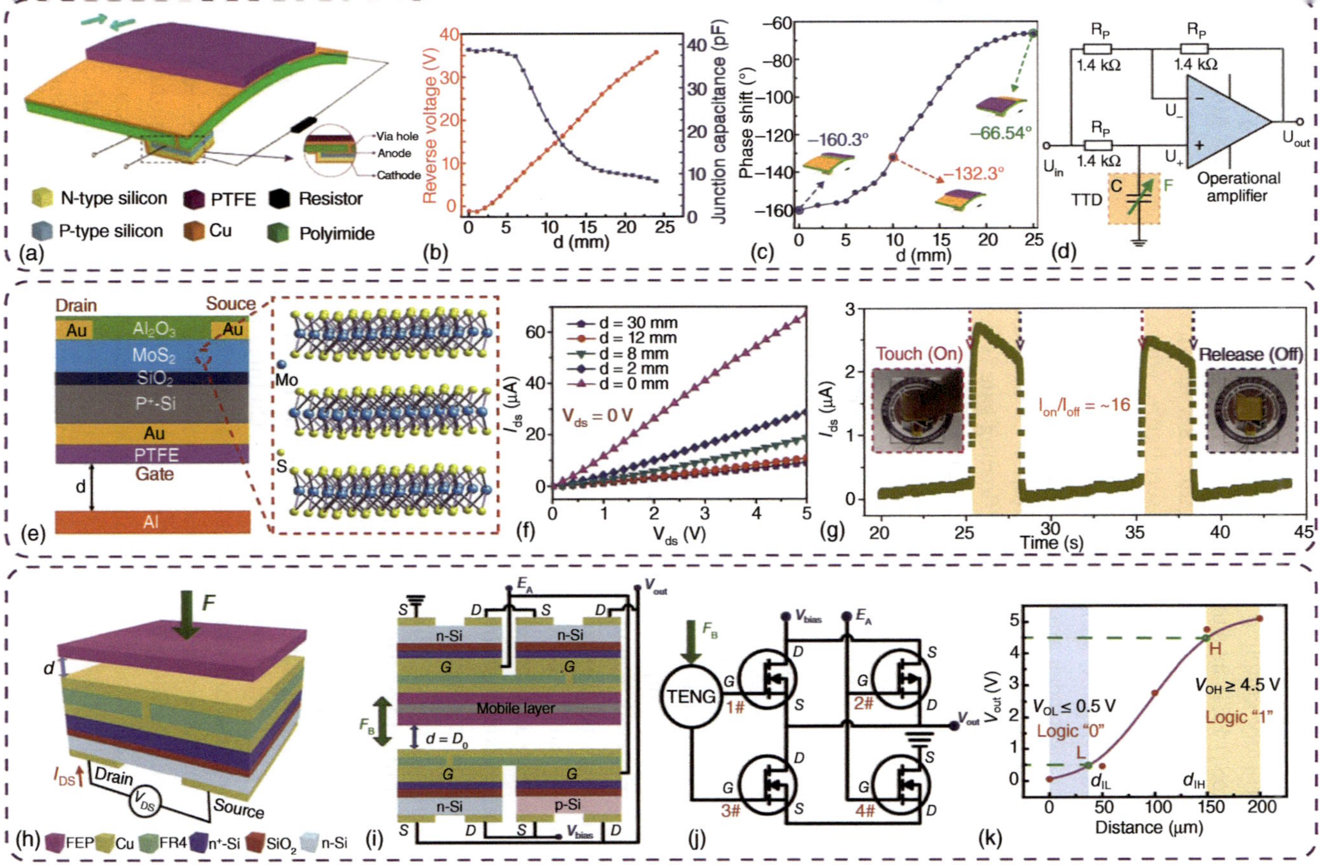

Figure 9.8 Tribotronics for active control. (a) Schematic illustration of the TTD. Inset: Cross-section configuration of the TTD. (b) Output reverse voltage from the TENG and the junction capacitance as a function of the sliding distance of the PTFE film. (c) Phase shift of the output signal as the sliding distance varied from 0 to 25 mm. (d) Schematic circuit diagram of the phase modulation by TTD in conjunction with an operational amplifier. Reproduced with permission of Ref. [41], 2017 © American Chemical Society. (e) Structure of the MoS_2 tribotronic transistor. (f) I_{DS}-V_{DS} characteristics with the PTFE and aluminum separated at different distances. (g) Finger-triggered response of I_{DS}. Reproduced with permission of Ref. [23], 2016 © John Wiley & Sons. (h) Structure of the FGTT. (i) Structure of the tribotronic NAND gate. (j) Equivalent circuit of the tribotronic NAND gate with mechanical electrical coupled inputs. (k) V_{out}−d transfer characteristic of the tribotronic NAND gate at the input E_A of 5 V. Reproduced with permission of Ref. [42], 2017 © Springer Nature.

shown great perspectives in electronic tuning circuits, active controlling sensors, intelligent sensing networks, smart wearable devices, and so on [41].

Recently, two-dimensional materials have been widely investigated as the semiconductor layer of the transistor. As a result, a novel tribotronic transistor has been proposed by vertically coupling a single-electrode mode TENG and a Molybdenum sulfide (MoS_2) FET, as shown in Figure 9.8e–g. The structure of the MoS_2 tribotronic transistor is illustrated in Figure 9.8e, in which the semiconducting MoS_2 is three-layered, as shown in the insert figure, which is obtained from bulk MoS_2 by using mechanical exfoliation method, followed by being transferred to the highly doped P-type Si substrate with 300 nm thick SiO_2. Additionally, the source and drain electrodes are precisely patterned on MoS_2 flakes by using UV lithography followed by successive evaporation of Cr and Au. A conductive and charge-induced layer of 50 nm Cu as the floating gate is deposited on the bottom surface of the Si substrate, underneath which attached a thin PTFE film as the triboelectric layer of the TENG. When the movable Al layer is triggered by external force, a relative movement between the PTFE film and the Al layer will lead to electrical output, serving as the gate voltage to tune the behavior of charge carriers, thus the I_{DS} can be modulated. Furthermore, the output characteristics I_{DS}-V_{DS} with different separation distance between the PTFE film and Al layer is shown in Figure 9.8f. It is observed that the increase in separation distance results in the decreased drain-source current, indicating that the TENG can effectively work as a gate to tune/control charge carrier transport in the transistor. Consequently, the as-fabricated device is employed as a finger-triggered active smart tactile switch and the finger-triggered response of I_{DS} is shown in Figure 9.8g, indicating that the finger-triggered on/off ratio can reach as high as 16, which is higher than the case with the Al plate. In conclusion, this work has extended the new field that may have innovative and important potential applications in the fields of wearable healthcare, human-computer interfacing, and artificial intelligence [23].

Furthermore, a logic circuit is an ideal physical unit to implement discrete signal transmission and operation, which has been widely used in digital control and communications. In order to realize direct control from the environment, a floating-gate tribotronic transistor (FGTT) has been proposed by combining a traditional Si-based FET and a contact-separation mode TENG as shown in Figure 9.8h–k. The basic structure of the FGTT is schematically shown in Figure 9.8h, consisting of an inverted SOI MOSFET, which employs a floating Cu integrated into the glass fiber epoxy (FR4) layer as the gate electrode. On the top of the FR4 layer, another Cu film serving as the triboelectric layer was sputtered and connected with the Cu below through the via hole, while the other triboelectric layer is FEP which realizes the periodical contact and separation with the Cu layer. Moreover, two Cu pads are deposited above the N-type semiconductor layer as source and drain electrodes, respectively. An external force is employed to control the periodical contact and separation between the triboelectric layer and the top gate, which can generate the edge electric field leakage effect. That's where the periodical expansion and shrink of the enhancement zone comes from. As a result, the drain current can be effectively modulated by external stimulus. By further coupling, two opposite

FGTTs and a couple of traditional N-type and P-type MOSFET, the tribotronic negated AND (NAND)gate has been fabricated as illustrated in Figure 9.8i, and the corresponding equivalent circuit is shown in Figure 9.8j. One FGTT is integrated into the lower FR4 layer with the P-type MOSFET, while the other one is integrated into the upper FR4 layer with the N-type MOSFET. The drain electrode of the former FGTT and the source electrode of the P-type MOSFET are connected to the bias voltage. Meanwhile, the source electrode of the former FGTT and the drain electrode of the P-type MOSFET are connected to the drain electrode of the latter FGTT, all of which serve as the output voltage. Moreover, the source electrode of the N-type MOSFET is connected to the drain electrode of the latter FGTT, and the source electrode of it is grounded. E_A is defined as the logic electric level input port connected by the gate electrode of both N-type and P-type MOSFET. An external force F_B is engaged to control the two mobile FEP layers, which offers periodical relative movement between the two triboelectric layers. Based on that, the transfer characteristic of the tribotronic NAND gate when the input E_A is 5 V has been studied, which is shown in Figure 9.8k, indicating that $V_{out} \geq 4.5$ V is defined as logic "1", and $0 \leq V_{out} \leq 0.5$ V is defined as logic "0". In addition, the blue region represents logic "0" while the yellow region represents logic "1" as is shown in the figure. In conclusion, the as-proposed logical device has shown excellent potential in micro-nanosensors, electromechanical storage, interactive control, and intelligent instrumentation [42].

9.5.3 Tribotronics for Artificial Synapse

Developing multifunctional and diversified artificial neural systems is an important task toward the emulation of neuromorphic computation. As an emerging field, tribotronics can also be used for bioinspired neuromorphic sensing. Inspired by somatosensory signal generation and neuroplasticity-based signal processing, intelligent neuromorphic tactile sensors with learning and memory based on the principle of a TENG are demonstrated [43]. As shown in Figure 9.9a, the device consists of a well-designed single-electrode-based triboelectric nanogenerator (SE-TENG) and the PI and reduced graphene oxide (PI:rGO) hybrid layer. Due to the employment of TENG technology, it can actively produce pressure-triggered electric signals without an external power supply. Furthermore, reduced graphene oxides (rGOs) embedded in the friction layer are employed to act as electron-body traps to obtain neuroplasticity. The significant feature of the intelligent neuromorphic sensor is the remarkable output, which increases with increasing number of presses. As shown in Figure 9.9b, for the device with rGO, the rGO sheets can act as electron traps in the PI layer to promote the charge transfer process from PI to rGO sheets. When the Al film is in contact with the PI:rGO layer, electrons in the Al film hop up into the surface traps of the PI:rGO layer. Then, the occupied energy levels in the surface trap increase to the Fermi level of Al, and the triboelectrification process stops. After the separation of the Al film from the PI:rGO layer, most electrons in the surface traps will subsequently transfer to the body traps. After enough training, the triboelectric electrons captured in the PI:rGO layer reach a

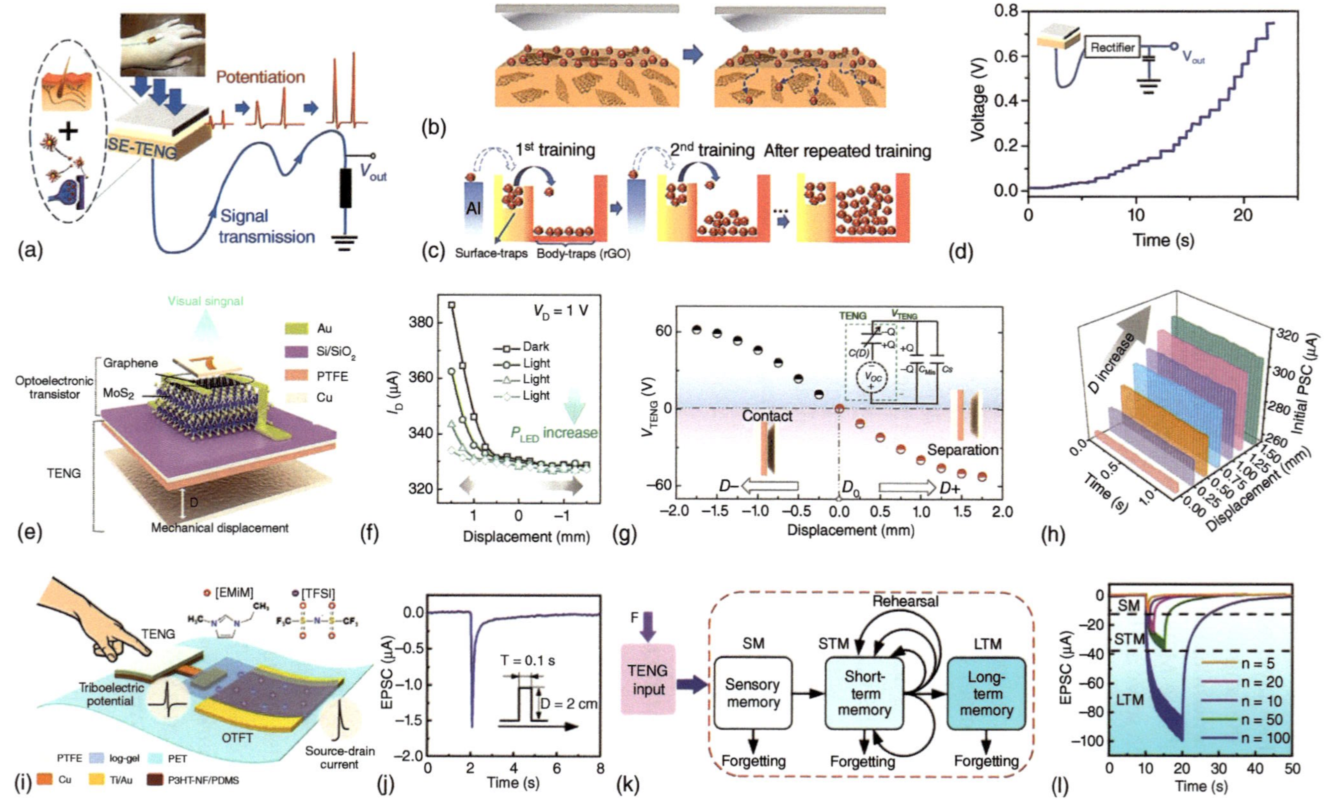

Figure 9.9 Tribotronics for artificial synapse. (a) The intelligent neuromorphic sensor based on SE-TENG technology can functionally emulate combined mechanoreceptor and neuromorphic systems. (b) Schematic process showing the triboelectric electrons transferring from the surface of the PI:rGO to the rGO sheet. (c) Schematic process showing the triboelectric electrons transferring from the surface traps to the body traps. (d) Charging process by using a capacitor as the load device. Inset is the schematic diagram of the system. Reproduced with permission of Ref. [43], 2020 © American Chemical Society. (e) Schematic diagram of the mechano-photonic artificial synapse based on graphene/MoS_2 (Gr/MoS_2) heterostructure. (f) Transfer curves (I_D versus D) in the dark and under different green light power intensities (P_{LED}). (g) The TENG output voltage (V_{TENG}) versus displacement (D). Inset: Equivalent circuit diagram for V_{TENG} characterization. (h) Real-time evaluation of the initial PSC in the dark; D varies from 0 to 1.5 mm [44]. (i) Schematic illustration of the TAS device. (j) EPSC response of the TAS device under single mechanical stimuli ($T = 0.1$ second, $D = 2$ cm). (k) Multistore memory model with TENG as input. (l) EPSC response of the TAS under multiple continuous mechanical stimuli ($T = 0.1$ second, $\Delta T = 0.1$ s, $D = 2$ mm). Reproduced with permission of

maximum, and the output voltage becomes saturated, as shown in Figure 9.9c. When using a capacitor as the load device, voltage across the capacitor can directly reflect the charge quantity. As shown in Figure 9.9d, under a series of pressures, the capacitor accumulates the charges and is charged to about 0.7 V gradually. In addition, the stacked structure in the friction layer was able to mediate the retention time of the information received. This highly integrated tactile sensor can serve as a functional element suitable for use in a pressure sensor network of the artificial intelligence system.

Inspired by the perception and cognition ability of human brain for acquiring somatosensory and visual information, a mechano-photonic artificial synapse with synergistic mechanical and optical plasticity is demonstrated [44]. The schematic illustration of the mechano-photonic artificial synapse is shown in Figure 9.9e. It includes an optoelectronic transistor based on Gr/MoS_2 heterostructure and an integrated TENG. The integrated TENG component is composed of Cu/polytetrafluoroethylene (PTFE)/Cu in contact-separation mode, with one friction layer (PTFE/Cu) connected to the transistor gate and another Cu electrode as the movable counter friction layer. The mechanical displacement (D) between two friction layers of TENG can induce triboelectric potential coupling to the transistor, which can directly affect the charge transfer/exchange in the transistor channel (i.e. Gr/MoS_2 heterostructure) and modulate the photocurrent of the optoelectronic synapse. As shown in Figure 9.9f, mechano-optoelectronic transistor based on Gr/MoS_2 heterostructure, the I_D markedly decreases with the increased P_{LED} in the region of $D > -0.1$ mm. However, it has no obvious change in the region of $D < -0.1$ mm. The optoelectrical performances under mechanical displacement are consistent with the situation under applied V_G. The TENG displacement of −0.1 mm is equivalent to the MoS_2 conduction threshold of ~5 V according to Figure 9.9g. Mechano-photonic artificial synapse based on Gr/MoS_2 heterostructure, the channel conductivity (i.e. synaptic weight) can be readily modulated by both mechanical displacement and light illumination, which is the fundamental of multimodal plasticity in the mechano-photonic artificial synapse. As shown in Figure 9.9h, the basal postsynaptic current (PSC, i.e. the I_D of the Gr/MoS_2 transistor under a dark state) represents a clear variation trend from 260 to 308 μA with the displacement changed from 0 to 1.5 mm, matching well with the transfer characteristics in Figure 9.9f. The PSCs can also be maintained stable at different levels under different displacement states, while the light illumination can lead to dynamic variation of PSC. This means that the triboelectric potential–modulated electrical behaviors and the optoelectrical behaviors can synergistically affect the PSC, the long-term memory and consecutive neural facilitation are also realized. In conclusion, the mechano-photonic artificial synapse is highly promising for implementing mixed-modal interaction, emulating complex biological nervous systems, and promoting the development of interactive artificial intelligence.

To satisfy the mechanical compatibility requirements of curved and dynamic surfaces, a flexible tribotronic artificial synapse (TAS) is demonstrated [45]. The structure is shown in Figure 9.9i, the Ti/Au electrodes of OTFT are sputtered on the PET film. Then, the Poly(3-hexylthiophene-2,5diyl) (P3HT-NF)/Poly(dimethylsiloxane) (PDMS) semiconductor solution is spun coated on the channel area of the source-drain electrodes. Subsequently, the ion-gel dielectric film is aligned and heat-laminated on the device. Finally, the Cu and PTFE films were assembled on the PET film in sequence, in which the Cu film was connected to the OTFT as the gate electrode. In this flexible TAS, the triboelectric potential generated by the external contact electrification is regarded as the action potential, and the source-drain current (I_{DS}) of the OTFT is analogous to the postsynaptic current. During the whole measurement of the electrical characteristics, the displacement distance (D) of TENG is 2 mm, and the V_d applied to the EDL OTFT is 0.5 V. A typical excitatory postsynaptic current (EPSC)of the TAS activated by a single mechanical stimulus with a duration time of 0.1 second is shown in Figure 9.9j. It can be observed that the EPSC rapidly rises to the peak current (1.6 μA), and then gradually decays back to the initial current. This process is similar to an expression of excitatory behavior in biological synapses, which is a basic characteristic of neurosensory behavior and can also be considered a sensory memory (SM) mode in the multistore memory model. The TAS also has synapse plasticity behavior. Inspired by the multistore model of memory, the simplified memory model corresponding to the bioinspired neurosensory behavior of the TAS under mechanical stimuli is shown in Figure 9.9k. The simplified memory model explains the hierarchical memory process of the TAS from SM to short-term memory and long-term memory under external mechanical stimuli. To verify the hierarchical memory process, the EPSC response of the TAS under multiple continuous mechanical stimuli has been studied, as shown in Figure 9.9l. When five times of the mechanical stimuli ($T = 0.1$ second, $\Delta T = 0.1$ second, $D = 2$ mm) are performed, the EPSC of the TAS gradually increases to a peak value of 12.8 μA, and then decay to the initial value at the end of the mechanical stimuli. As the number of mechanical stimuli increases, the output EPSC peak value of the TAS increases continuously. Until the number of the mechanical stimulus increases to 100, the highest EPSC peak value of the TAS can increase to 99.9 μA, and it can be observed that the decay of the EPSC returns to the initial value after about 50 seconds. Moreover, the TAS still exhibits excellent durability even after 1000 bending tensile and compressive strain cycles with a bending radius of 20 mm. This work will help to facilitate the development of artificial afferent nervous systems, which has great significance to the practical application of artificial limbs, robotics, and bionics in future.

9.6 Conclusion

This chapter systematically summarizes the field of tribotronics from a theoretical basis, primary element, functional devices, as well as the applications in information sensing, active control and artificial synapse. First, the working mechanism of tribotronics was introduced, in which the tribotronic potential can be used to control electrical transport in semiconductors. Second, the contact electrification field-effect transistor (CE-FET) was demonstrated, which is the fundamental

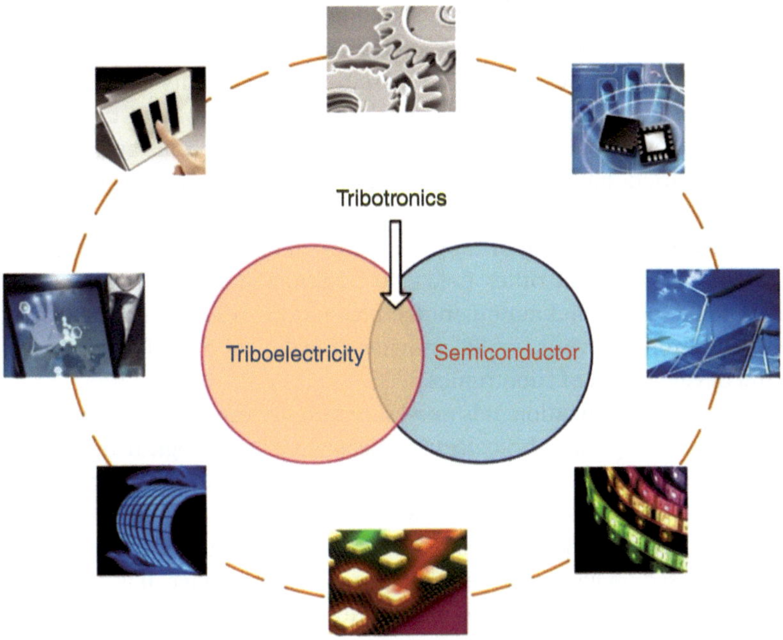

Figure 9.10 Application prospect of triboelectronics. Reproduced with permission of Ref. [21], 2016 © Elsevier.

unit of tribotronics. On this basis, several tribotronic functional devices have been generalized for active control and information sensing, demonstrating triboelectricity-controlled electronics. Based on the coupling of triboelectricity and semiconductor, tribotronics adjust electrical output and transfer characteristics of semiconductor through triboelectric charges generated by mechanical motion, establishing a direct interaction mechanism between external environment and semiconductor devices and realizing active mechanical perception of the external environment. Tribotronics provides new ideas and methods for applications in different fields such as human-machine interaction, micro- and nanoelectromechanical systems, and sensing and self-powered systems (Figure 9.10) [21].

Despite great progress, tribotronics still faces various challenges as an emerging field of research. There is still much to be done to fully understand and realize the application potential of tribotronics. First of all, although theoretical simulations have been carried out for single tribotronic transistor and logic devices, the corresponding theoretical research on the tribotronic functional arrays is relatively insufficient. Secondly, for semiconductor materials with special optical, magnetic and thermal properties, the modulation mechanism on coupling of triboelectric and semiconductor field effect remains to be studied. In addition, the working principles of tribotronics for artificial synapses also need to be further investigated. The above researches are expected to elucidate the coupling and controlling mechanism of triboelectricity and semiconductor field effect and provide a theoretical basis for the preparation and application of tribotronic functional devices.

In terms of materials, most of the materials used in the preparation of tribotronic functional devices are traditional inorganic materials, which limits the application and development of tribotronics. Therefore, based on the diversity of tribotronic material selection, research on the application of organic, nano, and other new materials in the field of tribotronics can greatly expand the functions and applications of tribotronic functional devices. For example, the use of new materials with stretchable, self-healing and other properties gives the tribotronic functional device skin-like characteristics, promoting the application in human-machine interaction, smart skin, bionic devices and other fields. In addition, it is also necessary to develop standardized micro- and nanomanufacturing processes to achieve efficient preparation of tribotronic devices, which provides a technical basis for the future application and development of tribotronics.

In terms of multifunctionalization, it is necessary to study various new functional devices such as light emitting, photoelectric, electromagnetic, logical operation, storage and MEMS regulated by triboelectricity, such as tribotronic memory, tribotronic light-emitting diode, tribotronic MEMS switch, etc. It greatly expands the application scenarios of tribotronics in storage, display, computing and other fields. In addition, based on verifying the feasibility of a single device, it is necessary to study the array extension of tribotronic multifunctional devices through integration technology to explore large-scale integrated tribotronic array devices and highly integrated MEMS for large area array human-machine interface.

In the future, the development of tribotronics will reflect the deep coupling of triboelectricity and semiconductors, whose connotation will be further expanded. On the one hand, tribotronics focuses on the electronic problems of interfacial triboelectric systems. In addition to realizing active mechanical sensing by triboelectric regulating semiconductor field effects, the tribovoltaic effect of direct triboelectricity acting on semiconductor interfaces to generate DC power generation is also included, thus developing new tribotronic devices for energy conversion, active sensing and control. On the other hand, tribotronics utilizes the theories and technologies of electronics to solve triboelectric problems, including energy management of semiconductor electronic devices for triboelectrification, energy storage, etc. It enables efficient collection of micromechanical energy, providing distributed sensing of microenergy solutions and developing self-powered/self-awakening intelligent microsystems. It is expected that tribotronics will make important breakthroughs in the establishment and development of basic theories, material systems and micro-nanomanufacturing as well as technical application of new tribotronic functional devices. It will deepen and expand emerging disciplines such as nanoenergy, nanotribology, and MEMS, showing great value in emerging applications such as smart sensing, energy science, human-machine interface, and so on.

References

1 Someya, T., Bao, Z., and Malliaras, G.G. (2016). The rise of plastic bioelectronics. *Nature* 540 (7633): 379–385.

2 Chortos, A., Liu, J., and Bao, Z. (2016). Pursuing prosthetic electronic skin. *Nat. Mater.* 15 (9): 937–950.

3 Lim, S., Son, D., Kim, J. et al. (2015). Transparent and stretchable interactive human machine interface based on patterned graphene heterostructures. *Adv. Funct. Mater.* 25 (3): 375–383.

4 Jeong, J.W., Yeo, W.H., Akhtar, A. et al. (2013). Materials and optimized designs for human-machine interfaces via epidermal electronics. *Adv. Mater.* 25 (47): 6839–6846.

5 Levin, E., Pieraccini, R., and Eckert, W. (2000). A stochastic model of human-machine interaction for learning dialog strategies. *IEEE Trans. Speech Audio Process.* 8 (1): 11–23.

6 Liu, Y., Norton, J.J.S., Qazi, R. et al. (2016). Epidermal mechano-acoustic sensing electronics for cardiovascular diagnostics and human-machine interfaces. *Sci. Adv.* 2 (11): e1601185.

7 Cao, R., Pu, X., Du, X. et al. (2018). Screen-printed washable electronic textiles as self-powered touch/gesture tribo-sensors for intelligent human–machine interaction. *ACS Nano* 12 (6): 5190–5196.

8 Kang, D., Pikhitsa, P.V., Choi, Y.W. et al. (2014). Ultrasensitive mechanical crack-based sensor inspired by the spider sensory system. *Nature* 516 (7530): 222–226.

9 Xu, S., Zhang, Y., Jia, L. et al. (2014). Soft microfluidic assemblies of sensors, circuits, and radios for the skin. *Science* 344 (6179): 70–74.

10 Zhao, S. and Zhu, R. (2017). Electronic skin with multifunction sensors based on thermosensation. *Adv. Mater.* 29 (15): 1606151.

11 Yick, J., Mukherjee, B., and Ghosal, D. (2008). Wireless sensor network survey. *Comput. Netw.* 52 (12): 2292–2330.

12 Wang, Z.L. (2017). On Maxwell's displacement current for energy and sensors, the origin of nanogenerators. *Mater. Today* 20 (2): 74–82.

13 Gao, W., Emaminejad, S., Nyein, H.Y.Y. et al. (2016). Fully integrated wearable sensor arrays for multiplexed in situ perspiration analysis. *Nature* 529 (7587): 509–514.

14 Alwan, A.M. and Jabbar, A.A. (2011). Design and fabrication of nanostructures silicon photodiode. *Modern Appl. Sci.* 5 (1): 106.

15 Fan, F.R., Tian, Z.Q., and Wang, Z.L. (2012). Flexible triboelectric generator. *Nano Energy* 1 (2): 328–334.

16 Jiang, D., Liu, G., Li, W. et al. (2020). A leaf-shaped triboelectric nanogenerator for multiple ambient mechanical energy harvesting. *IEEE Trans. Power Electron.* 35 (1): 25–32.

17 Liu, W., Xu, L., Bu, T. et al. (2019). Torus structured triboelectric nanogenerator array for water wave energy harvesting. *Nano Energy* 58: 499–507.

18 Liu, G.X., Li, W.J., Liu, W.B. et al. (2018). Soft tubular triboelectric nanogenerator for biomechanical energy harvesting. *Adv. Sustain. Syst.* 2 (12): 1800081.

19 Xu, L., Bu, T.Z., Yang, X.D. et al. (2018). Ultrahigh charge density realized by charge pumping at ambient conditions for triboelectric nanogenerators. *Nano Energy* 49: 625–633.

20 Guo, T., Liu, G., Pang, Y. et al. (2018). Compressible hexagonal-structured triboelectric nanogenerators for harvesting tire rotation energy. *Extreme Mech. Lett.* 18: 1–8.
21 Zhang, C. and Wang, Z.L. (2016). Tribotronics—a new field by coupling triboelectricity and semiconductor. *Nano Today* 11 (4): 521–536.
22 Bhardwaj, A., Kaur, J., Wuest, M. et al. (2017). In situ click chemistry generation of cyclooxygenase-2 inhibitors. *Nat. Commun.* 8 (1): 1.
23 Xue, F., Chen, L., Wang, L. et al. (2016). MoS_2 tribotronic transistor for smart tactile switch. *Adv. Funct. Mater.* 26 (13): 2104–2109.
24 Zhang, C., Zhang, L.M., Tang, W. et al. (2015). Tribotronic logic circuits and basic operations. *Adv. Mater.* 27 (23): 3533–3540.
25 Zhang, C., Zhang, Z.H., Yang, X. et al. (2016). Tribotronic phototransistor for enhanced photodetection and hybrid energy harvesting. *Adv. Funct. Mater.* 26 (15): 2554–2560.
26 Khan, U., Kim, T.H., Ryu, H. et al. (2017). Graphene tribotronics for electronic skin and touch screen applications. *Adv. Mater.* 29 (1): 1603544.
27 Sun, Y., Zheng, X., Yan, X. et al. (2017). Bioinspired tribotronic resistive switching memory for self-powered memorizing mechanical stimuli. *ACS Appl. Mater. Interfaces* 9 (50): 43822–43829.
28 Yang, B.Z., Lin, Y.S., and Wu, J.M. (2017). Flexible contact-electrification field-effect transistor made from the P3HT, PCBM conductive polymer thin film. *Appl. Mater. Today* 9: 96–103.
29 Jin, I.K., Park, J.Y., Lee, B.H. et al. (2018). Self-powered data erasing of nanoscale flash memory by triboelectricity. *Nano Energy* 52: 63–70.
30 Zhu, G., Pan, C., Guo, W. et al. (2012). Triboelectric-generator-driven pulse electrodeposition for micropatterning. *Nano Lett.* 12 (9): 4960–4965.
31 Peng, W., Yu, R., He, Y. et al. (2016). Theoretical study of triboelectric-potential gated/driven metal-oxide-semiconductor field-effect transistor. *Acs Nano* 10 (4): 4395–4402.
32 Jiang, T., Zhang, L., Zhang, X. et al. (2018). Theoretical study of sliding-electrification-gated tribotronic transistors and logic device. *Adv. Electron. Mater.* 4 (1): 1700337.
33 Zhang, C., Tang, W., Zhang, L. et al. (2014). Contact electrification field-effect transistor. *ACS Nano* 8 (8): 8702–8709.
34 Bu, T., Xu, L., Yang, Z. et al. (2020). Nanoscale triboelectrification gated transistor. *Nat. Commun* 11 (1): 1054.
35 Zhao, J., Guo, H., Pang, Y.K. et al. (2017). Flexible organic tribotronic transistor for pressure and magnetic sensing. *Acs Nano* 11 (11): 11566–11573.
36 Pang, Y., Li, J., Zhou, T. et al. (2017). Flexible transparent tribotronic transistor for active modulation of conventional electronics. *Nano Energy* 31: 533–540.
37 Cao, Y., Bu, T., Fang, C. et al. (2020). High-resolution monolithic integrated tribotronic InGaZnO thin-film transistor array for tactile detection. *Adv. Funct. Mater.* 30 (35): 2002613.
38 Zhao, J., Zeng, J., Chen, L. et al. (2022). Intrinsically stretchable and self-healable tribotronic transistor for bioinspired e-skin. *Mater. Today Phys.* 28: 100877.

39 Li, J., Zhang, C., Duan, L. et al. (2016). Flexible organic tribotronic transistor memory for a visible and wearable touch monitoring system. *Adv. Mater.* 28 (1): 106–110.
40 Yang, Z.W., Pang, Y., Zhang, L. et al. (2016). Tribotronic transistor array as an active tactile sensing system. *Acs Nano* 10 (12): 10912–10920.
41 Zhou, T., Yang, Z.W., Pang, Y. et al. (2017). Tribotronic tuning diode for active analog signal modulation. *Acs Nano* 11 (1): 882–888.
42 Zhang, L.M., Yang, Z.W., Pang, Y.K. et al. (2017). Tribotronic triggers and sequential logic circuits. *Nano Res.* 10 (10): 3534–3542.
43 Wu, C., Kim, T.W., Park, J.H. et al. (2020). Self-powered tactile sensor with learning and memory. *ACS Nano* 14 (2): 1390–1398.
44 Yu, J., Yang, X., Gao, G. et al. (2021). Bioinspired mechano-photonic artificial synapse based on graphene/MoS_2 heterostructure. *Sci. Adv.* 7 (12): eabd9117.
45 Zeng, J., Zhao, J., Bu, T. et al. (2023). A flexible tribotronic artificial synapse with bioinspired neurosensory behavior. *Nano-Micro Lett.* 15 (1): 18.

10

Tribophotonics

10.1 Introduction

Considering the future lifestyle, the smart city has rapidly developed in these years, which collected data from daily life to improve the operations of the city and has been widely revealed in smart home, smart manufacturing, human healthcare, environment monitoring, building automation, smart traffic system, wearable electronics, implantable electronics, etc. [1, 2]. As an important component of a smart city, the Internet of Things (IoT) plays a significant role in information exchange. Especially with the development of fifth-generation mobile network (5G), the data transmission speed has been greatly improved, resulting in lower response time, which greatly improved the performance of IoT [3, 4]. As called the eyes and ears of IoT, there is a high demand for sensor nodes with the fast development of the IoT. In that, the distributed power supply and the wireless signal transmission become two major challenges due to such a high number of distributed sensor nodes [5].

To address the distributed power supply challenge, in traditional ways, the sensor nodes were usually powered by batteries or wires [6, 7]. However, the wires bring huge limitations to the application scenario, especially in the harsh environment and dynamic motion system [8, 9]. On the one hand, the wires affect the normal movement of the dynamic motion system and may bring signal crosstalk with complex wirings, especially for robotics and highly integrated systems [10]. On the other hand, the wires increase the cost of the whole system, especially the sensor networks in remote or harsh environments, and the maintenance difficulty [11]. The batteries may partially address the difficulties in wire-based power supply systems, while they bring the new challenges of potential environmental issues and limited lifetime [12, 13]. Besides, the batteries may still bring limitations to the system due to their relatively large weight and size, with difficulties in maintenance and replacement, such as that in the implanted systems and systems in harsh environments [6, 14]. Even though another potential solution – the wireless power supply solution – has been studied for decades, it still meets the challenge of limited transmission distance and low energy transfer efficiency, which cannot be widely applied to power devices and systems [15, 16].

Triboelectric Nanogenerators: Technology, Applications, and Commercialization, First Edition.
Yunlong Zi, Hengyu Guo, Jie Wang, Chi Zhang, Xiangyu Chen, and Qing Zhao.

Alternatively, the emerging nanogenerator technology, especially the triboelectric nanogenerator (TENG) [17, 18], can directly convert the ambient environmental energy into electrical energy, such as the energy from vibration [19, 20], wind [21, 22], rain [23, 24], human motion [25–27], sound [28, 29], water wave [30, 31], and ultrasound wave [32, 33]. Here, some mechanical energy harvesting technologies and TENG-based sensing systems toward smart cities have been discussed [34–36]. Based on the TENGs, there are two potential solutions to realize a self-powered sensing system. First, these TENGs can be applied to harvest environmental energy and work as a sustainable power source to drive the traditional sensing system [37, 38]. Second, by directly converting the energy of the target signal (i.e. the movement of ocean waves and the vibration of the machine) into electrical energy, TENG can directly work as a component of the sensor without additional power supply needed, known as self-powered sensing [9, 39–42].

To address the wireless signal transmission challenge, the signals of sensor are preferably transmitted by the wireless transmission modules, such as Bluetooth, Zigbee, and Wi-Fi. The traditional wireless transmission solution still needs to be powered by wires or batteries, which is also facing the difficulties described above. The TENG may provide a new method to power wireless transmission modules [25, 43–47]. However, due to the high-power demands of the traditional wireless transmission modules and impedance mismatch issues, relatively large-size TENG and power management units are usually required, limiting the application scenarios [25, 44, 46–50]. An alternative way is still required to achieve unique wireless transmission methods that can be directly triggered by the high-voltage output from TENGs, enabling TENG output as the power and wireless signal source.

As the common media of signal transmission, photons or light are naturally transmitted in space without the need for wires, which sets the foundation for wireless sensing and transmission. Thus, TENG has been coupled with devices for photon generation or modulation toward novel fully self-powered wireless systems, which can be concluded as the field of tribophotonics. Tribophotonics can be considered an emerging research direction of TENG field as the coupling of triboelectricity with photonics. Compared with traditional self-powered systems, tribophotonic systems can achieve self-powered wireless sensing and transmission without any additional cable connections required for signal processing, transmission, or power supply. Tribophotonics can perfectly solve the distributed power supply and wireless signal transmission challenges existing in the traditional sensing networks, which can stimulate the corresponding research in signal processing, modulation, and application systems, facilitate the development of IoT and smart city, with potential applications in electronic skin, wearable and implantable electronics, smart buildings, smart manufacturing, human–machine interface (HMI), etc.

This chapter intends to present a comprehensive view of TENG-based self-powered wireless systems established by the concept of tribophotonics, including the principles, technologies, and application scenarios. The concept, origin, characteristics, and potential applications of tribophotonics were first introduced [51]. And then, the four major types of technological routes of tribophotonics, based on tribo-induced EM-wave generation (TIEG), tribo-induced light propagation tuning (TILPT), triboelectrification-induced electroluminescence (TIEL), and tribo-assisted spectrometry (TAS), are discussed. These methods with different working mechanisms, working spectrum, effective transmission distance, and transmission medium can realize fully self-powered wireless sensing toward different application scenarios. Challenges and perspectives of tribophotonics are discussed at the end of this chapter. These self-powered wireless solutions based on tribophotonics will greatly facilitate the general developing trends of electronics in functionality, miniatured integrability, wireless portability, and self-power-ability [52], toward the IoT and smart city.

10.2 Tribophotonics: Concept, Origin, Characteristics, and Potential Applications

Triboelectricity indicates the generation of surface charge with opposite polarity upon contact of two surfaces [53–56]. TENG based on the working principle of triboelectricity and electrostatic induction was invented in 2012 [17]. As an effective method to convert mechanical energy into electrical energy, especially low-frequency mechanical input, the TENG plays an important role in mechanical energy harvesting and self-powered sensing. However, considering the high-voltage output from the TENG, it is difficult to be directly applied to traditional electronic systems, where the power management circuit, including charge pump circuit, voltage and current controlling circuit, impendence matching circuit, etc., were studied to adjust the output of TENG [48–50, 57, 58]. Particularly in previous self-powered wireless sensing systems, TENG was combined with a commercial wireless signal transmission circuit, microcontrol unit (MCU), and sensor to realize the function. In 2019, Khan et al. reported a self-powered multibroadcasting wireless sensing system, where a TENG was applied to collect the vibration energy to power the radio frequency (RF) module to realize wireless sensing and transmission [44]. In 2020, Lu et al. reported a wind-driven TENG-based self-powered wireless sensing system, where the TENG structure worked both as the sensor part and power source, and the MCU and Bluetooth low-energy (BLE) modules were applied for the wireless signal transmission [46]. In 2020, Song et al. reported a self-powered wireless wearable sweat sensing system, which was powered by a human motion-driven TENG [25]. In these wireless sensing systems, the power management circuit and other components are still needed, which results in a complex system and high energy consumption.

As an alternative solution, the unique high-voltage output characteristic of TENG is perfectly suitable to stimulate physical processes to realize signal conversion for photonics-based wireless sensing, such as electric field-induced electrowetting to realize tunable liquid lens, breakdown discharge for electromagnetic (EM) wave generation, and tribo-induced electroluminescence. Through these mechanisms, the high-voltage electric signals from TENGs were directly converted into optical/EM signals, reducing complex intermediate steps, which may bring additional energy consumption. These studies were summarized as the concept of "tribophotonics" toward addressing the distributed energy supply and wireless signal transmission challenges in the IoT system.

To introduce the tribophotonics, we first illustrate the triboelectric effect. With the contact or sliding of two materials with different polarities, the charges are transferred between surfaces due to mechanisms of electron transfer, ion transfer, or mass transfer, resulting in surface static charges [56, 59–61]. When the two surfaces separate, due to electrostatic induction, an electric field is created with the direction from the positively charged surface to the negatively charged surface, generating voltage between electrodes. The triboelectricity-induced electric field or voltage can trigger related physical processes to generate the photons or tune the optical signal transmission, making the fundamental principle of tribophotonics coupling the triboelectricity with photonics. The general illustration of the working process of tribophotonics is shown in Figure 10.1a.

In addition, with the different signal transmission forms and triggering mechanisms, tribophotonics can be realized by four major technological routes as shown in Figure 10.1b, where the detailed mechanisms are discussed below:

(1) TIEG: The induced voltage can trigger discharge, time-varying displacement current or magnetic field, which further generates photons in EM-wave spectrum for wireless transmission.

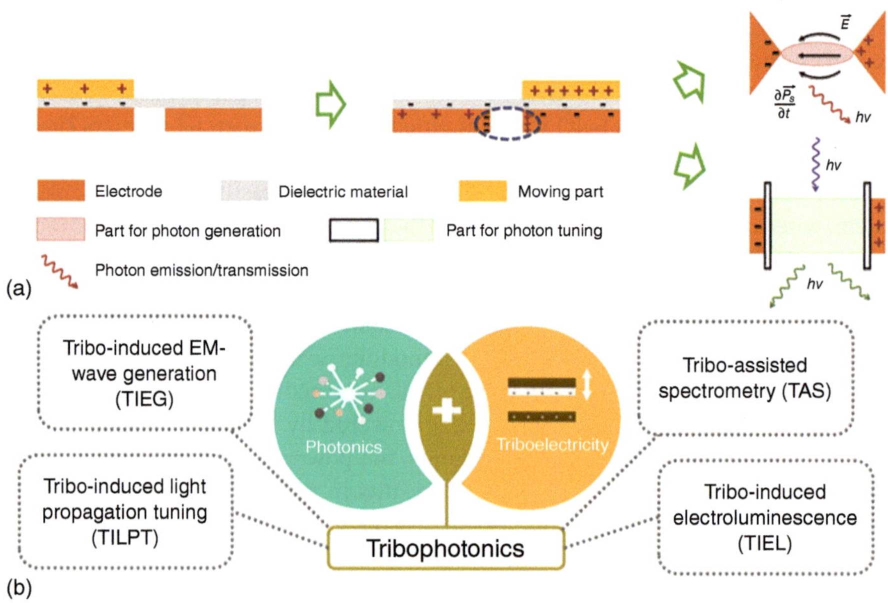

Figure 10.1 The overall illustration of tribophotonics. (a) The general illustration of working process of tribophotonics. (b) The scheme of tribophotonics and the principle of signal conversion methods. (i) Tribo-induced EM-wave generation (TIEG). (ii) Tribo-induced light propagation tuning (TILPT). (iii) Triboelectrification-induced electroluminescence (TIEL). (iv) Tribo-assisted spectrometry (TAS).

(2) TILPT: The induced electric field is applied to the optical modulation devices, where the photons can be modulated or tuned while passing through these devices.
(3) TIEL: The induced electric field excites and modulates the photon emission of luminescence materials.
(4) TAS: The induced electric field generates breakdown discharge or microplasma, where the photons are excited in this process involving the environment information in corresponding spectra.

As triggered by TENG, such wireless sensing methods can be conducted in a fully self-powered manner. The technological significances of tribophotonics are concluded below:

Fully self-powered: By directly using the output of TENG to tune or generate the optical signal or EM wave signal, the tribophotonics can directly convert the target signal into a wireless signal, where the TENG usually works as both power source and sensor. Hence, the sensing and transmission part can be triggered solely by the TENG, working in a fully self-powered manner, with no battery or any other additional power source needed, considered the fully self-powered system.

Wireless: Introduction of photonics brings the natural way of wireless signal transmission through photons. Besides the power-cable-free feature brought by the TENG as the power supplier, the tribophotonics-based system may work completely remotely and wirelessly for broad applications.

To conclude, tribophotonics features the self-power-ability enabled by the TENG and the wireless transmission enabled by the photonics, which can be widely used in harsh environment detection, wearable and implantable systems, smart homes, etc.

10.3 Tribo-Induced EM-Wave Generation (TIEG)

In physics, light refers more broadly to EM radiation of any wavelength, which is transmitted through photons. In this case, TIEG is the tribo-induced photon generation in EM radiation for self-powered wireless transmission, which satisfies the definition of tribophotonics. Here, the EM wave consists of both the electric field and magnetic field oscillating in phase, transferring energy and momentum in space, which has been widely applied in communication, wireless detection, wireless power transfer, etc. [70]. At the same time, the TENG can convert the input mechanical energy to high-voltage output electrical energy based on the displacement current, where the electricity contains both the mechanical input information and energy, which can be further used for sensor and power source. By combining TENG with corresponding EM radiation generation structure, it can generate an alternating displacement current or magnetic field to emit EM wave, which contains the sensing information, to realize wireless power transfer and signal transmission. In addition, by using a coil, metal electrode, or antenna to detect the field or wave, the transmitted signal can be received and further processed by the processing system. Among them, based on different working mechanisms including intrinsic displacement current, time-varying magnetic field, and discharge-induced displacement current, the fully self-powered systems can be realized, as stated below.

10.3.1 Intrinsic-Displacement-Current-Induced Fully Self-Powered Wireless Sensing System

The intrinsic-displacement-current-induced self-powered wireless sensing and transmission method can be realized by the time-varying electric field or the intrinsic displacement current generated in TENG itself. Considering the working principle of TENG, with the mechanical energy input and relative motion between two triboelectric materials, the triboelectric charge on the dielectric surface induces the electric field, and the electric displacement D can be added with the additional polarization P_S from the triboelectric charge and the input mechanical stimulation, the induced time-varying electric field can further generate the additional term of nondischarge displacement current J_D in D, as given by equation below [52, 71–73]:

$$\vec{J_D} = \frac{\partial \vec{P_S}}{\partial t} \tag{10.1}$$

Then, based on the Maxwell equation:

$$\nabla \times \vec{H} = \vec{J} + \frac{\partial \vec{D}}{\partial t} \tag{10.2}$$

Considering the displacement current induced by TENG, we can get the equation below:

$$\nabla \times \vec{H} = \vec{J_D} = \frac{\partial \vec{P_s}}{\partial t} \tag{10.3}$$

$$\nabla \times \vec{E} = -\frac{\partial \vec{B}}{\partial t} \tag{10.4}$$

Thus, Eqs. (10.3) and (10.4) show that the time-varying electric and magnetic fields can be generated in space successively. Hence, the EM wave can be generated by the TENG-induced displacement current. With such a changing displacement current, the energy induced by TENG can be transmitted into external space, which can be applied for wireless sensing and transmission.

Based on this working principle, in 2018, Cao et al. reported an electrodeless TENG, which can realize wireless power delivery and wireless sensing with mechanical energy input [74]. The reported electrodeless TENG consisted of a dielectric turntable, where the polypropylene (PP) worked as the stator and the polymethyl methacrylate (PMMA) worked as the rotator and an electrode which worked as the energy collector was placed wirelessly. With the mechanical motion input, based on the principle of triboelectrification, the PP got the negative charges, while PMMA got the positive charges, which led to the time-varying electric field and displacement current with the rotation. Then, the energy transmitted through the displacement current can be collected by the metal collector wirelessly, with an effective transmission distance of around 18 cm. In addition, the output current density was positively related to the rotating speed, which can be further applied for motion and rotating speed sensing. At the same time, the wireless power delivered by the integrated rotating TENG has also been discussed [75], which can also be applied for self-powered wireless sensing.

In 2018, Mallineni et al. reported a wireless TENG (W-TENG), which can directly convert the input mechanical energy to wireless electrical energy for wireless charging and sensing in an integrated TENG structure [76]. The reported W-TENG consisted of a TENG structure and an additional copper ribbon. The graphene polylactic acid (gPLA) nanocomposite and Teflon were applied in the TENG structure, while the gPLA electrode was connected with a copper ribbon as an antenna structure. With a motion loaded, the high voltage generated by W-TENG can further generate a strong time-varying electric field and displacement current, which can be received by the receiver system around 3 m away. The applications of wireless and secure actuation of smart home, such as smart windows (SWs), security alarms, and temperature sensors, have been demonstrated.

Specifically, based on the principle of electric field resonance (EFR), the high-frequency oscillating signal can be transmitted. In 2021, Kuang et al. reported EFR-based wireless TENG (EFR-TENG), which connected the TENG with EFR circuit [77]. The output of TENG can be converted to the high-frequency oscillating signal, and further generate alternating electric field and displacement current to realize wireless power delivery and sensing, where the signal can be adjusted by the connected capacitor and inductor. The wireless power delivery via EFR-TENG achieved an energy transfer efficiency of 67.8% with a transmission distance of 5 cm. At the same time, the EFR-TENG can also be applied for wireless sensing, with an effective transmission distance of up to 2.3 m.

10.3.2 Time-Varying Magnetic Field-Based Fully Self-Powered Wireless Sensing System

Compared with the intrinsic-displacement-current-based system, TENG combined with the inductor-capacitor oscillation circuit may generate a time-varying magnetic field for EM emission with controllable frequency. By connecting the TENG (which is considered a capacitor with capacitance C_T) with a coil structure with an inductance L, an electrical oscillation happened when there was a mechanical stimulation applied to TENG, with the oscillation frequency f given by:

$$f = \frac{1}{2\pi\sqrt{LC_T}} \tag{10.5}$$

In this case, a high-frequency alternating current (AC) is generated in the circuit loop between the TENG and the coil, where a time-varying magnetic field across the coil structure was given by [62]

$$\nabla \times \vec{B} = \mu_0 \left(\vec{J} + \varepsilon_0 \frac{\partial \vec{E}}{\partial t} \right) \tag{10.6}$$

where the μ_0 present vacuum permeability. Then a magnetic flux φ_B can be calculated by

$$\phi_B = \int_{\Sigma} \vec{B} \cdot d\vec{A} \tag{10.7}$$

where dA is an element of the surface Σ enclosed by the wire loop. Here, another coil with the inductance L_2 was used to collect the transmitted energy or signal, and the changing magnetic flux can induce the voltage potential in the receiver coil to realize wireless transmission, where the induced electromotive force ε can be calculated by

$$\varepsilon = -\frac{d\phi}{dt} \tag{10.8}$$

At the same time, the inductive coupling solution can also be applied to explain the working principle, where a mutual inductance M between the two coil structures can be calculated by

$$M = k\sqrt{L_1 L_2} \tag{10.9}$$

where k presents the coupling coefficient. Then the induced voltage potential v_2 in the second coil can be calculated by

$$v_2 = M\frac{di_1}{dt} \tag{10.10}$$

where i_1 is the current in the first coil. With the changing magnetic flux, the electrical energy or signal generated by TENG can be wirelessly transmitted. Then, by combining the TENG with the EM oscillation circuit, the self-powered wireless sensing system can be realized.

Based on this working principle, in 2020, Zhang et al. reported a time-varying magnetic field-based TENG wireless power transmission and self-powered wireless sensing system via coil structure (MR-WTENG) [62]. The reported MR-WTENG consisted of a transmit part and a receive part. For the transmit part, a TENG was connected with a coil and a tuning capacitance batch, which can convert the electrical output of TENG into a changing magnetic field. The receive part consisted of a coil, a tuning capacitor, and the load, which can receive the changing magnetic field and convert it into the induced voltage potential. This work schematically studied the influence of different parameters on the received signal, including the magnetic resonant coupling system and inductive coupling system with an effective transmission distance of over 2 m. Among them, the magnetic resonant coupling system contained the same natural frequency between the transmit part and receive part, which can be adjusted by the tuning capacitor, demonstrating higher energy transmission efficiency and longer effective transmission distance. To demonstrate the potential application of the MR-WTENG, wirelessly charging the watch and powering the lighting system were demonstrated for the ability of wireless power transmission, and the self-powered wireless force sensing and pressure sensing were demonstrated for the ability of wireless sensing. Then, in 2021, Xu et al. reported a fully self-powered instantaneous wireless humidity sensing system [78]. With the same working principle and wireless signal transmission system, the reported

TENG-based self-powered wireless-sensor system (WS-TENG) was connected with a capacitive humidity sensor, when there was a mechanical stimulation loaded to the TENG, the receiver coil can catch the transmit signal. By analyzing the oscillation frequency of the spectrum of the received signal, the capacitance value of the humidity sensor can be calculated to reveal the humidity. In addition, in 2021, based on the reported WS-TENG, Tang et al. reported a fully self-powered instantaneous wireless traffic monitoring system [79]. Here, a capacitive pressure sensor was integrated with WS-TENG in the sidewalk, where the movement of humans and other kinds of transportation can excite the WS-TENG and transmit signal wirelessly. By analyzing the waveform of different received signals, the passed objects and moving direction can be distinguished. Besides, the speed of the passed objects can be calculated by two TENG with the distance s, where the time interval between the two corresponding received signals was Δt, and then the instantaneous speed v can be calculated by

$$v = \frac{s}{\Delta t} \tag{10.11}$$

Then, the proposed system realized the traffic monitoring in fully self-powered manner, which can be applied in the IoT system. After that, in 2021, a real-time metal height and position recognition system was designed based on pumping switched TENG (PSW-TENG) [80]. Here, the PSW-TENG was connected with a coil structure to realize the resistor-inductor-capacitor (RLC) resonant circuit, where the changing magnetic field was generated under mechanical stimulation. The coil structure (probe) was set above the target object, where the metal object can generate the opposite magnetic field. Then, when a metal object passes the region, it can induce the net flux linkage between the probe and the object to further change the inductance of the coil structure through mutual induction. By analyzing the resonant frequency of the RLC resonant circuit, the system can sense the vertical height and horizontal position of the metal object. In 2022, the electronic switch and impedance-modulated signal analysis method were further studied toward the fully self-powered instantaneous wireless sensing system [81]. The reported system not only contained high-precision-sensing-ability and long-distance-transmission-ability, but also realized the signal modulation toward the resistive, capacitive, and inductive sensors. Based on the system, the fully self-powered wireless bicycle tire pressure and speed sensing system was demonstrated.

In 2020, Wen et al. reported a short-range self-powered wireless sensor network (SS-WSN) [82]. With the same working principle, the textile TENG was connected to a coil to convert the output to a time-varying magnetic field, where the electrical signal generated by TENG can be transmitted to the receiver coil wirelessly. Here, the flexible TENG consisted of nonconductive textile, nitrile, eco-flex-coated textile, and conductive textile. By applying the amplitude response and characteristic frequency response toward different input forces, the static force sensing system in fully self-powered wireless manner was realized. In addition, when the textile TENG was integrated into a sensing array, by analyzing the different characteristic frequencies and amplitudes, the HMI was realized, as demonstrated by a toy car control system and wireless 3D virtual reality drone control. The demonstrated work showed potential applications of SS-WSN in self-powered wireless static force, motion, and position sensing.

In 2019, Chen et al. reported an EM-triboelectric nanogenerator-based wireless power transmission system, which directly converts the input mechanical energy into electrical energy and transmits wirelessly [83]. The reported system consisted of the energy harvesting part and the wireless transmission part. The energy harvesting part was a hybrid nanogenerator with a combination of an EM generator (EMG) and TENG, which was disk-like, including a stator and a rotator. For the EMG part, the magnet and coils were placed on the disk. When the motion was loaded in the disk, the magnetic flux in the coil varied, which can convert the mechanical energy into electrical energy. For the TENG part, the Au electrode worked as the positive material, while the polytetrafluoroethylene (PTFE) worked as the negative material. When the input motion drove the rotator, triboelectrification was generated between the Au and PTFE, where the Au got the positive charges and PTFE got the negative charges, which converted the mechanical energy into electrical energy. Besides, the output current was positively related to the rotating speed, which can be further applied for the rotating speed sensing. The wireless transmission part consisted of two helical coil structures, which were connected with the energy harvesting part and the load, separately. When the energy is generated by the energy harvesting part, the AC voltage can be converted to a high-frequency changing current in the primary coil and then induce a time-varying magnetic field to realize wireless transmission. In addition, they have demonstrated the wireless transmission of over 1 m for charging smartphone. At the same time, this process can still be applied to self-powered wireless sensing.

10.3.3 Discharge-Induced Displacement-Current-Based Fully Self-Powered Wireless Sensing System

In Section 10.3.1, the intrinsic-displacement-current-based fully self-powered wireless sensing system was discussed. However, the electric field or displacement current decayed quickly in the environment, which limited the effective signal transmission distance. The operation frequency was also limited by the motion frequency as well. Alternatively, based on the triboelectric effect, a paradigm-shift mechanism of EM wave wireless signal generation can be realized by the additional rapidly varied term $\frac{\partial \vec{P_s}}{\partial t}$ as generated by the breakdown discharge, which was triggered by the high-voltage output characteristic of TENG. When the TENG-induced electric field was stronger than the threshold breakdown electric field of the air, as the threshold breakdown voltage was given by Paschen's law [84], the breakdown discharge happened between the two electrodes. With the process of breakdown discharge, the electron avalanche and plasma were generated, which induced electron and ion oscillation, generating the rapidly varied $\frac{\partial \vec{P_s}}{\partial t}$ and transmitting wireless signals. Then, the antenna can be applied to receive the discharge-induced wireless signal. Thus, the self-powered wireless sensing system was realized.

In 2021, our team reported a paradigm-shift fully self-powered long-distance wireless-sensing solution enabled by discharge-induced displacement-current-based self-powered wireless sensing e-sticker (SWISE), which can directly convert the input mechanical energy into EM wave, featuring the smallest system size with the longest effective signal transmission distance along with existing wireless sensing and transmission system, as shown in Figure 10.2 [63]. The SWISE consisted of a tribo-charge layer, electrode layer, and a substrate layer. When there was a motion input, the triboelectric effect happened between the tribocharge layer and contact objects, which further induced a strong electric field on the electrode layer with the principle of electrostatic induction. The strongest electric field was generated at the gap position of the electrode layer, where two tips were fabricated opposite to each other in the electrode layer. Once the generated electric field reached the threshold strength of the electric field, the breakdown discharge happened, which induced the EM wave signal with the principle discussed above. Besides, the SWISE featured the characteristics of small (down to 9 mm by 9 mm), thin (down to 95 μm), light (16 mg), soft and flexible. A schematic study of the major influencing

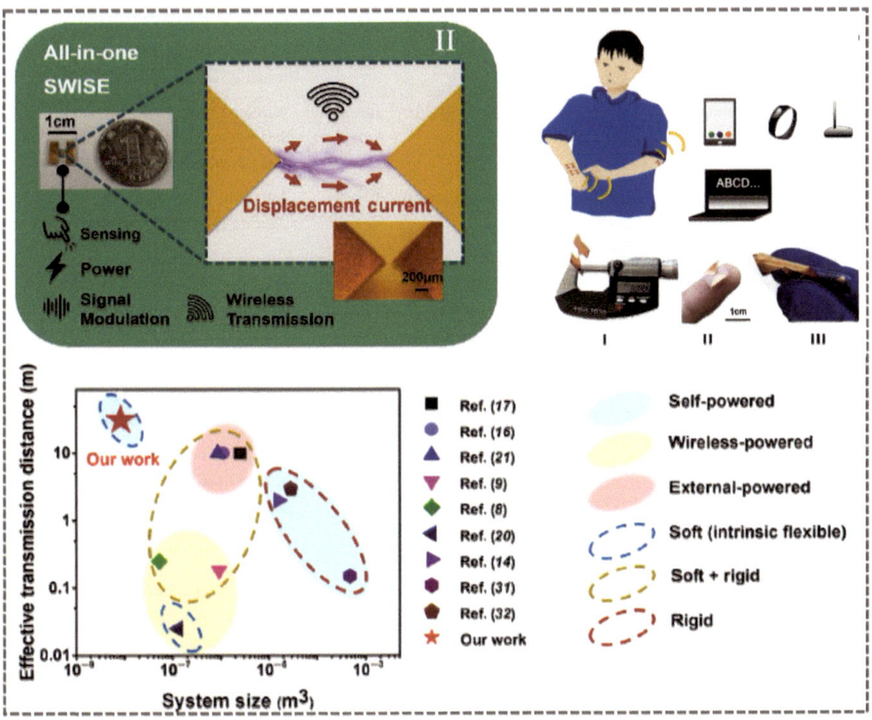

Figure 10.2 Discharge-induced displacement-current-based fully self-powered wireless sensing system. The discharge-induced displacement-current-based SWISE. Source: [63] / American Association for the Advancement of Science / CC-BY-4.0.

factors on the wireless signal was conducted. Based on the different wireless signal waveforms in different gas environments, the self-powered wireless gas-sensing ability of SWISE was demonstrated. In addition, the SWISE-based self-powered wireless motion-sensing e-skin system, the SWISE-based self-powered wireless soft keyboard, and the SWISE-based smart wristband system were demonstrated for potential applications.

For TIEG, since the sensing signal is transmitted by the time-varying EM wave/field, the transmitted signal can avoid the influence of opaque objects and

Table 10.1 Summary of TIEG devices.

Ref	Working mechanism	Effective transmission distance (m)	System size	Application
[74]	Intrinsic displacement current	0.18	~4.76×10^{-4} m^3	Wireless energy transmission
[75]	Intrinsic displacement current	0.1	~9.04×10^{-4} m^3	Wireless energy transmission
[76]	Intrinsic displacement current	3	~2.88×10^{-5} m^3	Wireless and secure actuation of smart-home applications
[77]	Intrinsic displacement current (electric field resonance)	2.3	~2.88×10^{-4} m^3	Wireless energy transmission; Self-powered real-time wireless sensing (liquid level sensing)
[62]	Time-varying magnetic field (magnetic resonance)	2	~1.59×10^{-5} m^3	Wireless energy transmission; Self-powered wireless sensing (force and pressure sensing)
[78]	Time-varying magnetic field	0.9 (20 cm diameter air-core coil pair)	~2.62×10^{-5} m^3	Self-powered instantaneous wireless sensing (humidity sensing)
[79]	Time-varying magnetic field	1	0.6 m × 1.5 m + 2.5 cm × 7.5 cm (Coil diameter = 20 cm)	Self-powered wireless sensing (traffic monitoring, including traffic recognition and speed monitoring)
[80]	Time-varying magnetic field	2	~1.19×10^{-4} m^3	Self-powered wireless sensing (metal tin height and position recognition and counting)
[81]	Time-varying magnetic field	0.12	~2.69×10^{-5} m^3	Self-powered wireless sensing system combined with resistive, capacitive, and inductive sensors (bicycle tire pressure and speed sensing)
[82]	Time-varying magnetic field	1	N/A	Wireless 2D/3D control; Self-powered wireless sensing (static force sensing)
[83]	Time-varying magnetic field	1	N/A	Wireless energy transmission
[63]	Discharge-induced displacement current	30	~7.70×10^{-9} m^3	Self-powered wireless sensing (motion sensing e-skin and gas environment recognition); Remote control (self-powered wireless soft keyboard and smart wristband)

transmit in the free space. The TIEG-based tribophotonic devices can realize both directional and omnidirectional transmission, which are suitable for different application scenarios. Additionally, by using discharge-induced displacement-induced EM wave, the TIEG-based tribophotonic devices can be fabricated with ultra-small system size with long effective transmission distance, which satisfies demands of HMI, such as wearable human motion sensing, and remote control. However, the TIEG devices may meet the issue of EM interference (EMI), which may raise the demands of a high signal-to-noise ratio on the transmitted signal. Currently, TIEG-based tribophotonic devices may be widely applied in different scenarios in smart cities, such as wireless motion sensing, structure/transportation monitoring, electronic skin, static force, and position sensing. The specifications of the reported TIEG devices are summarized in Table 10.1.

10.4 Tribo-Induced Light Propagation Tuning (TILPT)

The tribocharge can also be employed to modulate optical devices, such as liquid lenses, to realize self-powered wireless sensing. At the same time, this method can be considered one application scenario of free-space optical communication (FSO), which is an optical communication technology employing light propagation in free space as the information carrier [85]. Here, with the high-voltage output, the TENG can be applied to an optical device to tune the optical signal to realize wireless sensing via FSO. According to the modulation principle, we can divide the tribo-induced light-propagation-tuning methods into light ON/OFF control, light direction control, light intensity control, and light color control.

10.4.1 Tribo-Induced Light ON/OFF Control

As the fundamental operation of the TILPT, the light ON/OFF control can be realized by combining the electrostatic actuators or liquid crystal (LC) with TENG. Driven by the TENG, the actuators or LC can respond to block the light path, giving the status "ON", or allow the passing light, as considered the status "OFF". During this process, the mechanical–electrical–optical signal conversion can be realized, which can be considered a binary signal transmission method. By using the receiver to detect the binary optical signal, it can realize the simplest TILPT system for self-powered wireless sensing.

Based on the working principle, in 2020, our team reported a TENG-driven self-powered electrostatic force optical switch (EFOS) with the ability of ON/OFF mechanical–electrical-optical signal conversion and delay characteristic toward motion sensing [86]. The reported EFOS consisting of two electrode layers and a dielectric film worked as actuators, and the freestanding sliding TENG (FS-TENG) was series-connected with the EFOS. In the initial status, the dielectric film was in the middle position between the two electrodes or attached to the bottom layer, and the light could pass through the EFOS, which was defined as the OFF status. When there was an input motion loaded in the FS-TENG, the high voltage generated by FS-TENG induced a strong electric field between the top and bottom electrode layers of EFOS, which drove the dielectric film to be attached to the top electrode layer. In this case, the dielectric film blocked the light, which was defined as the ON status. After that, the generated charges gradually dissipated into the air, and then the EFOS returned to initial status to allow the light beam to pass through the EFOS again, returning to OFF status. Capable of signal transfer, the self-powered EFOS was demonstrated to be effective for the wireless motion-sensing application. By combining with the delay characteristic, the self-powered EFOS has been further demonstrated to be applied in motion-sensing delay lighting systems and FSO systems.

In 2021, our team reported a bistable SW with the modes of transparent and hazy [64]. The reported SW consisted of glass, indium tin oxide (ITO), silicon spacer, and cholesteric LC (CLC). In initial status, the SW was in transparent status with the haze ratio of 1.2%, which was induced by the orderly assembled helical structure. When there was mechanical stimulation applied to the FS-TENG, an electric field was generated by the tribocharge and loaded to the CLC film, which derived the SW conversion to hazy mode with the haze ratio of 63%, which was caused by electric field-induced helical structure randomly oriented. Here, the haze ratio of 1.2% can be considered the OFF status, when the light can directly pass through the structure, while the haze ratio of 63% can be defined as the ON status when the light is blocked by the SW. With these features, the CLC and FS-TENG-based SW realized the mechanical–electrical–optical signal conversion, which can be applied in wireless sensing by detecting the ON/OFF status of the SW. In addition, the SW can return to its initial status with a gentle pressure load.

In 2021, Chen et al. reported a triboelectrification-driven CLC display, which can be applied for self-powered information securing, vision-correcting, and wireless sensing [87]. The reported CLC display contained three statuses – the planar (P) state, focal conic (FC) state, and homeotropic (H) state, which showed partially transparent, opaque, and fully transparent, respectively. In addition, the FC to H state-switching device was developed toward practical application. Similarly, the high-voltage output of TENG can adjust the state of CLC display to modulate the transmittance between transparent state and opaque state to realize light ON/OFF control. Thus, the self-powered wireless motion sensing system can be further realized through this system. With the same working principle, in 2021, Liu et al. reported an electroresponsive CLC for display panel, SW, and wireless alarm system [88]. Here, the reported CLC contained the three statuses as well, where the three statuses can be controlled by the applied electric field, which can affect the arrangement of the LC molecules. Toward wireless sensing, CLC, which can transform between FC to H state, was designed and driven by CS-TENG to realize the mechanically stimulated optical response. Based on the tribo-induced light ON/OFF control, the self-powered wireless sensing system was realized, which was further demonstrated by the remote alarm sensor.

10.4.2 Tribo-Induced Light Direction Control

By using the TENG-driven opaque actuators and LC structure, the ON/OFF control of the TILPT strategy was realized. However, in this way, the transmitted information is limited. As an alternative strategy, the direction of the light propagation can be controlled, which delivers higher information density in the transmitted optical signal.

Based on this strategy, in 2017, Chen et al. reported a dielectric elastomer actuator (DEA)-based tunable optical grating (TOG) [89]. The strain of the reported DEA can be controlled by applied voltage to compress or expand, where the movement of DEA can further control the TOGs to tune the light path (via the tuning of grating period). Here, by combining the TOG with a TENG, the TENG drove the decrease of grating period by 16.5% or the increase by 9.4%, resulting in the expansion or compression of the light spot array. Besides, the period change was positively related to the output of TENG, which can be used to reflect the applied voltage. Based on this principle, the designed system realized the mechanical–electrical–optical-position signal conversion, which realized the tribo-induced light propagation position control.

In 2021, Chen et al. reported the programmable compound prism (PCP) which consisted of ITO glass, Teflon, deionized (DI) water, and oil for solar beam steering, as another TILPT strategy [90]. When the light passed through the PCP, there were two oil–water interfaces, which can affect the light path and deflect the incident beam, whereas the oil–water interface can be controlled by the applied voltage. Here, a disc TENG was designed with a peak-to-peak open-circuit voltage of 530 V, short-circuit current of 1.2 μA, and transferred charge of 140 nC, and an electric circuit was applied to modulate the PCP of the refraction angle, where the refraction angle was related to the applied voltage. Then, the light path direction can be controlled, where the position of the received light spot can be controlled in the range of −18 mm–16 mm. In this way, the reported system can be applied for sensing, where the sensed signal can be revealed by the position of the received light spot.

10.4.3 Tribo-Induced Light Intensity Control

The above-discussed works have realized the ON/OFF control and light propagation direction control. Besides, it is also possible to design the optical device to tune the light intensity in propagation path to further increase the information density. By analyzing the light intensity variation in the receiver, self-powered wireless sensing can be realized. Due to the relative lower integration currently, these technical solutions can be considered tribophotonic hybrid systems, which hopefully are to be further developed and also be highly integrated tribophotonic devices in the future.

In 2017, Chen et al. reported a smart optical modulator (SOM) system, which was a combination of a single-electrode TENG and a DEA [91]. As a combination of the elastomer with nanowire electrode, the DEA was kept in transparent status (high transmittance) in initial status. When there is an input mechanical stimulation, the high voltage generated by TENG can be applied across the dielectric elastomer, resulting in Coulombic attraction and then inducing spatially uniform Maxwell stress. Thus, many localized deformations generated by the Maxwell stress and the generated nano wrinkles can refract the light path to adjust the optical transmittance, where the transmittance was negatively relevant to the input voltage. Then, the reported SOM showed potential toward the self-powered wireless motion or force sensing, by detecting the light intensity variation of the passed light beam to reveal the input signal (i.e. motion speed, separation height).

In 2019, our team reported a self-powered electrowetting optical switch (EOS) for light-intensity tuning [65]. The reported EOS was fabricated by the ITO glass, olive oil, and 10% NaCl solution, where ITO glass served as the electrode to tune the interface between the oil and solution, working as a liquid lens. By combining the EOS and an FS-TENG, the self-powered EOS system was built to realize the mechanical–optical–electrical signal conversion to realize self-powered wireless sensing. Initially, there was no voltage applied to EOS, and the interface between the oil and solution was flat, where the light could directly pass through the structure. With the mechanical input applied to the TENG, there was a strong electric field generated between the two ITO glasses, resulting in the electrowetting effect. Hence, the curvature of the interface between the oil and solution varied, where the light was scattered by the interface. A photoresistor was applied as the light detector, where the change of the light intensity could be detected, which was related to the input voltage. The self-powered wireless motion-sensing system was demonstrated based on the designed self-powered EOS.

In 2020, our team reported a rotary freestanding sliding TENG (RFS-TENG) and a polymer network LC (PNLC) cell-based SW [92]. The alignment of LC molecules can be controlled by the alternating applied electric field, where the electric field can realign the LC molecules to induce the multiple-scattering effect, resulting in the low transmittance of the PNLC. Based on this, the RFS-TENG was designed to generate the alternating electric field, which could be potentially driven by wind energy, water waves, etc. Here, the PNLC was connected and powered by the reported RFS-TENG, where the output of RFS-TENG could control the alignment of LC molecules to affect the transmittance of the SW. In initial status, the SW was in high transparent status. When there was the mechanical input applied to the RFS-TENG, the charge induced by TENG affected the transmittance of the SW, where the transmittance of SW was negatively relevant to the applied voltage. Among them, the lowest transmittance of the SW was below 20%. In this case, the SW could realize the control of transmitted light intensity.

With a similar working principle, in 2021, our team reported an aerodynamics-driven TENG and a tribo-induced smart reflector for self-powered wireless sensing [63]. Here the reported smart reflector was fabricated by the silver-coated PNLC, where the transmittance of PNLC can be controlled by the applied voltage. Here, the aerodynamics-driven TENG was designed to convert the airflow input to electrical output to drive the PNLC. In initial status, the PNLC contained the transmittance of above 75%, where the light can pass through the PNLC, and be reflected and detected by the photodetector (PD). When there was a wind stimulation, the wind could drive the motion of fluorinated ethylene propylene (FEP) film to generate high-voltage output between the electrodes, where the transmittance was negatively relevant to the applied voltage. Especially, the PNLC with applied high voltage (above 30 V) was converted to the haze status with a transmittance of lower than 30%, where the light was scattered in the PNLC and almost could not be detected by the PD. With this working mechanism, our team realized the mechanical–electrical–optical signal conversion, which was demonstrated for the self-powered wireless sensing of airflow.

10.4.4 Tribo-Induced Light Color Control

The tribo-induced light intensity and direction control may be used for self-powered wireless sensing with high information density. However, the precise position measurement system or light-intensity detector to receive and detect the optical signal is still required, raising the high demand on the receiver system. As an alternative solution, the tribo-induced light color control may be achieved by tuning the wavelength of the received light. In this way, the precise optical detector can be replaced by the camera, closed circuit television (CCTV), smartphone, or eyes, which can be applied to monitor the color change in a convenient way.

Based on this working principle, in 2021, our team reported a tribo-induced color tuner, which could control the light spectrum via the mechanical input, as shown in Figure 10.3 [93]. Here the tribo-induced color tuner mainly consisted of a concentric color-conversion plate (CCP), tribo-induced light lens, connector, and RFS-TENG, where a blue light diode worked as the light source. The middle part of the CCP was filled with TiO_2, where the blue photons were scattered while reaching TiO_2. The red phosphor was placed around the TiO_2, where the wavelength can be red-shifted when the blue photons reached the phosphor. When there was no mechanical input on the TENG structure, the liquid lens was a concave lens with a dispersive light spot, while a majority of blue photons were lighted to the phosphor, resulting in pink color. When there was a mechanical input loaded on the TENG, the liquid lens was changed to a convex lens with a focused light spot. In this case, most of blue photons reached TiO_2 while a few reached the phosphor, resulting in purple color. Using this way, the light spectrum can be adjusted and the corresponding input mechanical signal can be sensed.

For TILPT, since the signal is transmitted by light, the TILPT device contains the strong ability to avoid EMI, and the transmitted signal is also easily detected by the receiver system (by photodiode, photoresistor, camera, cellphone, etc.). In addition, the TILPT devices can also be designed with both directional and omnidirectional transmissions, which fit in different application scenarios. However, in long-distance and outdoor applications, such as bridge structure monitoring, environment airflow sensing, etc., the transmitted light signal is easily affected by environmental factors, such as fog and structure vibration. Therefore, a relatively accurate light path is usually needed, and further research is needed in standardized signal calibration and reference status detection. The TILPT device can be applied for the detection of structural and environmental factors toward the applications in smart cities, such as infrastructure vibration sensing and wind speed sensing.

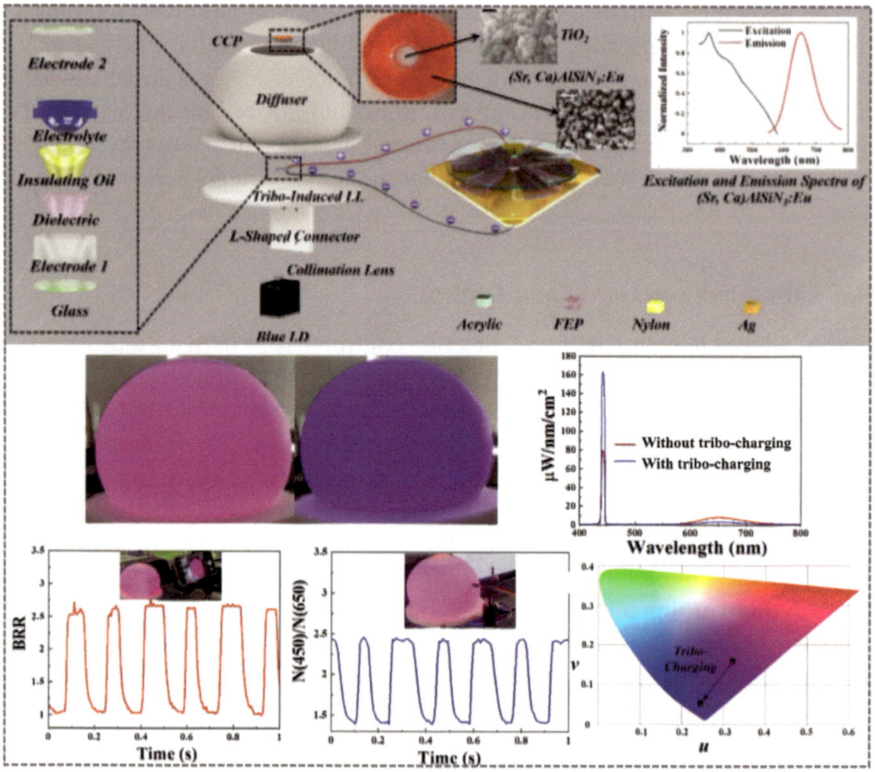

Figure 10.3 Tribo-induced light color control through tribo-induced color tuner. Source: [93] / John Wiley & Sons / CC-BY-4.0.

The specifications of the reported TILPT devices/hybrid systems are summarized in Table 10.2.

10.5 Triboelectrification-Induced Electroluminescence (TIEL)

As another solution for light-based self-powered wireless sensing or FSO, the light can be directly generated by the TIEL through the coupling effect of triboelectrification and electroluminescence. TIEL-based fully self-powered wireless sensing has been applied to different kinds of sensing and applications with all-in-one designs. Currently, the TIEL device has been widely demonstrated for applications such as motion trajectory sensing, water droplet dynamic motion monitoring, vertical and sliding force detection, stress and strain sensing, and realization of the visualization system.

As an initial design, TENG was applied to an electroluminescence device to realize the self-powered lighting system for the application of visualized motion sensing and lighting. In 2015, Wei et al. reported a self-powered electroluminescent

Table 10.2 Summary of TILPT devices/hybrid systems.

Ref	Modulation principle	Modulation mechanism	Threshold voltage (V)	Output signal	Application
[86]	Tribo-induced light ON/OFF control	EFOS	2500	Transmittance: N/A	Delay characteristic wireless motion sensing
[64]		CLC	12	Haze ratio: 1.2% and 63%	Mechanically triggered smart window; Self-powered information covering and retrieval system; (Can also be applied for light signal-based motion sensing)
[87]		CLC	400	Transmittance: ~78% and ~20%	Automotive glass; Self-powered webcam and cellphone security; Occlusion therapy device; (Can also be applied for light signal-based motion sensing)
[88]		CLC	60	Transmittance (400–700 nm): Lower than 40% and ~80%	Self-powered bistable smart window and E-paper; Self-powered wireless remote control/motion wireless sensing
[89]	Tribo-induced light direction control	DEA-based TOG	0–1400 (compression mode); 0–2800 (expansion mode).	Grating period:16.5% decrease to 9.4% increase	Display system, optical sensors, optical communications
[90]		PCP	V_R: ~90–~195	Light spot position: −18 mm–16 mm	Deflect oblique incident laser beam toward the output performance boosting of solar cell; (Can also applied for light signal-based motion sensing)
[91]	Tribo-induced light intensity control	DEA-based SOM	0–1800	Transmittance: from 72% to 40%	MEMS/NEMS, human–robots interaction, and IoT; (Can also be applied for light signal-based motion sensing)
[65]		Self-powered EOS	200–1100	Aspect ratio: from 0 to ~0.25	Wireless hand-sliding and stepping sensing
[92]		PNLC	0–68	Haze ratio: from 2.28% to 80.3%	Smart window, self-powered sunroofs, wind-driven smart systems, etc.
[63]		Silver-coated PNLC	10–30	Transmittance: from above 75% to lower than 30%	Self-powered wind speed sensing and breath monitoring
[93]	Tribo-induced light color control	Tribo-induced color tuner	0–140	Color: from pink to purple	Wireless rotation, wind speed, and vehicle speed sensing

system that can directly convert the input mechanical signal into an optical signal to emit light [94]. The reported self-powered electroluminescent system was composed of a TENG and an alternating-current thin-film electroluminescent (ACTFEL) lamp. The ACTFEL lamp can be excited by the electric field generated by TENG to generate luminescence, as triggered by motion such as footsteps. As a result, the mechanical–electrical–optical signal conversion was realized, which can further be applied for self-powered wireless motion sensing. Besides, they designed a footstep-triggered area-scalable electroluminescent system, which can also be considered self-powered footstep-sensing system.

Then, various TIEL devices by integrating the TENG part and the electroluminescent part into the all-in-one style were designed and fabricated. Among them, the TIEL devices were widely applied for visualized motion trajectory sensing. In 2016, Wei et al. proposed an integrated TIEL device, which can convert the dynamic motion signal to optical signal, realizing the self-powered wireless visualized motion sensing [95]. In initial status without dynamic motion input, the device was in non-luminous status. A dynamic motion input, such as the finger or pen sliding through the electrification layer, can generate a tribocharge-induced electric field based on the triboelectric effect between the electrification layer and the contact object. Then, the induced strong electric field can further excite the luminescent layer consisting of PMMA matrix and ZnS:Cu phosphor particles, to realize light-emitting. The loaded motion can be sensed with the different luminescent light intensities and areas. Besides, they fabricated a high-resolution segmented structure TIEL device for real-time self-powered position sensing and trajectory detection. In 2019, Wang et al. presented a self-powered high-resolution triboelectrification-induced electroluminescence (HR-TIEL) sensor, which can convert the input motion signal (including the contact profile and dynamic trajectory) into optical signal through TIEL [96]. Similarly, once there was a motion stimulation, the triboelectrification generated between the contact object and the electrification layer could produce charges to further activate the electric field and excite luminescence, where the light intensity was related to the stress. Additionally, the transparent conductive layer was fabricated by the silver nanowire, which had high transmittance and guided the electric field, resulting in an increase in the generated light intensity and the lateral spatial resolution. Here, the visualized surface texture of the nitrile glove can be revealed, providing the potential for detection of the motion and the profile of the contact objects. In 2020, Zhao et al. designed a self-powered, user-interactive e-skin (SUE-skin). As an integration TIEL device, the SUE-skin can convert the input dynamic touch signal to optical signal and corresponding electrical signal [97]. The programmable touch operation platform further demonstrated the good performance of SUE-skin in the HMI field. In 2019, Chen et al. established a stretchable hybrid luminescent composite (HLC), which can convert the input sliding motion signal or mechanical strain to luminescence through TIEL and mechanoluminescence (ML), respectively [98]. The light intensity was positively related to the parameters of the input motion (i.e. strain, stress, velocity). Besides, the phenomenon of TIEL and ML can be combined to enhance the light intensity of electroluminescence.

In addition to these sensing systems, in 2021, Song et al. established flexible motion-driven trajectory storage and display functional film, which can convert the input motion signal to optical signal and store the motion trajectory information to be re-excited by applying a polar-liquid mist [99]. The TIEL was generated in the functional film with an input motion, where the excited light intensity was related to the sliding velocity, sliding force, and rubbing times. The emitting light can only be maintained for a while and then is recorded by the functional film, which can be re-excited by the polar-liquid mist to show the recorded information once. Thus, the functional film can be employed to detect motion trajectory and realize sensing and handwriting recording in a self-powered wireless manner.

Except for the solid–solid interface, the TIEL-based motion trajectory sensing system can be further applied for liquid–solid interface. In 2020, Zhao et al. reported a water droplet-driven electricity-free electroluminescence device, which can convert the motion of water droplets into an optical signal in the solid–liquid interface [100]. Here, the TIEL can be excited at the solid–liquid interface of the reported device, where the generated light area and intensity were related to the types and falling height of water droplets. In the process, the demonstrated device realized the signal conversion from slight mechanical motion to optical with multidimensional information for calibration, which can be further applied to liquid leak detection and liquid motion sensing.

Besides, the TIEL devices can be applied for display and wearable electronics. In 2019, Wei et al. designed a motion-driven electroluminescence (MDEL) device, which can convert the human motion signal to optical signal toward the applications of display [101]. The reported MDEL was a combination of a lateral electroluminescence device and a human motion energy harvester. The electrical signal generated by the energy-harvesting structure can further excite the lateral electroluminescence device to emit an optical signal. The customized EL display can be realized by adjusting the design of the electrode. In 2019, Park et al. put forward a textile motion-driven electroluminescent (TMEL) device, which was fabricated by fibers and can convert multiple types of motion signals to optical signals through TIEL [102]. The generated light intensity was associated with the input force and the device material, which further improved the information density of the TMEL. On this basis, the TMEL device was designed to detect the input motion, presenting the potential in smart cloth and wearable applications.

The TIEL devices also played an important role in intelligent sensing systems toward detecting various physical signals, such as wind speed, pressure, and ultrasound. In 2019, Su et al. proposed a TIEL-based wind speed sensor with a low detection limit and high sensitivity, which can convert the input wind energy to optical signal through TIEL [66]. Here, the wind speed sensor was fabricated by combining the TIEL part with a designed cylinder support structure and perovskite-based PD. When there was a wind with a wind speed larger than 5 m/s, the wind energy was collected by the structure and induced the electroluminescence in the TIEL part, which can be identified by human eyes or detected by the perovskite-based PD. In addition, the light intensity increased with the increasing input wind speed, enabling wind speed sensing. In 2021, our team reported a self-powered visualized flexible pressure sensor (SP-VFPS), which could map local pressure information and generate color-adjustable visual optical signals without any additional data processing modules [103]. Based on the compression property of the porous structure, the ratio of green light to red light transmitted by the device gradually increased with the applied pressure. In addition, the device can be further integrated with silver nanowires/PDMS electrodes into a single-electrode TENG matrix, which generates electrical and optical responses in dual modes. With the capability of self-powered wireless pressure sensing through this color-tunable all-in-one TIEL device, the SP-VFPS showed potential in intelligent sensing

Table 10.3 Summary of TIEL devices.

Ref	Working mechanism	Target input signal	Detect limit	Responsivity (kPa^{-1})	Application
[94]	TENG driven ACTFEL lamp	Diverse mechanical sources	100 V	N/A	Self-powered illumination, display, and footsteps sensing
[95]	TIEL	Dynamic motions	10 kPa	0.03	Self-powered wireless position locating and motion tracking
[96]	TIEL	Dynamic motions	10 kPa	0.035	Self-powered wireless contact profile and dynamic trajectory visualized sensing
[97]	TIEL (optical signal) + TENG (electrical signal)	Touch stimuli	20 kPa	0.014	Programmable touch operation platform
[98]	TIEL + ML	Mechanical strain or contact object sliding	5 kPa	0.046	Self-powered lighting system and dynamic motions and interaction sensing
[99]	TIEL	Contact object sliding	0.5 N	N/A	Self-powered contact trajectory visualized sensing and storage
[100]	Solid–liquid interfacing triboelectric field based electroluminescence (SL-TEL)	Motion of liquid	Drop falling height = 0.25 m	N/A	Self-powered visualized water droplet sensing and leakage detection
[101]	MDEL	Dynamic motions	20 kPa	0.015	Self-powered customized display
[102]	TIEL	Various movements	N/A	N/A	Self-powered luminescent textile system
[66]	TIEL	Wind	5 m/s	N/A	Self-powered wind speed sensing
[103]	TIEL	Vertical pressure	10 kPa	N/A	Self-powered visualized flexible pressure sensing
[67]	TIEL	Ultrasound	0.2 W/cm^2	N/A	Self-powered wireless underwater ultrasound sensing and ultrasound source detecting

networks and HMI. In 2021, our team reported an underwater self-powered all-optical wireless ultrasonic sensor (SAWS), which can realize self-powered ultrasonic sensing and positioning through TIEL [67]. When there is an ultrasonic signal, it can excite the SAWS to emit light, where the light intensity is relevant to the design parameters of SAWS, input ultrasonic power density, and receiver distance. In addition, a three-SAWS-based positioning system was demonstrated, which can further be applied for underwater signal sensing, showing great potential toward the applications of ocean sensing systems.

For TIEL, similar to the TILPT device, since the transmitted signal is visible light, the TIEL device also contains the strong ability to avoid EMI. In addition, the TIEL device can be fabricated with all-in-one and flexible characteristics, which is suitable for different kinds of application scenarios. The TIEL devices have high sensitivity toward various kinds of slight movements (i.e. solid–solid contact, liquid–solid contact). Thus, the contact profile and trajectory can also be recorded by TIEL devices, which can be widely applied for trajectory sensing, writing feature recognition, and liquid leakage detection, especially in dark environments. However, since the light intensity of TIEL devices is usually limited, the TIEL devices are difficult to be applied in high-light-intensity environments. Thus, the TIEL devices are considered the sensitivity motion detection methods in dark environments, such as deep ocean sensing and sealed environment sensing. The specifications of the reported TIEL devices are summarized in Table 10.3.

10.6 Tribo-Assisted Spectrometry (TAS)

The high-voltage output characteristics of TENG, which can cause discharge internally or drive external discharge devices, as well as the photon emission generated by the interface interatomic electron-transition in triboelectrification (or contact electrification, CE) process, are able to generate spectra for wireless detection of the samples, known as TAS. Here, among various wireless sensing and detecting methods, spectrometry is a critical one, especially in component analysis and detection of the physical process in microworld. It is an indispensable tool in many disciplines not only for determining the presence or absence of specific components but also for accurately measuring their contents. In our previous studies about TENG-triggered nanoelectrospray ionization, we have studied and tailored the TENGs with high-voltage output characteristics for highly sensitive mass spectrometry [104]. By applying TENG to work as the high-voltage power source to create the breakdown discharge or microplasma, the generated light spectra may contain the target information, which can be detected by the receiver wirelessly. Additionally, the CE process can also generate photon emission, in which the information can be detected wirelessly through the light spectra as well. Thus, the TAS can be realized.

TENG has been developed for detecting material type directly by plasma spectra through discharge. In 2018, Cheng et al. reported a TENG-based microplasma system toward emission spectroscopy, as shown in Figure 10.4a [68]. As shown in

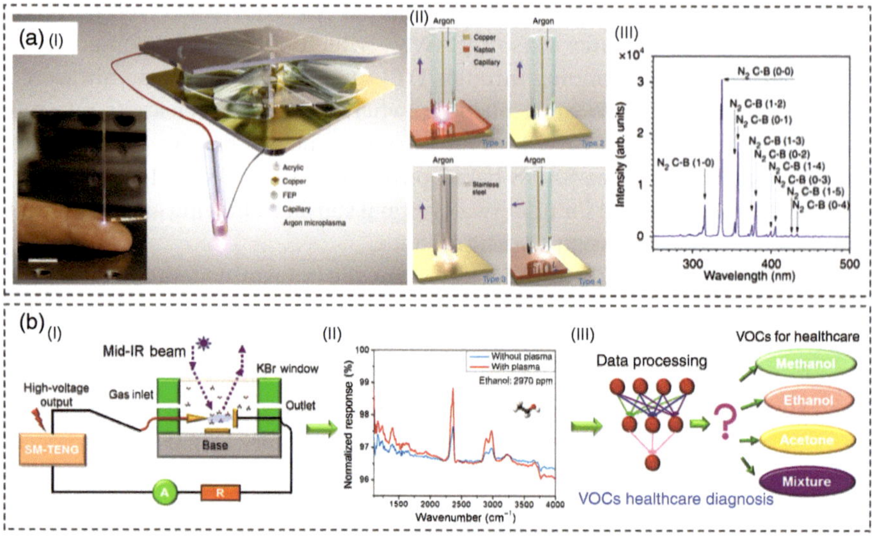

Figure 10.4 Tribo-assisted spectrometry (TAS). (a) Microplasma driven by TENG. I. The schematic diagram of argon microplasma directly driven by rotation-freestanding TENG. II. Four different types of microplasma through changing material and moving mode. III. Emission spectra of N_2 corona discharge. Source: [68] / Springer Nature / CC-BY-4.0. (b) A fast, accurate and sensitive plasma-enhanced VOC detector based on TENG. I. The mechanism of plasma-enhanced mid-IR VOCs detector. II. The normalized response spectrum of detectors with (red line) and without (blue line) plasma. III. Data processing step assisted by machine learning. Source: Reproduced with permission of [105], 2021 © American Chemical Society.

Figure 10.4a-I, emission spectra were generated from argon discharging in a capillary. Due to its stable and high-voltage output characteristics, a rotation-freestanding TENG worked as the power source. This device can achieve different discharge modes, as shown in Figure 10.4a-II. Moreover, in addition to the differences between these discharge modes mainly in electrical signals and the morphology discussed in the article, their optical emission spectra, such as Figure 10.4a-III (emission spectra of N_2 corona discharge) also contained the physical information, as shown by the characteristic spectral lines.

In addition to emission spectrum, reflection spectrum can also achieve wireless substance detection. In 2021, Zhu et al. reported a plasma discharge enhancement of mid-infrared (IR) system toward the application of volatile organic compound (VOC) sensing, as shown in Figure 10.4b [105]. The mechanism of this system is based on the response of the vibrational modes of molecular bonds in the range of 2–25 μm, leading to absorption and reflection of the subsequent mid- IR beam. To solve the weak coupling problem and improve the sensitivity, the reported system introduced a triboelectrification-microplasma to enhance their recognition, as shown in Figure 10.4b-I. Figure 10.4b-II demonstrated the obviously increased sensitivity of this device, compared to that without plasma. With the help of machine learning, the system's recognition capabilities can be further improved, as shown in Figure 10.4b-III.

In 2021, Li et al. reported the CE-induced interface photon emission spectroscopy (CEIIPES), where the induced photon emission spectra revealed the physical process of electron transfer between two solids, as shown in Figure 10.4c [69]. Here, a hollow cylinder structure was designed to drive the rotatory movement between the fans and the wall surface, where the CE occurred at the interface between the FEP and quartz/acrylic with the motion input. A spectrometer and charge-coupled device (CCD) detector were applied to detect the generated light spectrum. Accompanied by the input motion, the light was generated at the interface between the FEP and acrylic, where several sharp lines were recorded at the light spectrum, where the peak positions and the electron transitions of corresponding element atoms were accordant. Thus, the spectrum detected the details process in CE, which can further be applied for the detailed sensing in the charge transfer process in different material interfaces, such as solid–liquid interface, liquid–liquid interface, and liquid–gas interface.

10.7 Potential Applications and Perspectives

Toward the rapid development of IoT and smart cities, the next generation of sensing systems and networks are being rapidly developed, where the tribophotonics-enabled self-powered wireless systems provide a potential solution toward future distributed systems. This chapter presents and concludes the emerging tribophotonics, as a self-powered wireless solution by coupling triboelectricity and photonics. The tribophotonics can collect environmental mechanical information and energy, acting as both sensors and power sources to tune light propagation or generate photons, enabling the direct conversion of sensed signals into wireless signals. Thus, tribophotonics shows the potential in applications toward IoT, such as sensing for human motion, traffic, wind, rain, and ocean waves. This chapter discusses the major four types of technological routes of tribophotonics, based on TIEG, TILPT, TIEL, and TAS. The discussed tribophotonics features advantages and shows great potential in the application of distributed sensing systems, as discussed below. The fully self-powered wireless sensing ability allows tribophotonics to realize the function of sensing, signal modulation, and wireless transmission without any additional power source or wires. Compared with wired or battery-powered sensing systems (including the traditional sensing systems and some nanogenerator-powered sensors), tribophotonics brings fewer limitations in applications, especially in harsh environments, dynamic motion systems, and implantable systems, such as systems for deep ocean sensing, robotics force sensing, and heartbeat sensing. Besides, as the tribophotonics can directly convert the target signal into a wireless signal without complex intermediate steps, it can precisely sense and transmit each stimulation, working continuously for real-time sensing of random signals (i.e. wind speed sensing, robotic dog plantar pressure sensing). In addition, as various materials can be chosen for tribophotonics, such as biocompatible materials and high wear-resistant materials, tribophotonics-based systems can be easily applied in different kinds of application scenarios. Some tribophotonic devices such as SWISE can be fabricated with soft, flexible, thin, light, and small characteristics, which shows great potential in wearable and implantable sensing. Through tribophotonics, we can also avoid the EMI, which makes tribophotonic devices suitable for slight signal and long-distance signal transmission. More importantly, the cost of tribophotonics is usually low, and the materials and fabrication process are environmentally friendly, which shows the potential for large-scale fabrication and application in widely distributed sensing systems.

10.8 Challenges and Summary

In the meanwhile, there are still some challenges in the development of tribophotonics, considering the perspective research in the future, as summarized below:

1) For TIEG, the time-varying electric field and magnetic field decay with the transmitted distance in the environment, and thus it is required to study how to greatly promote the wireless transmission distance. At the same time, the EM-wave transmission is easily affected by the environment (i.e. metal and conductive liquid), where the device packaging, signal transmission routes, and signal processing algorithm still need to be further researched to enhance the compatibility and robustness.
2) For TILPT, the system usually contains rigid optical components with a relatively large system size. Some designs suffer from the fixed transmission path, which makes the system easily affected by some environmental factors (i.e. microvibration and fog), and the receiving system is also required. Thus, the systems need to be further optimized to avoid interference toward different application scenarios.
3) For the TIEL and TAS, there are demands for the receiving system and the signal transmission path as well. Besides, the TIEL device still meets the challenges of limited resolution, crosstalk induced by the tribocharges, limited electroluminescent color, etc., which still needs to be improved in future research through multiple methods, including lithography-based design, electrode design, hybridizing fluorescent materials, and doping of electroluminescence materials. In the TAS system, the threshold discharge voltage may be a limitation for detecting very slight signals, where the electrode shape, breakdown discharge gap distance, and detection process can be further researched to improve the sensitivity.
4) The environmental factors (i.e. humidity) may impact the operation and the stability of TENG and photonic devices, and thus further studies are needed toward precise sensing (i.e. force, displacement, and velocity).

Thus, studies on tribophotonics are still required on materials, design, system integration, fabrication process, etc., including the suitable device design toward various types of application scenarios. At the same time, efforts may also be required to ensure repeatability and long-term stability. There is still a lot of research work to do toward achieving mature products and large-scale applications for tribophotonics.

In summary, by combining triboelectricity with photonics, the emerging field of tribophotonics recently put forward a self-powered wireless solution toward the IoT and smart city. The appealing advantages of tribophotonics, including fully self-powered wireless manner, real-time response, low cost, etc., reveal the brilliant future in the market of wireless solutions toward the rapid development of IoT and smart city.

References

1 Al-Fuqaha, A., Guizani, M., Mohammadi, M. et al. (2015). Internet of things: a survey on enabling technologies, protocols, and applications. *IEEE Commun. Surv. Tutor.* 17 (4): 2347–2376.

2 Su, K., J. Li, and H. Fu 2011. Smart city and the applications. in 2011 International Conference on Electronics, Communications and Control (ICECC).

3 Akpakwu, G.A., Silva, B.J., Hancke, G.P. et al. (2018). A survey on 5G networks for the internet of things: communication technologies and challenges. *IEEE Access* 6: 3619–3647.

4 Li, S., Xu, L.D., and Zhao, S. (2018). 5G internet of things: a survey. *J. Ind. Inf. Integration* 10: 1–9.

5 Wang, Z.L. (2019). Entropy theory of distributed energy for internet of things. *Nano Energy* 58: 669–672.

6 Gao, W., Emaminejad, S., HYY, N. et al. (2016). Fully integrated wearable sensor arrays for multiplexed in situ perspiration analysis. *Nature* 529 (7587): 509–514.

7 Kim, S.Y., Choo, Y., Bilodeau, R.A. et al. (2020). Sustainable manufacturing of sensors onto soft systems using self-coagulating conductive Pickering emulsions. *Sci. Robot.* 5 (39): eaay3604.

8 Jin, T., Sun, Z., Li, L. et al. (2020). Triboelectric nanogenerator sensors for soft robotics aiming at digital twin applications. *Nat. Commun.* 11 (1): 5381.

9 Wen, F., Zhang, Z., He, T. et al. (2021). AI enabled sign language recognition and VR space bidirectional communication using triboelectric smart glove. *Nat. Commun.* 12 (1): 5378.

10 Yang, J.C., Mun, J., Kwon, S.Y. et al. (2019). Electronic skin: recent progress and future prospects for skin-attachable devices for health monitoring, robotics, and prosthetics. *Adv. Mater.* 31 (48): 1904765.

11 Fu, Y., Hoang, T., Mechitov, K. et al. (2018). Sudden event monitoring of civil infrastructure using demand-based wireless smart sensors. *Sensors* 18 (12): 4480.

12 Hu, X., Xu, L., Lin, X. et al. (2020). Battery lifetime prognostics. *Joule* 4 (2): 310–346.

13 Ciez, R.E. and Whitacre, J.F. (2019). Examining different recycling processes for lithium-ion batteries. *Nat. Sustain.* 2 (2): 148–156.

14 Ledet, E.H., D'Lima, D., Westerhoff, P. et al. (2012). Implantable sensor technology: from research to clinical practice. *J. Am. Acad. Orthop. Surg.* 20 (6): 383–392.

15 Assawaworrarit, S., Yu, X., and Fan, S. (2017). Robust wireless power transfer using a nonlinear parity–time-symmetric circuit. *Nature* 546 (7658): 387–390.

16 Niu, S., Matsuhisa, N., Beker, L. et al. (2019). A wireless body area sensor network based on stretchable passive tags. *Nat. Electron.* 2 (8): 361–368.

17 Fan, F.-R., Tian, Z.-Q., and Lin Wang, Z. (2012). Flexible triboelectric generator. *Nano Energy* 1 (2): 328–334.

18 Wu, C., Wang, A.C., Ding, W. et al. (2019). Triboelectric nanogenerator: a foundation of the energy for the new era. *Adv. Energy Mater.* 9 (1): 1802906.

19 Kim, W., Bhatia, D., Jeong, S. et al. (2019). Mechanical energy conversion systems for triboelectric nanogenerators: kinematic and vibrational designs. *Nano Energy* 56: 307–321.

20 Wu, H., Wang, Z., and Zi, Y. (2021). Multi-mode water-tube-based triboelectric nanogenerator designed for low-frequency energy harvesting with ultrahigh volumetric charge density. *Adv. Energy Mater.* 11 (16): 2100038.

21 Chen, B., Yang, Y., and Wang, Z.L. (2018). Scavenging wind energy by triboelectric nanogenerators. *Adv. Energy Mater.* 8 (10): 1702649.

22 Yong, S., Wang, J., Yang, L. et al. (2021). Auto-switching self-powered system for efficient broad-band wind energy harvesting based on dual-rotation shaft triboelectric nanogenerator. *Adv. Energy Mater.* 11 (26): 2101194.

23 Xu, W., Zheng, H., Liu, Y. et al. (2020). A droplet-based electricity generator with high instantaneous power density. *Nature* 578 (7795): 392–396.

24 Nie, J., Wang, Z., Ren, Z. et al. (2019). Power generation from the interaction of a liquid droplet and a liquid membrane. *Nat. Commun.* 10 (1): 2264.

25 Song, Y., Min, J., Yu, Y. et al. (2020). Wireless battery-free wearable sweat sensor powered by human motion. *Sci. Adv.* 6 (40): eaay9842.

26 Zou, Y., Xu, J., Fang, Y. et al. (2021). A hand-driven portable triboelectric nanogenerator using whirligig spinning dynamics. *Nano Energy* 83: 105845.

27 Jiang, Y., Dong, K., An, J. et al. (2021). UV-protective, self-cleaning, and antibacterial nanofiber-based triboelectric nanogenerators for self-powered human motion monitoring. *ACS Appl. Mater. Interfaces* 13 (9): 11205–11214.

28 Wang, Z., Wu, Y., Jiang, W. et al. (2021). A universal power management strategy based on novel sound-driven triboelectric nanogenerator and its fully self-powered wireless system applications. *Adv. Funct. Mater.* 31 (34): 2103081.

29 Qiu, W., Feng, Y., Luo, N. et al. (2020). Sandwich-like sound-driven triboelectric nanogenerator for energy harvesting and electrochromic based on Cu foam. *Nano Energy* 70: 104543.

30 Liang, X., Jiang, T., Liu, G. et al. (2020). Spherical triboelectric nanogenerator integrated with power management module for harvesting multidirectional water wave energy. *Energy Environ. Sci.* 13 (1): 277–285.

31 Wang, Y., Liu, X., Wang, Y. et al. (2021). Flexible seaweed-like triboelectric nanogenerator as a wave energy harvester powering marine internet of things. *ACS Nano* 15 (10): 15700–15709.

32 Hinchet, R., Yoon, H.-J., Ryu, H. et al. (2019). Transcutaneous ultrasound energy harvesting using capacitive triboelectric technology. *Science* 365 (6452): 491–494.

33 Deng, W., Libanori, A., Xiao, X. et al. (2022). Computational investigation of ultrasound induced electricity generation via a triboelectric nanogenerator. *Nano Energy* 91: 106656.

34 Li, J., Wu, C., Dharmasena, I. et al. (2020). Triboelectric nanogenerators enabled internet of things: a survey. *Intell. Converg. Netw.* 1 (2): 115–141.

35 Liu, L., Guo, X., and Lee, C. (2021). Promoting smart cities into the 5G era with multi-field Internet of Things (IoT) applications powered with advanced mechanical energy harvesters. *Nano Energy* 88: 106304.

36 Jin, L., Zhang, B., Zhang, L. et al. (2019). Nanogenerator as new energy technology for self-powered intelligent transportation system. *Nano Energy* 66: 104086.

37 Xu, L., Xu, L., Luo, J. et al. (2020). Hybrid all-in-one power source based on high-performance spherical triboelectric nanogenerators for harvesting environmental energy. *Adv. Energy Mater.* 10 (36): 2001669.

38 Ren, Z., Liang, X., Liu, D. et al. (2021). Water-wave driven route avoidance warning system for wireless ocean navigation. *Adv. Energy Mater.* 11 (31): 2101116.

39 Liu, C., Wang, Y., Zhang, N. et al. (2020). A self-powered and high sensitivity acceleration sensor with V-Q-a model based on triboelectric nanogenerators (TENGs). *Nano Energy* 67: 104228.

40 Wang, L., Liu, Y., Liu, Q. et al. (2020). A metal-electrode-free, fully integrated, soft triboelectric sensor array for self-powered tactile sensing. *Microsyst. Nanoeng.* 6 (1): 59.

41 Yu, J., Hou, X., He, J. et al. (2020). Ultra-flexible and high-sensitive triboelectric nanogenerator as electronic skin for self-powered human physiological signal monitoring. *Nano Energy* 69: 104437.

42 Fu, K., Zhou, J., Wu, H. et al. (2021). Fibrous self-powered sensor with high stretchability for physiological information monitoring. *Nano Energy* 88: 106258.

43 Tang, X., Wang, X., Cattley, R. et al. (2018). Energy harvesting technologies for achieving self-powered wireless sensor networks in machine condition monitoring: a review. *Sensors* 18 (12): 4113.

44 Khan, A.A., Mahmud, A., Zhang, S. et al. (2019). A self-powered multi-broadcasting wireless sensing system realized with an all-in-one triboelectric nanogenerator. *Nano Energy* 62: 691–699.

45 Chen, X., Gao, L., Chen, J. et al. (2020). A chaotic pendulum triboelectric-electromagnetic hybridized nanogenerator for wave energy scavenging and self-powered wireless sensing system. *Nano Energy* 69: 104440.

46 Lu, S., Gao, L., Chen, X. et al. (2020). Simultaneous energy harvesting and signal sensing from a single triboelectric nanogenerator for intelligent self-powered wireless sensing systems. *Nano Energy* 75: 104813.

47 Rui, P., Zhang, W., Zhong, Y. et al. (2020). High-performance cylindrical pendulum shaped triboelectric nanogenerators driven by water wave energy for full-automatic and self-powered wireless hydrological monitoring system. *Nano Energy* 74: 104937.

48 Liu, W., Wang, Z., Wang, G. et al. (2020). Switched-capacitor-convertors based on fractal design for output power management of triboelectric nanogenerator. *Nat. Commun.* 11 (1): 1883.

49 Xia, X., Wang, H., Basset, P. et al. (2020). Inductor-free output multiplier for power promotion and management of triboelectric nanogenerators toward self-powered systems. *ACS Appl. Mater. Interfaces* 12 (5): 5892–5900.

50 Rawy, K., Sharma, R., Yoon, H.-J. et al. (2020). A triboelectric nanogenerator energy harvesting system based on load-aware control for input power from 2.4 μW to 15.6 μW. *Nano Energy* 74: 104839.

51 Wang, H. et al. (2023). A general self-powered wireless sensing solution based on triboelectric-discharge effect. *Nano Energy* 105: 107982.

52 Wang, Z.L. (2017). On Maxwell's displacement current for energy and sensors: the origin of nanogenerators. *Mater. Today* 20 (2): 74–82.

53 Xia, X., Wang, H., Guo, H. et al. (2020). On the material-dependent charge transfer mechanism of the contact electrification. *Nano Energy* 78: 105343.

54 Xu, C., Zi, Y., Wang, A.C. et al. (2018). On the electron-transfer mechanism in the contact-electrification effect. *Adv. Mater.* 30 (15): 1706790.

55 McCarty, L.S. and Whitesides, G.M. (2008). Electrostatic charging due to separation of ions at interfaces: contact electrification of ionic electrets. *Angew. Chem. Int. Ed.* 47 (12): 2188–2207.

56 Baytekin, H.T., Patashinski, A.Z., Branicki, M. et al. (2011). The mosaic of surface charge in contact electrification. *Science* 333 (6040): 308–312.

57 Harmon, W., Bamgboje, D., Guo, H. et al. (2020). Self-driven power management system for triboelectric nanogenerators. *Nano Energy* 71: 104642.

58 Zhang, H., Marty, F., Xia, X. et al. (2020). Employing a MEMS plasma switch for conditioning high-voltage kinetic energy harvesters. *Nat. Commun.* 11 (1): 3221.

59 Liu, C. and Bard, A.J. (2008). Electrostatic electrochemistry at insulators. *Nat. Mater.* 7 (6): 505–509.

60 Wiles, J.A., Grzybowski, B.A., Winkleman, A. et al. (2003). A tool for studying contact electrification in systems comprising metals and insulating polymers. *Anal. Chem.* 75 (18): 4859–4867.

61 Kim, W.-G., Kim, D.-W., Tcho, I.-W. et al. (2021). Triboelectric nanogenerator: structure, mechanism, and applications. *ACS Nano* 15 (1): 258–287.

62 Zhang, C., Chen, J., Xuan, W. et al. (2020). Conjunction of triboelectric nanogenerator with induction coils as wireless power sources and self-powered wireless sensors. *Nat. Commun.* 11 (1): 58.

63 Wang, H., Wang, J., Yao, K. et al. (2021). A paradigm shift fully self-powered long-distance wireless sensing solution enabled by discharge-induced displacement current. *Sci. Adv.* 7 (39): eabi6751.

64 Wang, J., Meng, C., Wang, C.-T. et al. (2021). A fully self-powered, ultra-stable cholesteric smart window triggered by instantaneous mechanical stimuli. *Nano Energy* 85: 105976.

65 Wang, J., Wang, H., Li, X. et al. (2019). Self-powered electrowetting optical switch driven by a triboelectric nanogenerator for wireless sensing. *Nano Energy* 66: 104140.

66 Su, L., Wang, H., Tian, Z. et al. (2019). Low detection limit and high sensitivity wind speed sensor based on triboelectrification-induced electroluminescence. *Adv. Sci.* 6 (23): 1901980.

67 Tian, Z., Su, L., Wang, H. et al. (2021). Underwater self-powered all-optical wireless ultrasonic sensing, positioning and communication with ultrafast response time and ultrahigh sensitivity. *Adv. Opt. Mater.* 10 (5): 2102091.

68 Cheng, J., Ding, W., Zi, Y. et al. (2018). Triboelectric microplasma powered by mechanical stimuli. *Nat. Commun.* 9 (1): 3733.

69 Li, D., Xu, C., Liao, Y. et al. (2021). Interface inter-atomic electron-transition induced photon emission in contact-electrification. *Sci. Adv.* 7 (39): eabj0349.

70 Ishimaru, A. (2017). Electromagnetic wave propagation, radiation, and scattering: from fundamentals to applications. In: *The IEEE Press Series on Electromagnetic Wave Theory*. Piscataway, NJ: Wiley-IEEE Press.

71 Wang, Z.L. (2021). From contact electrification to triboelectric nanogenerators. *Rep. Prog. Phys.* 84 (9): 096502.

72 Wang, Z.L. (2020). On the first principle theory of nanogenerators from Maxwell's equations. *Nano Energy* 68: 104272.

73 Shao, J., Jiang, T., and Wang, Z. (2020). Theoretical foundations of triboelectric nanogenerators (TENGs). *Sci. China Technol. Sci.* 63 (7): 1087–1109.

74 Cao, X., Zhang, M., Huang, J. et al. (2018). Inductor-Free wireless energy delivery via Maxwell's displacement current from an electrodeless triboelectric nanogenerator. *Adv. Mater.* 30 (6): 1704077.

75 Jie, Y., Ma, J., Chen, Y. et al. (2018). Efficient delivery of power generated by a rotating triboelectric nanogenerator by conjunction of wired and wireless transmissions using Maxwell's displacement currents. *Adv. Energy Mater.* 8 (31): 1802084.

76 Mallineni, S.S.K., Dong, Y., Behlow, H. et al. (2018). A wireless triboelectric nanogenerator. *Adv. Energy Mater.* 8 (10): 1702736.

77 Kuang, H., Huang, S., Zhang, C. et al. (2022). Electric-field-resonance-based wireless triboelectric nanogenerators and sensors. *ACS Appl. Mater. Interfaces* 14 (1): 794–804.

78 Xu, L., Xuan, W., Chen, J. et al. (2021). Fully self-powered instantaneous wireless humidity sensing system based on triboelectric nanogenerator. *Nano Energy* 83: 105814.

79 Tang, Y., Xuan, W., Zhang, C. et al. (2021). Fully self-powered instantaneous wireless traffic monitoring system based on triboelectric nanogenerator and magnetic resonance coupling. *Nano Energy* 89: 106429.

80 Chen, J., Liao, Z., Wu, Y. et al. (2021). Self-powered pumping switched TENG enabled real-time wireless metal tin height and position recognition and counting for production line management. *Nano Energy* 90: 106544.

81 Zhang, C., Chen, J., Xuan, W. et al. (2022). Triboelectric nanogenerator-enabled fully self-powered instantaneous wireless sensor systems. *Nano Energy* 92: 106770.

82 Wen, F., Wang, H., He, T. et al. (2020). Battery-free short-range self-powered wireless sensor network (SS-WSN) using TENG based direct sensory transmission (TDST) mechanism. *Nano Energy* 67: 104266.

83 Chen, Y., Cheng, Y., Jie, Y. et al. (2019). Energy harvesting and wireless power transmission by a hybridized electromagnetic–triboelectric nanogenerator. *Energy Environ. Sci.* 12 (9): 2678–2684.

84 Xia, X., Fu, J., and Zi, Y. (2019). A universal standardized method for output capability assessment of nanogenerators. *Nat. Commun.* 10 (1): 4428.

85 Chan, V.W.S. (2006). Free-space optical communications. *J. Light. Technol.* 24 (12): 4750–4762.

86 Wang, H., Wang, J., Yao, K. et al. (2020). Multifunctional self-powered switch toward delay-characteristic sensors. *ACS Appl. Mater. Interfaces* 12 (20): 22873–22880.

87 Chen, Y.-H., Lin, P.-Y., Wang, T.-W. et al. (2021). Dynamics of electrically driven cholesteric liquid crystals by triboelectrification and their application in self-powered information securing and vision correcting. *ACS Energy Lett.* 6 (9): 3185–3194.

88 Liu, H., Guo, Z.H., Xu, F. et al. (2021). Triboelectric-optical responsive cholesteric liquid crystals for self-powered smart window, E-paper display and optical switch. *Sci. Bull.* 66 (19): 1986–1993.

89 Chen, X., Wu, Y., Yu, A. et al. (2017). Self-powered modulation of elastomeric optical grating by using triboelectric nanogenerator. *Nano Energy* 38: 91–100.

90 Chen, G., Wang, Y., Jiang, D. et al. (2021). A programmable compound prism powered by triboelectric nanogenerator for highly efficient solar beam steering. *Nano Energy* 80: 105524.

91 Chen, X., Pu, X., Jiang, T. et al. (2017). Tunable optical modulator by coupling a triboelectric nanogenerator and a dielectric elastomer. *Adv. Funct. Mater.* 27 (1): 1603788.

92 Wu, F., Zhao, S., Yu, B. et al. (2020). A new coronavirus associated with human respiratory disease in China. *Nature* 579 (7798): 265–269.

93 Wang, J., Wang, H., Yin, K. et al. (2021). Tribo-induced color tuner toward smart lighting and self-powered wireless sensing. *Adv. Sci.* 8 (12): 2004970.

94 Yan Wei, X., Kuang, S.Y., Yang Li, H. et al. (2015). Interface-free area-scalable self-powered electroluminescent system driven by triboelectric generator. *Sci. Rep.* 5 (1): 13658.

95 Wei, X.Y., Wang, X., Kuang, S.Y. et al. (2016). Dynamic triboelectrification-induced electroluminescence and its use in visualized sensing. *Adv. Mater.* 28 (31): 6656–6664.

96 Wang, Y., Wang, H.L., Li, H.Y. et al. (2019). Enhanced high-resolution triboelectrification-induced electroluminescence for self-powered visualized interactive sensing. *ACS Appl. Mater. Interfaces* 11 (14): 13796–13802.

97 Zhao, X., Zhang, Z., Liao, Q. et al. (2020). Self-powered user-interactive electronic skin for programmable touch operation platform. *Sci. Adv.* 6 (28): eaba4294.

98 Chen, Y., Wei, X., Li, H. et al. (2019). Stretchable hybrid bilayered luminescent composite based on the combination of strain-induced and triboelectrification-induced electroluminescence. *ACS Omega* 4 (24): 20470–20475.

99 Song, C., Zhu, S., Ma, L. et al. (2021). Complete contact trajectory visualization based on triboelectrification and droplet luminescence. *Appl. Mater. Today* 24: 101081.

100 Zhao, X.J., Kuang, S.Y., Wang, Z.L. et al. (2020). Electricity–free electroluminescence excited by droplet impact driven triboelectric field on solid–liquid interface. *Nano Energy* 75: 104823.

101 Wei, X.Y., Wang, H.L., Wang, Y. et al. (2019). Fully-integrated motion-driven electroluminescence enabled by triboelectrification for customized flexible display. *Nano Energy* 61: 158–164.

102 Park, H.-J., Kim, S., Lee, J.H. et al. (2019). Self-powered motion-driven triboelectric electroluminescence textile system. *ACS Appl. Mater. Interfaces* 11 (5): 5200–5207.

103 Su, L., Jiang, Z., Tian, Z. et al. (2021). Self-powered, ultrasensitive, and high-resolution visualized flexible pressure sensor based on color-tunable triboelectrification-induced electroluminescence. *Nano Energy* 79: 105431.

104 Li, A., Zi, Y., Guo, H. et al. (2017). Triboelectric nanogenerators for sensitive nano-coulomb molecular mass spectrometry. *Nat. Nanotechnol.* 12 (5): 481–487.

105 Zhu, J., Ren, Z., and Lee, C. (2021). Toward healthcare diagnoses by machine-learning-enabled volatile organic compound identification. *ACS Nano* 15 (1): 894–903.

11

TENG-Based Wearable Biomechanical Sensors and Human–Machine Interface

11.1 Introduction

With the growing emphasis on adopting healthy lifestyles, there has been a surge in the importance of tracking and managing personal well-being [1–3]. Consequently, self-monitoring of health at home has gained substantial traction. Wearable devices have revolutionized the ability to monitor various facets of physical health, including exercise routines, metabolic indicators, and dynamic and static physiological traits [4, 5]. Furthermore, these devices can be worn on various body parts, facilitating the continuous real-time collection of data on biomechanical information [6]. By integrating wearable devices with cutting-edge medical technologies, the reception and analysis of bodily information for early detection of potential ailments and medical diagnostics becomes plausible [7, 8]. This underscores the significant potential of portable, versatile wearable sensor devices in domains such as disease diagnosis, rehabilitation treatment, and routine health evaluation [9]. Moreover, wearable sensors play a crucial role in achieving precise control over devices, robots, or virtual environments, acting as a link between humans and machines, as well as driving forward the progress of artificial intelligence and the Internet of Things (IoT) [10–12]. Striving for effective human–machine interface (HMI) experiences necessitates a focus on enhancing the sensitivity and detection range of these sensors, enabling them to detect subtle stimuli and a wider range of versatile application scenarios [9, 13–15].

Among the array of advanced wearable sensors, triboelectric nanogenerator (TENG) has recently gained prominence as a compelling technology, leveraging the synergy between the triboelectric effect and electrostatic induction [16, 17]. TENG has the capability to harness the biomechanical motions of human body to construct self-powered systems, liberating them from the constraints of external power sources [18, 19]. TENG devices can be crafted using an extensive range of biocompatible materials, encompassing both synthetic and natural options, thereby enabling the creation of flexible, soft, lightweight, and biocompatible solutions [20, 21]. In addition, the simple structure and small size of TENG make it easier to increase the portability of wearable sensors, which facilitates enhanced integration and miniaturization of TENG sensors [22]. Moreover, four distinct working modes of TENG empower the sensor structure to be fine-tailored to the target, such as a

Triboelectric Nanogenerators: Technology, Applications, and Commercialization, First Edition.
Yunlong Zi, Hengyu Guo, Jie Wang, Chi Zhang, Xiangyu Chen, and Qing Zhao.

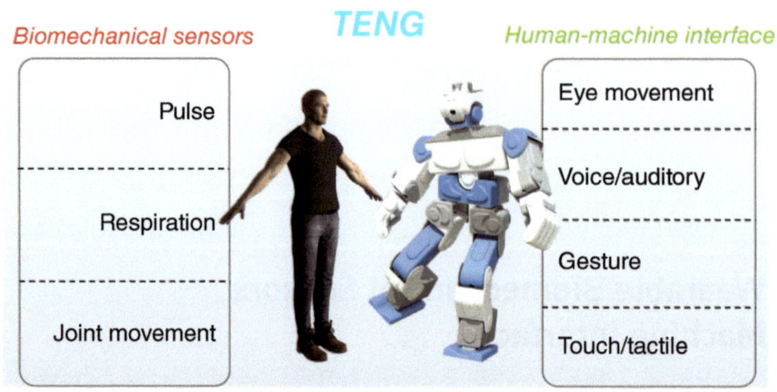

Figure 11.1 An overview of TENG-based wearable biomechanical sensors and HMIs.

flexible e-skin in single-electrode mode for temperature regulation [23], a smart mask in vertical contact-separation mode for respiratory monitoring [24], a conductive sponge brush in linear sliding mode for trajectory tracking [25], and a hybrid textile in freestanding mode for harvesting energy from human movement [26]. Collectively, these attributes position TENG as a robust contender for wearable biomechanical sensors and HMIs [27–29].

This chapter provides an overview of wearable biomechanical sensors founded on TENG technology, utilized for the surveillance of pulse, respiration, and joint movement, along with their role in HMI, encompassing eye movement, voice/auditory, gesture, and touch/tactile (Figure 11.1). Each section briefly introduces the practical necessity and references typical illustrative instances. The composition and materials of the devices, the operation principles, and the key application are succinctly elucidated. Furthermore, this chapter engages in a discourse concerning the potential development trends of wearable sensors dedicated to biomechanical sensing and HMIs.

11.2 TENG-Based Biomedical Sensing

Wearable electronic devices have gained significant popularity in the field of physical sign measurement and physiological information research owing to their portability and versatility [30]. TENG excels at converting external stimuli into electrical signals, rendering it an ideal candidate for the development of self-powered sensors, thus facilitating the creation of wearable sensors for biomedical applications. This section primarily focuses on TENG-based wearable sensors designed for monitoring pulse, respiration, and joint movement, while also discussing potential directions for future advancements.

11.2.1 Pulse

The human pulse, a typical physiological signal that provides valuable information for health assessment and disease diagnosis, necessitates the monitoring of these signals using wearable electronic sensing devices. TENGs can harvest pulsation resulting from skin surface and yield electric signals that carry individual information. This section presents three examples of biomechanical sensors in pulse, including an elastomer-encapsulated pulse monitoring and diagnostic system [31], a stretchable and washable gauze mesh-embedded sensor for measuring arterial pulse waves [32], and a no-spacer TENG fabricated using an environmentally friendly in-situ gap generation method for monitoring pulse signals [33], which are of great significance for monitoring these bodily signals for assessment and disease diagnosis.

In 2017, Ouyang et al. proposed a flexible self-powered ultrasensitive pulse sensor (SUPS) based on TENG, as illustrated in Figure 11.2a. The sensor design incorporates a nanostructured Kapton (n-Kapton) film as the triboelectric layer. An ultrathin Cu layer is deposited on the back side of the Kapt-on film to function as the electrode. A nanostructured Cu film is employed as both a triboelectric layer and an electrode. The entire device is enclosed by a flexible polydimethylsiloxane (PDMS) layer to bolster its structural stability. The SUPS effectively captures pulse signals from various body regions, including the carotid, brachial, radial, finger, and ankle arteries. It can quantify the amplitude of different pulses and calculate pulse wave velocity (PWV). Moreover, the SUPS can be integrated with a Bluetooth chip, enabling wireless real-time monitoring of pulse signals from the cardiovascular system on smartphones and PCs.

In 2018, Dong et al. introduced a stretchable and washable skin-inspired triboelectric nanogenerator (SI-TENG) for biomechanical energy harvesting and skin surface pressure sensing, as shown in Figure 11.2b. The SI-TENG incorporates a silver-coated nylon yarn configured in a three-ply-twisted arrangement, chosen for its remarkable electrical conductivity, mechanical durability, and ease of use. The elastomeric dielectric is composed of silicone rubber, capitalizing on its inherent biocompatibility, superior mechanical properties, excellent flexibility/stretching, and strong affinity for acquiring electrons. Embedded within the silicone rubber elastomer, there exists a continuous, planar, and "chain-link" fence-shaped interlaced conductive network of zigzag distributed nylon yarn electrodes. This design confers the SI-TENG with adequate transparency, considerable stretchability, high-pressure sensitivity, and exceptional mechanical stability. The working principle of SI-TENG is rooted in the single-electrode mode, which exploits the combined effects of contact electrification and electrostatic induction. In this mode, human skin serves as another triboelectric electrode. The highly sensitive SI-TENG is capable of detecting the subtle physiological signals emanating from arterial pulses and differentiating between real-time arterial pulses under normal and exercise conditions. The amplified waveforms yield typical arterial pulse signals, including the percussion wave (P-wave), tidal wave (T-wave), and diastolic wave (D-wave). By analyzing these waveforms, the wearer's systolic and diastolic blood pressure, ventricular pressure, and heart rate can be extracted. Consequently, the SI-TENG displays excellent biomechanical energy harvesting capability and multifunctional pressure sensing properties, rendering it eminently suited for applications involving wearable artificial skin and humanoid robotics.

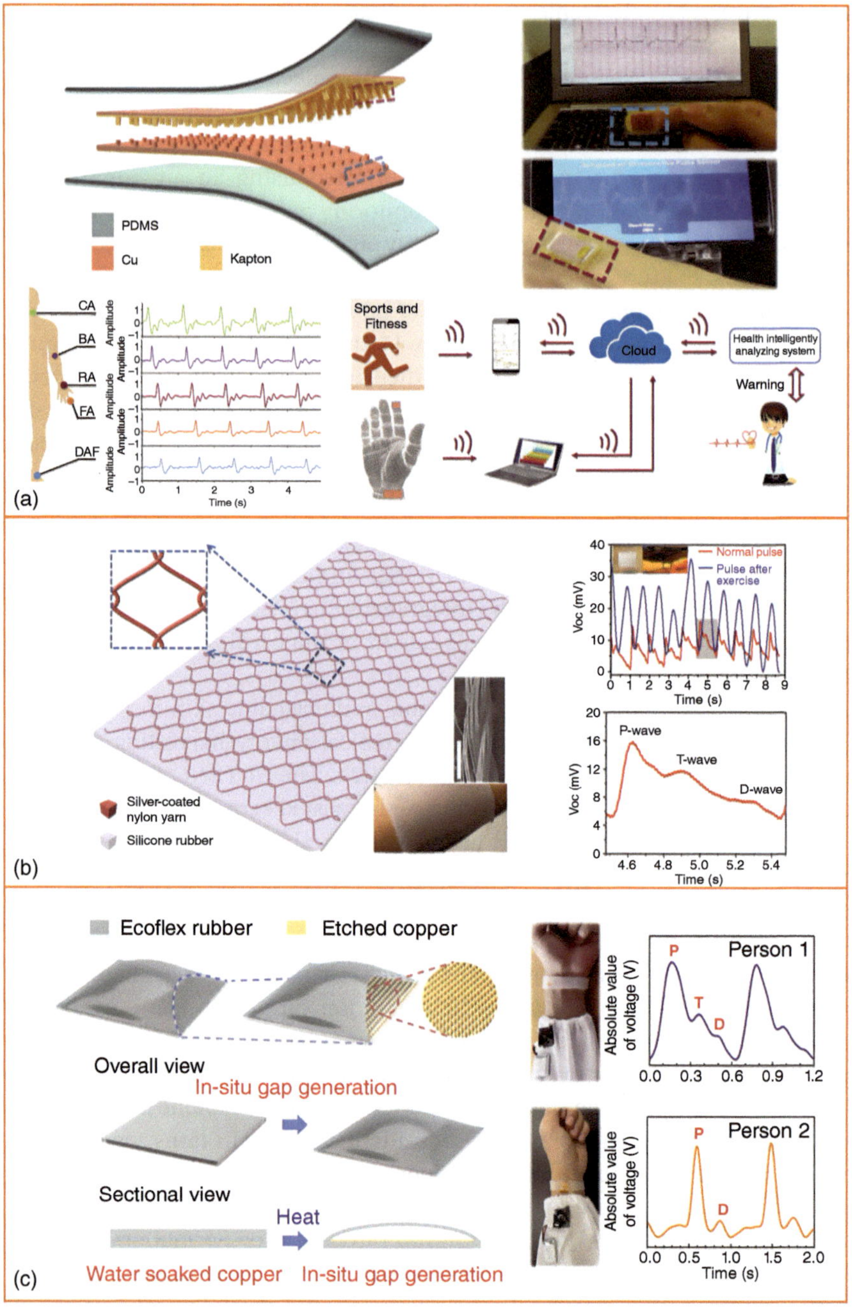

Figure 11.2 TENG-based biomechanical sensors in pulse monitoring. (a) Self-powered pulse sensor for anti-diastole of cardiovascular disease. Reproduced with permission of Ref. [31] © 2017, John Wiley & Sons. (b) A stretchable yarn-embedded TENG for biomechanical energy harvesting and multifunctional pressure sensing. Reproduced with permission of Ref. [32] © 2018, John Wiley & Sons. (c) No-spacer TENG for monitoring cardiovascular activities. (Reproduced with permission of Ref. [33], 2021 © Elsevier).

In 2021, Zhao et al. developed a novel eco-friendly in-situ method for generating gaps by vaporizing soaked distilled water, enabling the fabrication of a no-spacer triboelectric nanogenerator (NS-TENG), as depicted in Figure 11.2c. The structure of NS-TENG comprises a layer of Cu serving as an electrode and an outer rubber layer, with an internally situated gap. When the heart contracts, resulting in artery dilation, localized pressure is exerted on the NS-TENG, generating relative motion between the triboelectric layers and facilitating charge transfer at the interface. Characterized by uniform stress/strain distributions, excellent mechanical stability, and high sensitivity, the NS-TENG accurately detects arterial pulse signals in real-time. These signals are subsequently visualized on a mobile phone screen, accomplished by quantifying voltage changes triggered by arterial dilation and contraction. Beyond pulse monitoring, the NS-TENG is adept at detecting respiration and arrhythmia, thereby supplying vital insights for the monitoring and early detection of cardiovascular ailments.

11.2.2 Respiration

Breathing is a fundamental physiological activity, carries profound significance in daily human life and stands as a vital indicator for health monitoring. In addition to providing rich physiological insights and potential disease markers, respiration can provide a continuous and stable energy source for devices that utilize airflow or vibrations. Using TENG to formulate wearable respiratory sensors, human breathing patterns can be monitored in real-time in the most comfortable manner. A selection of TENG-based respiratory sensors include: an integrated sensor comprising a metallic yarn braided chest strap for detecting changes in respiration across different regions [34], an arch-structured penetration sensor integrated into a face mask to capture airborne vibrations caused by respiration [35], a textile sensor array-based monitoring system for assessing cardiovascular and respiratory syndromes [36], and a pneumatic elastic lamellae embedded sensor specifically designed to monitor the respiratory status of gas mask users [37], which are capable of monitoring the body's respiration and providing a timely response.

In 2016, Zhao et al. presented a meticulously designed textile-based triboelectric nanogenerator (T-TENG) with exceptional washing durability, as illustrated in Figure 11.3a. The T-TENG comprises 2-ply Cu-coated polyethylene terephthalate (Cu-PET) yarns and polyimide (PI)-coated Cu-PET (PI-Cu-PET) yarns, woven as the vertical (warp) and horizontal (weft) components, respectively. This single electrode T-TENG obviates the necessity for a secondary substrate to generate triboelectricity, with each yarn intersection at the crisscross considered as an individual TENG. The output hinges on the alteration of contact area at these crisscross junctions under mechanical force, culminating in a short-circuit current density of 15.5 mA/m^2. Integrating the T-TENG into a chest strap paves the way for real-time monitoring of human respiration parameters, including rate and depth. Moreover, the T-TENG's sensitivity to even minor body movement positions it as an exceedingly promising technology, suitable for a wide range of applications in sports, healthcare, and various other fields.

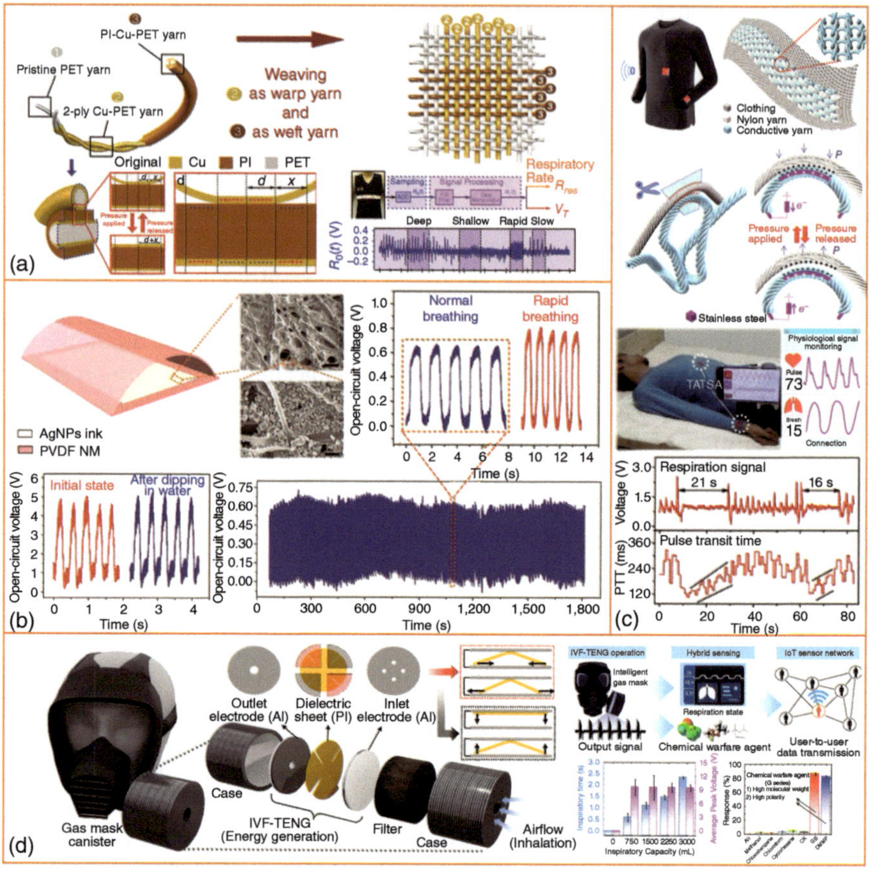

Figure 11.3 TENG-based biomechanical sensors in respiration monitoring. (a) Machine-washable textile TENG for effective human respiratory monitoring. Reproduced with permission of Ref. [34] © 2016, John Wiley & Sons. (b) Self-powered nanofiber-based screen-print biomechanical sensors for respiratory monitoring. Reproduced with permission of Ref. [35] © 2018, Springer Nature. (c) Machine-knitted washable sensor array textile for precise epidermal physiological signal monitoring. Reproduced with permission of Ref. [36] © 2020, American Association for the Advancement of Science – AAAS. (d) Inhalation-driven vertical flutter TENG as a gas mask. Reproduced with permission of Ref. [37], 2022 John Wiley & Sons.

In 2017, Cao et al. developed a self-powered nanofiber-based triboelectric sensor (SNTS) for respiratory monitoring, as depicted in Figure 11.3b. The SNTS fabrication involves electrospinning and screen-printing techniques, scaled for batch production. Comprising two layers, the SNTS features a polyvinylidene fluoride (PVDF) nanofiber membrane (NM) serving as the triboelectric material. Concurrently, Ag nanoparticles (AgNPs) are deposited on the PVDF NM substrate, fulfilling the roles of opposing triboelectric material and electrode. Operating under a single-electrode mode, equivalent positive and negative triboelectric charges are generated on the surfaces of PVDF NM and AgNPs, respectively. Facilitating the passage of gases, a conductive network comprised of PVDF NMs and AgNPs equips the SNTS with breathability, capable of accommodating airflow up to 6.16 mm/s. With integration into a mask, the SNTS becomes a respiratory monitoring tool. Notably, the SNTS demonstrates consistent voltage levels over a span of more than 30 minutes of usage, underscoring its favorable stability and moisture-proof characteristics.

In 2020, Fan et al. reported a triboelectric all-textile sensor array (TATSA) with exceptional pressure sensitivity and wearer comfort, as shown in Figure 11.3c. The TATSA's construction employs a full cardigan stitch to knit the conductive yarn and commercial nylon yarn. This stitch transforms the contact between the loop head of the conductive yarn and the adjacent tuck stitch head of the nylon yarn from a point contact to a surface contact. The transformation yields two contact surfaces between the interlocked yarns, both of which adapt in response to external forces, facilitating the transfer of triboelectric charge. A standout feature of the TATSA is sensitivity to subtle epidermal pressure, highlighted by its 7.84 mV/Pa sensitivity and a rapid 20 ms response time. Furthermore, this textile can withstand machine washing for more than 40 cycles. Seamlessly integrating into clothing, the TATSA can be positioned on various areas of a garment, enabling the transmission of measured physiological parameters to a mobile phone. This capability empowers the TATSA to capture and assess respiratory and pulse signals in patients with sleep apnea syndrome, exemplifying its outstanding performance in real-time and remote health monitoring scenarios.

In 2022, Heo et al. designed an inhalation-driven vertical flutter triboelectric nanogenerator (IVF-TENG) with enhanced current output, as illustrated in Figure 11.3d. The IVF-TENG comprises an inlet Al electrode, an aeroelastic PI sheet, and an outlet Al electrode. The two Al sheets serve as both triboelectric layers and electrodes, featuring four inlet holes at the inlet electrode and one outlet hole at the outlet electrode. This design allows for the generation of vertically applied airflow upon the inhalation of a gas mask user, prompting the pneumatic elastic sheet to undergo a vertical churning motion. Consequently, voltage and current outputs are generated upon contact with the Al sheets, resulting in maximum peak voltage and current outputs of 456 V and 288 mA, respectively. The IVF-TENG sensor exhibits the ability to detect various gases such as air, methanol, chlorobenzene, chloroform, and cyclohexane. This attribute makes it suitable for monitoring the wearer's respiratory status, while also holding potential in the identification of chemical warfare agents. Beyond its respiratory monitoring applications, the IVF-TENG holds promise in the fields of hybrid sensing and IoT sensor networks.

11.2.3 Joint Movement

The acquisition of biological information pertaining to human joint movement states holds utmost importance in healthcare, facilitating health assessment and early diagnosis, particularly for elderly individuals and those with injuries. This section highlights two TENG-based sensors that contribute to joint movement monitoring. The first sensor is a shape-adaptive design that incorporates rubber-covered conductive fluid electrodes, ensuring flexibility for monitoring joint movements [38]. The second example pertains to a joint monitoring sensor for robotic arms,

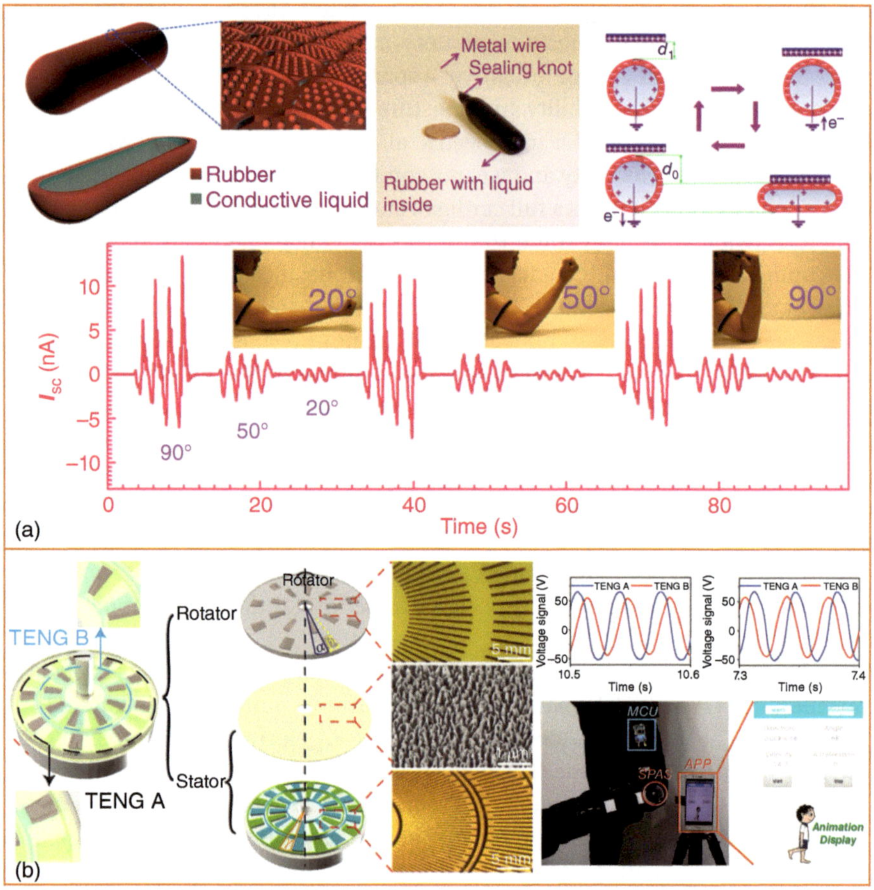

Figure 11.4 TENG-based biomechanical sensors in joint movement monitoring. (a) A highly shape-adaptive design based on conductive liquid for self-powered biomechanical monitoring. Reproduced with permission of Ref. [38] © 2016, American Association for the Advancement of Science – AAAS. (b) A self-powered angle sensor at nanoradian resolution for robotic arms and personalized medicare. Reproduced with permission of Ref. [39] © 2020, John Wiley & Sons.

which utilizes a rotary grating structure angle sensor to count pulses and detect phase difference [39]. Both of these examples are instrumental in tracking body joint motions, with particular relevance in areas such as athletic training, medical diagnosis and rehabilitation, as well as ensuring safety measures.

In 2016, Yi et al. presented a scalable approach for creating highly deformable and stretchable energy harvesters and self-powered sensors, as portrayed in Figure 11.4a. The shape-adaptive triboelectric nanogenerator (SATENG) consists of an inner conductive liquid electrode and an outer rubber layer. NaCl solution and water are selected as the electrode materials due to their renewability, low cost, and eco-friendliness. By connecting the conductive liquid electrode to the ground through a load, the SATENG operates in the single-electrode mode, effectively extracting energy from mechanical motion. Consequently, a bracelet-like SATENG can be fabricated and worn on the upper arm to detect the bending angle of the forearm. This is accomplished by tracking the change in current resulting from the temporary deformation of the biceps muscle during elbow flexion.

In 2020, Wang et al. reported a highly sensitive triboelectric self-powered angle sensor (SPAS) with the highest resolution of 2.03 nano-radians, as shown in Figure 11.4b. The SPAS comprises a rotator and a stator part, which integrates two rotary contact-sliding mode TENGs. The SPAS serves as an energy-efficient angle sensing system, generating sensing signals without the need for external power and offering a signal magnitude sufficient to obviate amplification. With its light weight, high sensitivity, and high signal-to-noise (SNR) ratio, the SPAS finds utility as a high-resolution and low-energy sensor in conjunction with medical scaffolds. This enables the creation of an active flexion measurement system capable of proactively recording the daily flexion and extension movements of a joint. Such a system is well-suited for personalized rehabilitation of individuals with arthritis or limited joint mobility.

11.3 TENG-Based Human–Machine Interface

The rapid evolution of AI and IoT has been further accelerated with the emergence of the fifth-generation mobile network (5G), which has intensified the demand for adept control of intelligent external devices. Consequently, intuitive and efficient human–machine interactions are crucial, especially in enabling wearable devices as lightweight HMI [40]. The integration of TENG-based wearable sensors has significantly expanded the range of HMI applications, exhibiting considerable potential across various domains. This section primarily concentrates on TENG-driven HMIs that facilitate the control of external devices through actions like eye movements, voice/auditory, gestures, touch/tactile interactions.

11.3.1 Eye Movement

To enhance the quality of life for patients and streamline their daily routines, the development of eye movement-based HMIs that do not rely on noticeable speech or hand/arm actions is essential. The inherent capability of TENG to detect micromotion, even pulses, grants it advantageously for HMIs. Thus, three examples of eye movement-based HMIs are presented here, a micromotion sensor utilizing a tactile TENG for controlling electrical appliances and communication [41], an eye-motion detection system based on triboelectric interaction and near-field electrostatics [42], and an active eye tracking (AET) system relying on the electrostatic induction effect [43], which implement blink control electronics while monitoring eye movements.

In 2017, Pu et al. introduced a novel micromotion sensor based on TENG (msTENG) that can be affixed to eyeglasses. This technology facilitates smart home control systems and hands-free wireless typing, as illustrated in Figure 11.5a. Operating in the single-electrode mode, the sensor boasts a multilayer membrane structure. The supporting substrate comprises a thin tadpole-shaped layer of PET.

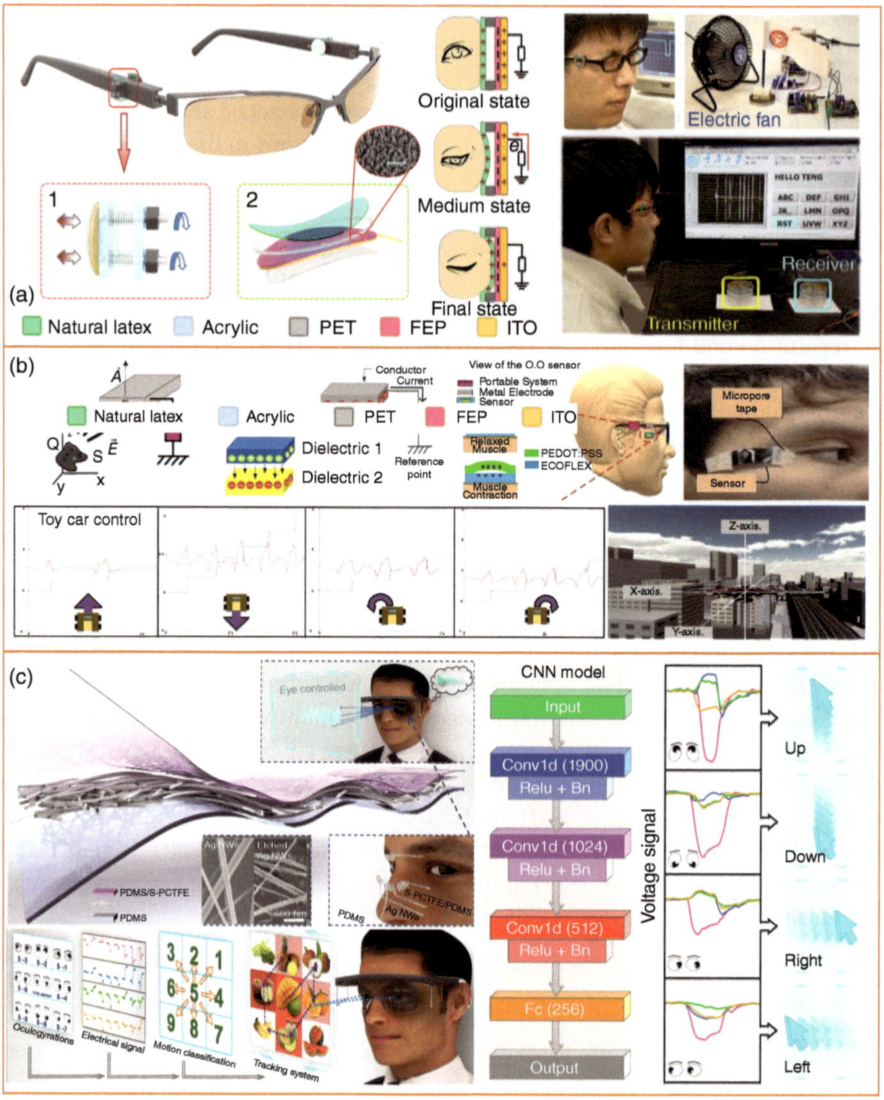

Figure 11.5 TENG-based HMIs in eye movement monitoring. (a) Eye motion triggered a self-powered mechnosensational communication system using TENG. Reproduced with permission of Ref. [41] © 2017, American Association for the Advancement of Science – AAAS. (b) Self-powered eye motion sensor for wearable assistive technologies. Reproduced with permission of Ref. [42] © 2020, Elsevier. (c) Eye tracking and eye expression decoding based on electrostatic interface. [43] / Springer Nature / CC-BY-4.0.

The triboelectric layer consists of natural latex and fluorinated ethylene propylene (FEP) film, with the natural latex positioned above, in direct contact with the skin. A slender acrylic ring functions as a spacer, separating the two triboelectric layers. For the back electrode, indium tin oxide is laminated onto the PET substrate. This flexible, transparent sensor excels in detecting blinking movements, yielding strong signals (~750 mV) and surpassing conventional electrooculography methods. Moreover, the msTENG-equipped eyeglasses demonstrate their proficiency in controlling home appliances with simple signal processing circuits. By integrating a wireless module, the eyeglasses enable the development of a hands-free virtual keyboard typing system. These cost-effective msTENGs hold promise for mass production and cater specifically to individuals reliant on ambient intelligence.

In 2020, Anaya et al. conducted a study on a novel configuration termed the nonattached electrode-dielectric triboelectric sensor (NEDTS). This sensor is designed for monitoring both voluntary and involuntary blinking, as shown in Figure 11.5b. When the triboelectric and dielectric layers of the NEDTS move in relation to each other, an electrostatic induction effect triggers the generation of charge distribution and potential difference within the electrode layer. Experimental findings demonstrate the NEDTS's capability to detect voltage output corresponding to muscle movement, with a peak voltage range of 300–600 mV. Through tracking the motion of eyelid muscles, NEDTS can effectively steer cursor movement and execute left-click actions. Additionally, the NEDTS exhibits potential for utilization in tasks such as remote car control, drone control, and driver fatigue monitoring. The integration of the NEDTS in specialized HMIs aims to provide support for individuals with disabilities in their daily lives.

In 2023, Shi et al. proposed an advanced and versatile electrostatic sensing interface for the implementation of an AET system, as illustrated in Figure 11.5c. This interface is transparent, flexible, and highly durable, allowing for ophthalmic detection with an angular resolution of 5°. The electrostatic sensing interface comprises a precharged composite dielectric bilayer and a stretchable back electrode made of Ag NWs, adhered to a PDMS substrate. Through noncontact electrostatic interactions with the human skin around the eyes, the precharged interface detects eye movements and translates them into corresponding signals. The design of dielectric bilayer and electrode enhances charge storage capacity, culminating in a surface density of 1671.10 mC/m^2. After 1000 cycles of noncontact operation, the system retains 96.91% of the initial charge. Leveraging deep learning techniques, the AET system showcases its potential in real-time decoding of eye movements and gaze playback, thereby enabling analysis of user preferences in business settings.

11.3.2 Voice/Auditory

Verbal communication stands as the predominant means for conveying information among humans. The intelligent auditory and voice recognition systems based on self-powered auditory sensors contribute significantly to enhancing the convenience and sensitivity of HMIs. Furthermore, self-powered auditory sensors exhibit promising prospects in various fields, including smart homes, unmanned surveillance, and the IoT. In this segment, two exemplary TENG-based acoustic sensors are outlined for a comprehensive understanding. First, the circular single-channel triboelectric auditory sensor (TAS) is discussed, catering to applications in robotic systems and human hearing aids [44]. Additionally, the introduction of a near-zero static power TAS, specifically designed for sound recognition systems [45], holds the capacity to elevate speech communication in HMIs while offering enhanced convenience to individuals with hearing impairments.

In 2018, Shi et al. presented a self-powered auditory sensor called TAS, as depicted in Figure 11.6a. The functionality of the sensor relies on the vibration of Kapton and its contact separation with the upper FEP, resulting from the air vibrations induced by external sounds. By implementing a boundary structure in the form of a ring or sector, along with system optimization, a broad frequency response ranging from 100 to 5000 Hz is achieved. This feature effectively emulates the human auditory system, thus facilitating accurate speech recognition and music recording in intelligent HMIs. Furthermore, the resonant frequency of the TAS can be adjusted through geometric design modifications of the inner boundary structure, enabling the amplification of specific sound waves. This attribute further positions it as an ideal solution for amplifying sound in hearing aids within a specific frequency range. With its outstanding qualities including high sensitivity, wide response range, and self-powered operation, the TAS paves the way for efficient and precise interaction between intelligent robots and humans, while also providing a novel approach in the field of hearing aid technology.

In 2023, Wang et al. introduced a near-zero quiescent power sound wake-up and identification system (NP-SWS). This system utilizes the potential of a sound triboelectric nanogenerator (S-TENG) to realize the recognition of various sound sources, such as drilling holes, kids playing, barking dogs, and street music, as shown in Figure 11.6b. The NP-SWS consists of an S-TENG, a wake-up circuit, a microcontroller unit (MCU) with a wireless transmitter, a microelectromechanical system microphone, and a computer with a receiver. Functioning in a vertical contact-separation mode, the S-TENG collects energy from sound-induced vibration. Additionally, both the open-circuit voltage and short-circuit current of the S-TENG increase proportionally with the intensity of the captured sound. Through the integration of deep learning techniques, specifically convolutional neural networks, the NP-SWS achieves an 98% accuracy in recognizing sound sources.

11.3.3 Gesture

Utilizing the bending and contact motions of finger joints presents a straightforward and intuitive approach to achieving machine control. In order to attain precise and real-time control of machinery, especially when merging reality and the virtual world, it becomes necessary to ensure the completeness, accuracy, and quantification of motion-related information. Therefore, it is worthwhile to explore in depth the use of TENG to create gesture sensors and use them to control external electronic devices. This section delves into three TENG-based gesture sensing and interaction systems. These include a rotary grating structure sensor that quantifies joints through positive/negative impulse counts and enables transient control of a robotic hand [46]. Another system integrates nanophotonic readout technology for sensing intermediate states [47]. Lastly, a stretchable sensor array designed for a wearable sign language interpreter system [48], enables precise machine control through gestures and interaction. The amalgamation of these technologies facilitates

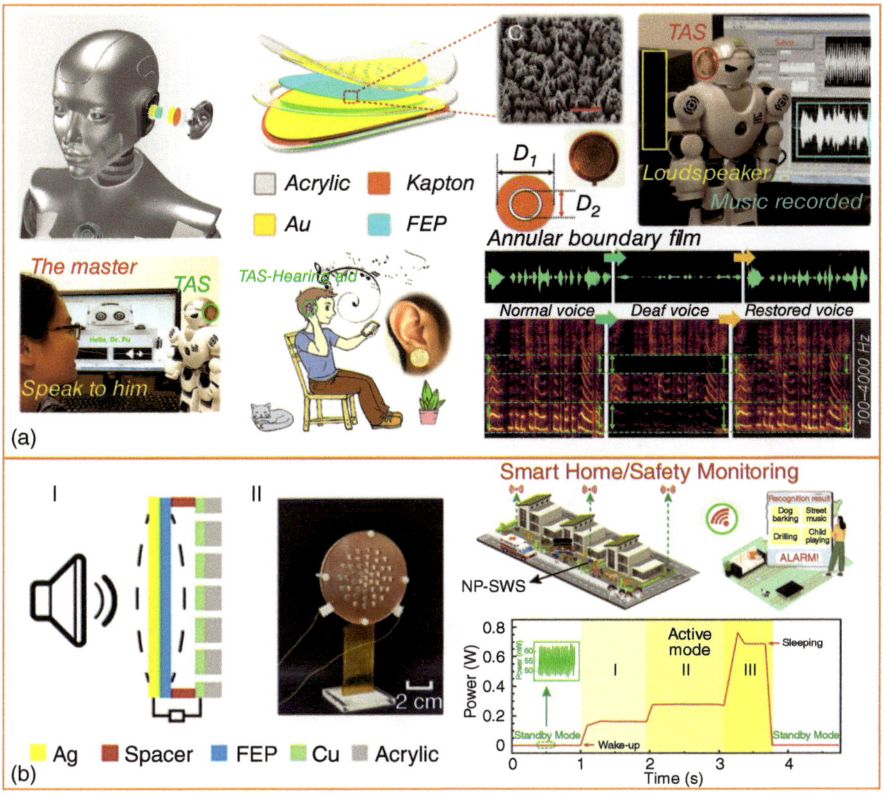

Figure 11.6 TENG-based HMIs in voice/auditory recognition. (a) A self-powered triboelectric auditory sensor for social robotics and hearing aids. Reproduced with permission of Ref. [44] © 2018, American Association for the Advancement of Science – AAAS. (b) A near-zero quiescent power sound wake-up and identification system. Reproduced with permission of Ref. [45] © 2023, American Chemical Society.

effective interaction with machines through gestures, ushering in new dimensions of control and manipulation.

In 2018, Pu et al. developed a joint movement triboelectric quantization sensor (jmTQS) to create a synchronous control system for a robotic hand, as illustrated in Figure 11.7a. The jmTQS comprises an acrylic rectangular chamber and a slider. The slider's movement, induced by finger flexion or extension, triggers relative displacement between the slider and the two interdigitated electrodes. This motion leads to periodic separation from one electrode and contact with the other electrode. Through the combined effect of contact initiation and in-plane displacement-induced charge transfer, alternating electrical signals can be detected between the interdigitated electrodes. The number of pulses in a given time interval and the positive or negative peak of the signal indicate the extent of finger flexion or extension. These signals effectively determine the angular position of the finger joint, establishing an intuitive correspondence between human hand movement and robotic hand control. Additionally, the jmTQS finds applications in

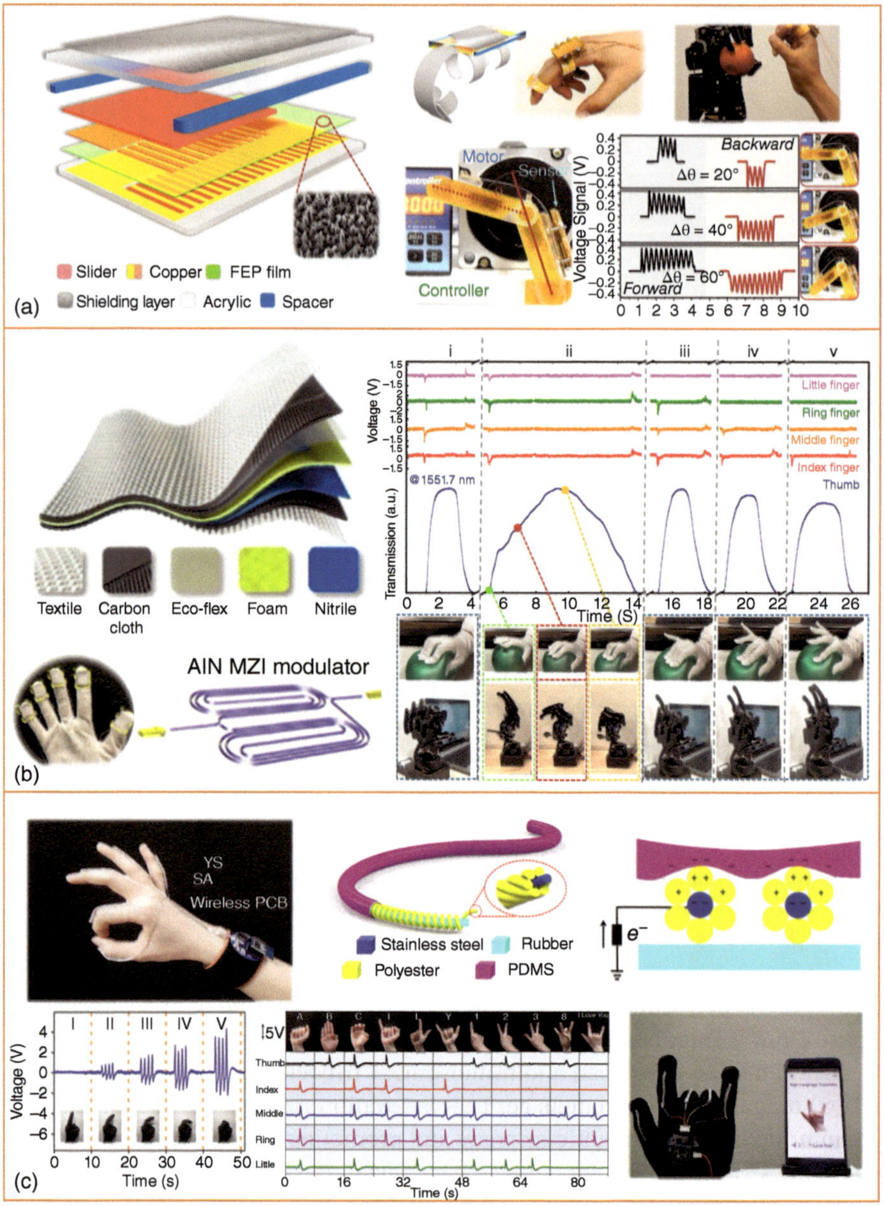

Figure 11.7 TENG-based HMIs in gesture recognition. (a) Rotation sensing and gesture control of a robot joint via triboelectric quantization sensor. Reproduced with permission of Ref. [46] © 2018, Elsevier. (b) A wearable triboelectric–human–machine interface using robust nanophotonic readout. Reproduced with permission of Ref. [47] © 2020, American Chemical Society. (c) A sign-to-speech translator using machine-learning-assisted stretchable sensor arrays. Reproduced with permission of Ref. [48] © 2020, Springer Nature.

virtual reality, augmented reality, and game control, enhancing the immersion and interactivity of the user experience.

In 2020, Dong et al. introduced a triboelectric HMI (THMI) utilizing nanophotonic aluminum nitride (AlN) modulators for readout. These modulators suppress electrical state transitions, resulting in a stable, information-lossless, and real-time HMI, as shown in Figure 11.7b. Two wearable THMI–nanophotonics systems are available in this study. The first system integrates a spacer THMI (STHMI) with a short AlN Mach-Zehnder interferometer (MZI) modulator, obtaining a linear force sensitivity of 0.659 mV/N within the range of 35–60 N. The second system features a textile THMI (T-THMI) coupled with a longer AlN MZI modulator. The adaptability of THMI design enables researchers to achieve diverse linear force sensitivities across various ranges by optimizing the THMI and corresponding nanophotonic readout circuit. Through the creation of a smart glove utilizing THMI-nanophotonic technology, the researchers effectively demonstrate continuous real-time robot control and virtual/augmented reality interaction. The proposed THMI-nanophotonic technology system offers a versatile solution for HMI applications in wearable devices.

In 2020, Zhou et al. reported a wearable sign-to-speech translation system aimed at translating American Sign Language hand gestures into spoken language using machine learning techniques, as depicted in Figure 11.7c. The wearable sign language interpretation system comprises yarn-based stretchable sensor arrays (YSSA) and a wireless printed circuit board. The central component of the sensing unit incorporates a conductive yarn wound around a rubber microfiber, all enveloped within a PDMS sleeve. As the system undergoes cycles of stretching and releasing, different levels of axial tensile strain impact in both the PDMS sleeve and the coiled conductive yarn, resulting in a continuous change in the contact area between these elements. This leads to the accumulation of electrical potential due to the electron affinity difference between the materials. The YSSA can detect positional and shape changes in hand gestures and convert them into electrical signals. By converting these gestures into spoken language, the system facilitates communication for individuals with hearing impairments. Additionally, the integration potential of this technology into gloves and garments enhances its practicality, with the capability to significantly enhance effective communication between sign language users and those who do not use sign language. This innovation marks a significant step forward in bridging the communication gap and promoting inclusivity among diverse communities.

11.3.4 Touch/Tactile

Tactile sensors with high performance find applications in various fields such as AI, prosthetics, health monitoring, and HMIs. As the evolution of HMIs continues, the demand for flexible, sensitive, efficient, and multifunctional tactile sensors is on the rise.

This section provides a review of four TENG-based tactile sensor applied in HMIs. These include a transparent and self-cleaning pressure sensor that utilizes an ionic hydrogel for coded communication [49], a user-interactive electronic skin that employs a triboelectric-optical model for touch trajectory mapping [50], a flexible touchpad with segmentation units enabling haptic XY localization [51], and a tactile sensor that integrates triboelectric-optical sensing arrays and a smart finger for the precise recognition of material type and roughness [52], which are flexible, sensitive, efficient and versatile enough to serve as tactile HMIs.

In 2018, Lee et al. introduced a self-cleaning, transparent, and attachable ionic hydrogel-based ionic communicators (STAICs), capable of wireless communication, as shown in Figure 11.8a. These STAICs are crafted using polyacrylamide hydrogels with lithium chloride as the primary material. Operating in the single-electrode mode, STAICs maintain stable light transmittance and stretchability, making them ideal for thimble-type, wireless, real-time human-computer communication, efficiently detecting gentle finger touches. Self-cleaning property is achieved through surface treatment, which significantly improved transmittance (40.6% increase) and electrical output voltage (19.4% increase) during the water jet step. This performance was notably superior to untreated STAICs. Consequently, these STAICs have the potential to advance self-powered, transparent, and connectable HMIs for wireless sensor networks.

In 2020, Zhao et al. developed a self-powered, user-interactive electronic skin (SUE skin) designed to convert touch stimuli into electrical signals and visible light, without relying on an external power source, as illustrated in Figure 11.8b. When a finger makes contact with the SUE skin, it generates electrical signals that can be processed by an MCU as control commands for various consumer electronic devices. Simultaneously, the SUE skin emits light signals that are directly observable by the human eye. The unique attributes of this technology stem from the accumulation of positive and negative triboelectric charges on the user's skin and the dielectric binder, respectively, during contact. As the surfaces separate, the dielectric binder's top layer becomes negatively charged, thus engendering an induced current in the external circuit. Moreover, the four-channel SUE skin can visualize touch lighting and track mapping. The potential applications of SUE skin include wearable devices, military applications, artificial prostheses, and intelligent robots. As a high-performance and cost-effective solution, the SUE skin's practicality extends across a wide spectrum of applications.

In 2020, Pu et al. presented a flexible 3D triboelectric touch panel (3D-TTP) capable of haptic XY localization, as depicted in Figure 11.8c. This 3D-TTP consists of two essential components: the top triboelectric layer for XY localization, and the bottom triboelectric layer for pressure sensing. To achieve efficient haptic mapping, the researchers implement a matrix scheme that reduces the number of channels and enables a high-resolution wiring layout. Furthermore, an XY complementary subdivision pattern (XYCSP) based on two orthogonal electrodes is arranged to minimize crosstalk through the reduction of overlap area between the orthogonal electrode layers. This is attributed to the enhancement of object localization accuracy irrespective of object size or contact points. The integration of the pressure sensing layer at the bottom also enables the establishment of a touch code, making it applicable for anti-peep authentication purposes.

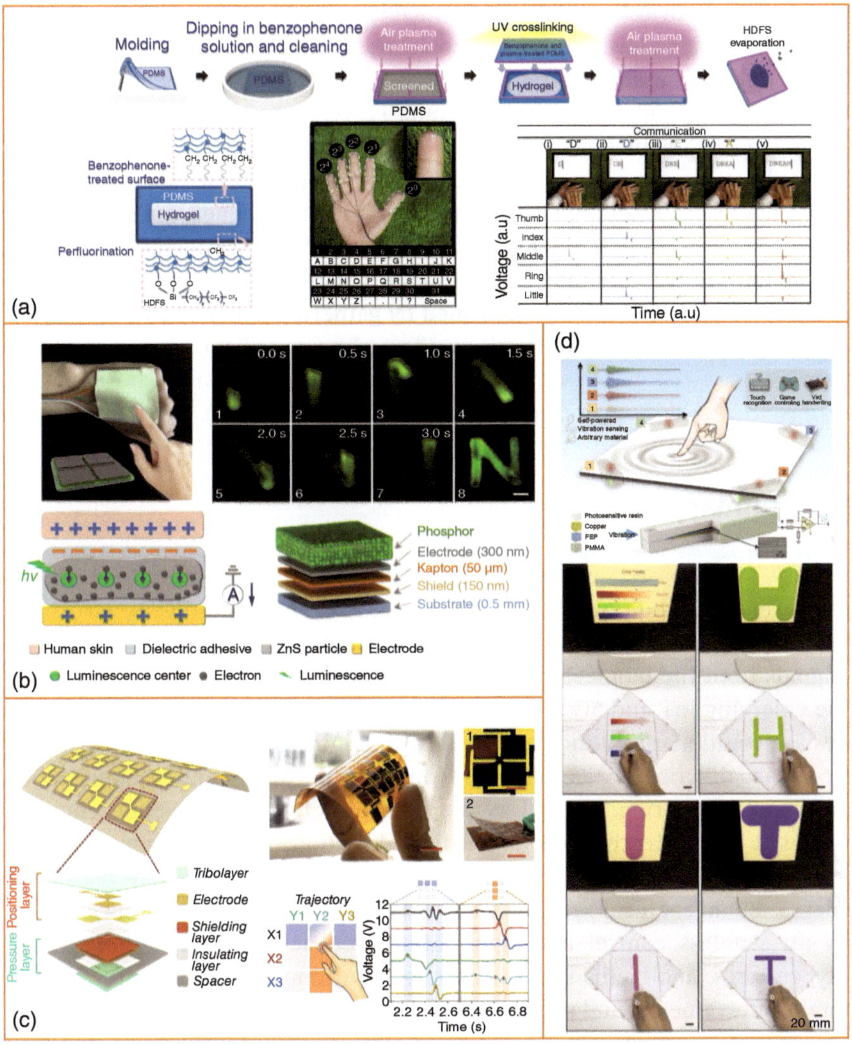

Figure 11.8 TENG-based HMIs in touch/tactile recognition. (a) Ionic communicators based on self-cleanable TENGs. [49] / Springer Nature / CC-BY-4.0. (b) Self-powered user-interactive electronic skin for programmable touch operation platform. Reproduced with permission of Ref. [50] © 2020, American Association for the Advancement of Science – AAAS. (c) Flexible triboelectric 3D touchpad for effective XY positioning and pressure sensing. Reproduced with permission of Ref. [51] © 2020, Elsevier. (d) Artificial tactile perception smart finger for material identification. Reproduced with permission of Ref. [52], 2024 © Elsevier.

In 2022, Qu et al. reported a smart finger with the ability to accurately identify the type of material and its roughness. This feat is realized by integrating triboelectric sensing and machine learning techniques, as depicted in Figure 11.8d. The smart finger comprises the sensor array, data acquisition and transmission module, and display module. The construction of the sensor array further not only facilitates the elimination of environmental interference but also contributes to more accurate and stable data acquisition. The triboelectric layer of each sensor consists of PI, PET, polystyrene, and polytetrafluoroethylene, complemented by an aluminum film functioning as the electrode layer, forming TENG in the single-electrode mode. Capitalizing on the distinct electron-capturing or releasing capacities of materials, TENG-based sensor arrays generate corresponding output signals when in contact with different objects. By employing machine learning techniques, the smart finger discerns key features such as maxima, minima, peak-to-valley intervals, number of zeros, number of inflection points, and absolute squared values. Classification and identification of materials are accomplished by gathering data from all four channels, resulting in a material identification and roughness prediction accuracy of 96% or higher.

11.4 Summary

In summary, TENG-based sensing technology offers a range of advantages, including self-powered capabilities, a wide selection of materials, and diverse structural designs. This innovation enables the harvesting of energy and generation of electrical signals from the biomechanical movements of the human body. Combined with the emerging IoT and 5G communication technologies, TENG-based sensing provides a revolutionary solution for medical monitoring systems and HMIs. This chapter aims to describe TENG wearable biomechanical sensor and its applications in biomedical monitoring, as well as the latest developments in biomechanical sensing-based HMIs. In the field of medical detection, wearable biomechanical sensors are primarily used for monitoring pulse, respiration, and various body joint movements. Biomechanical sensing-based HMIs employ wearable biomechanical sensors for wireless control and information feedback using eye movements, auditory, gestures, and tactile signals. Through continuous material optimization, structural improvements, and application innovations, biomechanical sensors play a pivotal role in active sensing, real-time information feedback, and intelligent HMIs. As a result, TENG-based devices have made substantial strides in enhancing human well-being and enabling sophisticated interactions between humans and computers. This trajectory is steering us towards an era characterized by heightened activity and intelligence in our technological landscape.

References

1 Wang, X., Liu, Z., and Zhang, T. (2017). Flexible sensing electronics for wearable/attachable health monitoring. *Small* 13: 1602790.

2 Liu, Y., Pharr, M., and Salvatore, G.A. (2017). Lab-on-skin: a review of flexible and stretchable electronics for wearable health monitoring. *ACS Nano* 11: 9614–9635.

3 Liu, L., Stroulia, E., Nikolaidis, I. et al. (2016). Smart homes and home health monitoring technologies for older adults: a systematic review. *Int. J. Med. Inform.* 91: 44–59.

4 Wang, Z.L., Zhu, G., Yang, Y. et al. (2012). Progress in nanogenerators for portable electronics. *Mater Today* 15: 532–543.

5 Ates, H.C., Yetisen, A.K., Der, F.G.U., and Dincer, C. (2021). Wearable devices for the detection of COVID-19. *Nat. Electron.* 4: 13–14.

6 Iqbal, S.M., Mahgoub, I., Du, E. et al. (2021). Advances in healthcare wearable devices. *NPJ Flex. Electron.* 5: 9.

7 Jin, H., Abu-Raya, Y.S., and Haick, H. (2017). Advanced materials for health monitoring with skin-based wearable devices. *Adv. Healthc. Mater.* 6: 1700024.

8 Takei, K., Honda, W., Harada, S. et al. (2015). Toward flexible and wearable human-interactive health-monitoring devices. *Adv. Healthc. Mater.* 4: 487–500.

9 Pu, X., An, S., Tang, Q. et al. (2021). Wearable triboelectric sensors for biomedical monitoring and human-machine interface. *iScience* 24: 102027.

10 Wang, M., Wang, T., Luo, Y. et al. (2021). Fusing stretchable sensing technology with machine learning for human–machine interfaces. *Adv. Funct. Mater.* 31: 2008807.

11 Jeong, J., Yeo, W., Akhtar, A. et al. (2013). Materials and optimized designs for human-machine interfaces via epidermal electronics. *Adv. Mater.* 25: 6839–6846.

12 Sun, Z., Zhu, M., and Lee, C. (2021). Progress in the triboelectric human–machine interfaces (HMIs)-moving from smart gloves to AI/haptic enabled HMI in the 5G/IoT era. *Nanoenergy Adv.* 1: 81–120.

13 Wang, L., Liu, W., Yan, Z. et al. (2021). Stretchable and shape-adaptable triboelectric nanogenerator based on biocompatible liquid electrolyte for biomechanical energy harvesting and wearable human–machine interaction. *Adv. Funct. Mater.* 31: 2007221.

14 Ding, W., Wang, A.C., Wu, C. et al. (2019). Human–machine interfacing enabled by triboelectric nanogenerators and tribotronics. *Adv. Mater. Technol.* 4: 1800487.

15 Lan, B., Wu, F., Cheng, Y. et al. (2022). Scalable, stretchable and washable triboelectric fibers for self-powering human-machine interaction and cardiopulmonary resuscitation training. *Nano Energy* 102: 107737.

16 Wang, Z.L. (2020). On the first principle theory of nanogenerators from Maxwell's equations. *Nano Energy* 68: 104272.

17 Choi, D., Lee, Y., Lin, Z. et al. (2023). Recent advances in triboelectric nanogenerators: from technological progress to commercial applications. *ACS Nano* 17: 11087–11219.

18 Shi, Q., He, T., and Lee, C. (2019). More than energy harvesting–combining triboelectric nanogenerator and flexible electronics technology for enabling novel micro-/nano-systems. *Nano Energy* 57: 851–871.

19 Wu, C., Wang, A.C., Ding, W. et al. (2019). Triboelectric nanogenerator: a foundation of the energy for the new era. *Adv. Energy Mater.* 9: 1802906.

20 Lin, Z., Yang, J., Li, X. et al. (2018). Large-scale and washable smart textiles based on triboelectric nanogenerator arrays for self-powered sleeping monitoring. *Adv. Funct. Mater.* 28: 1704112.

21 Wu, H., Su, Z., Shi, M. et al. (2018). Self-powered noncontact electronic skin for motion sensing. *Adv. Funct. Mater.* 28: 1704641.

22 Yang, Y., Guo, X., Zhu, M. et al. (2023). Triboelectric nanogenerator enabled wearable sensors and electronics for sustainable internet of things integrated green earth. *Adv. Energy Mater.* 13: 2203040.

23 Xiang, S., Liu, D., Jiang, C. et al. (2021). Liquid-metal-based dynamic thermoregulating and self-powered electronic skin. *Adv. Funct. Mater.* 31: 2100940.

24 Lu, Q., Chen, H., Zeng, Y. et al. (2022). Intelligent facemask based on triboelectric nanogenerator for respiratory monitoring. *Nano Energy* 91: 106612.

25 Cui, X., Zhao, T., Yang, S. et al. (2020). A spongy electrode-brush-structured dual-mode triboelectric nanogenerator for harvesting mechanical energy and self-powered trajectory tracking. *Nano Energy* 78: 105381.

26 Qiu, Y., Yang, D., Li, B. et al. (2018). Wearable triboelectric nanogenerators based on hybridized triboelectric modes for harvesting mechanical energy. *RSC Adv.* 8: 26243–26250.

27 Wang, H., Han, M., Song, Y., and Zhang, H. (2021). Design, manufacturing and applications of wearable triboelectric nanogenerators. *Nano Energy* 81: 105627.

28 Yang, Y., Zhang, H., Lin, Z. et al. (2013). Human skin based triboelectric nanogenerators for harvesting biomechanical energy and as self-powered active tactile sensor system. *ACS Nano* 7: 9213–9222.

29 Haghayegh, M., Cao, R., Zabihi, F. et al. (2022). Recent advances in stretchable, wearable and bio-compatible triboelectric nanogenerators. *J. Mater. Chem. C* 10: 11439–11471.

30 Heikenfeld, J., Jajack, A., Rogers, J. et al. (2018). Wearable sensors: modalities, challenges, and prospects. *Lab. Chip* 18: 217–248.

31 Ouyang, H., Tian, J., Sun, G. et al. (2017). Self-powered pulse sensor for antidiastole of cardiovascular disease. *Adv. Mater.* 29: 1703456.

32 Dong, K., Wu, Z., Deng, J. et al. (2018). A stretchable yarn embedded triboelectric nanogenerator as electronic skin for biomechanical energy harvesting and multifunctional pressure sensing. *Adv. Mater.* 30: 1804944.

33 Zhao, D., Zhuo, J., Chen, Z. et al. (2021). Eco-friendly in-situ gap generation of no-spacer triboelectric nanogenerator for monitoring cardiovascular activities. *Nano Energy* 90: 106580.

34 Zhao, Z., Yan, C., Liu, Z. et al. (2016). Machine-washable textile triboelectric nanogenerators for effective human respiratory monitoring through loom weaving of metallic yarns. *Adv. Mater.* 28: 10267–10274.

35 Cao, R., Wang, J., Zhao, S. et al. (2018). Self-powered nanofiber-based screen-print triboelectric sensors for respiratory monitoring. *Nano Res.* 11: 3771–3779.

36 Fan, W., He, Q., Meng, K. et al. (2020). Machine-knitted washable sensor array textile for precise epidermal physiological signal monitoring. *Sci. Adv.* 6: eaay2840.

37 Heo, D., Song, M., Chung, S. et al. (2022). Inhalation-driven vertical flutter triboelectric nanogenerator with amplified output as a gas-mask-integrated self-powered multifunctional system. *Adv. Energy Mater.* 12: 2201001.

38 Yi, F., Wang, X., Niu, S. et al. (2016). A highly shape-adaptive, stretchable design based on conductive liquid for energy harvesting and self-powered biomechanical monitoring. *Sci. Adv.* 2: e1501624.

39 Wang, Z., An, J., Nie, J. et al. (2020). A self-powered angle sensor at nanoradian-resolution for robotic arms and personalized medicare. *Adv. Mater.* 32: 2001466.

40 Ganesh, R.S., Yoon, H., and Kim, S. (2020). Recent trends of biocompatible triboelectric nanogenerators toward self-powered e-skin. *EcoMat* 2: e12065.

41 Pu, X., Guo, H., Chen, J. et al. (2017). Eye motion triggered self-powered mechnosensational communication system using triboelectric nanogenerator. *Sci. Adv.* 3: e1700694.

42 Anaya, D.V., He, T., Lee, C., and Yuce, M.R. (2020). Self-powered eye motion sensor based on triboelectric interaction and near-field electrostatic induction for wearable assistive technologies. *Nano Energy* 72: 104675.

43 Shi, Y., Yang, P., Lei, R. et al. (2023). Eye tracking and eye expression decoding based on transparent, flexible and ultra-persistent electrostatic interface. *Nat. Commun.* 14: 3315.

44 Guo, H., Pu, X., Chen, J. et al. (2018). A highly sensitive, self-powered triboelectric auditory sensor for social robotics and hearing aids. *Sci. Robot.* 3: eaat2516.

45 Wang, Z., Zhao, J., Gong, L. et al. (2023). Near-zero quiescent power sound wake-up and identification system based on a triboelectric nanogenerator. *ACS Appl. Mater. Interfaces* 15: 23328–23336.

46 Pu, X., Guo, H., Tang, Q. et al. (2018). Rotation sensing and gesture control of a robot joint via triboelectric quantization sensor. *Nano Energy* 54: 453–460.

47 Dong, B., Yang, Y., Shi, Q. et al. (2020). Wearable triboelectric–human–machine interface (THMI) using robust nanophotonic readout. *ACS Nano* 14: 8915–8930.

48 Zhou, Z., Chen, K., Li, X. et al. (2020). Sign-to-speech translation using machine-learning-assisted stretchable sensor arrays. *Nat. Electron.* 3: 571–578.

49 Lee, Y., Cha, S.H., Kim, Y. et al. (2018). Transparent and attachable ionic communicators based on self-cleanable triboelectric nanogenerators. *Nat. Commun.* 9: 1804.

50 Zhao, X., Zhang, Z., Liao, Q. et al. (2020). Self-powered user-interactive electronic skin for programmable touch operation platform. *Sci. Adv.* 6: eaba4294.

51 Pu, X., Tang, Q., Chen, W. et al. (2020). Flexible triboelectric 3D touch pad with unit subdivision structure for effective XY positioning and pressure sensing. *Nano Energy* 76: 105047.

52 Qu, X., Liu, Z., Tan, P. et al. (2022). Artificial tactile perception smart finger for material identification based on triboelectric sensing. *Sci. Adv.* 8: eabq2521.

12

TENG as the High-Voltage Source

12.1 Introduction

Triboelectric nanogenerator (TENG), a fundamental energy technology invented in 2012, has quickly become popular in the field of energy harvesting and self-powered devices [1]. TENG can convert various mechanical energies (vibrations, water waves, wind, and human motions) into electricity [2] using contact electrification and electrostatic induction [3]. Its attractive advantages lie in its ease of manufacturing [4], dependability, cost-effectiveness, and the abundance of low-frequency resources available. [5] TENG offers sustainable power supply through four operation modes [6]: vertical contact-separation mode [7], lateral sliding mode [8], single-electrode mode [9], and freestanding triboelectric-layer mode. [10] Optimization efforts have been made on TENG's materials and interfaces to increase the material factor of TENG, leading to improved performance with area power density exceeding 700 W/m^2 [11, 12], volume power density over 15 MW/m^3 [13], instantaneous conversion efficiency over 70% [14], and total efficiency up to 85% [15]. The development of TENG has impacted industries like wind [16] and ocean energy [17–19] harvesting, self-powered smart systems [20, 21], portable energy packages [22], human–machine interfacing (HMI) [23], and highly sensitive tactile sensors [24, 25]. TENG serves as an advanced and durable energy source, revolutionizing various sectors with efficient and sustainable energy utilization.

TENG has shown significant development in terms of applications and industrialization, particularly in the field of high-voltage sources [26]. With its output characteristic of high open-circuit voltage (V_{oc}) reaching a few thousand volts and low short-circuit current (I_{sc}) on a microampere scale [27], TENG is suitable for controlling capacitive devices with high internal resistance [28]. This capability allows TENG to directly drive electrically responsive materials and devices like dielectric elastomers [29], ferroelectric polymers [30], and field emission devices [31], enabling the creation of self-powered smart systems with various exciting capabilities. By selecting specific operational parameters, the output voltage and transmitted charges of TENG can be accurately controlled, making it parameterizable for managing and controlling these systems. Additionally, TENG acts as an interface between humans and machines, translating human gestures into

Triboelectric Nanogenerators: Technology, Applications, and Commercialization, First Edition.
Yunlong Zi, Hengyu Guo, Jie Wang, Chi Zhang, Xiangyu Chen, and Qing Zhao.

real-time electrical signals. This feature enables the use of both static and dynamic mechanical stimulation as trigger or control signals for many electromechanical systems [32]. Furthermore, TENG's restricted tribocharges are swiftly consumed when electrical breakdown is imminent, providing a sensitive self-protection mechanism that lowers the risk of electrical damage for high-voltage devices. These characteristics make TENG an efficient high-voltage source with excellent controllability and reliable safety for smart systems and devices. The advancements of TENG in high-voltage applications have paved the way for its integration into a wide range of smart systems, offering precise control, safety, and enhanced functionality [33].

Smart materials and electrostatic devices can be combined with TENG to create a composite system, achieving a self-powered smart system with various practical functions [34]. This chapter focuses on the latest advancements in self-powered systems by combining these elements and determining key factors for successful integration with TENG. Figure 12.1 illustrates the concept and provides numerous examples of utilizing TENG's high voltage. The review begins with an in-depth introduction to generating high voltage with TENG. Subsequently, the following sections discuss electrically responsive materials and electrostatic devices that can be effectively driven by TENG, such as dielectric elastomers, piezoelectric materials, electrostatic manipulators, electrostatic air cleaners, field emission devices, mass

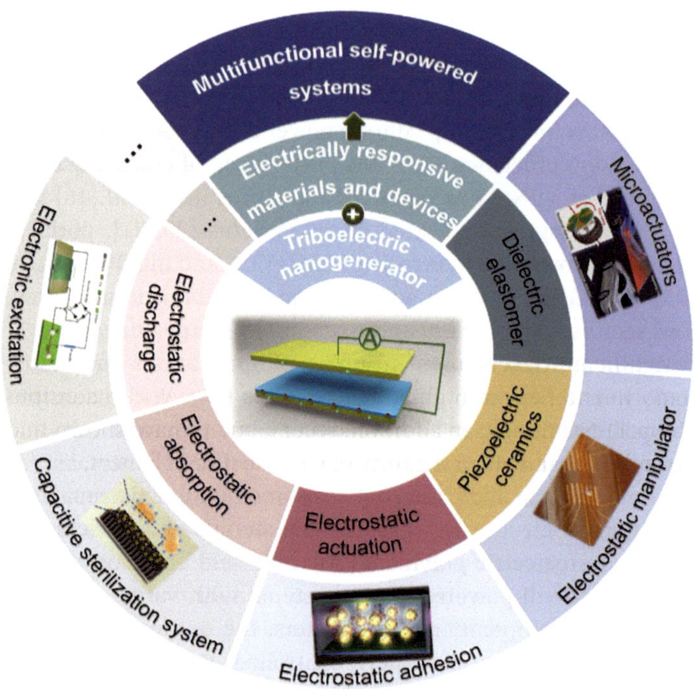

Figure 12.1 Combination of TENG and electrically response materials/devices for functional self-powered systems.

spectrometers, plasma sources, and electrospinning devices. Finally, the analysis offers a broader perspective and recommendations for future research directions.

12.2 Overview of Materials and Universal Methods for TENG's Performance Enhancement

To efficiently operate various functional systems based on TENG, it is crucial to maintain a high output voltage. TENG operates through the combined impact of triboelectrification and electrostatic induction [35]. There are two methods to optimize TENG's output voltage. The first approach involves reducing parasitic capacitance at TENG's maximum separated distance [36, 37]. Enhancing the capacitance change improves the amount of voltage and transferred charge. Researchers have developed four operation modes for TENG and established a structure analyzing method for optimizing its parameters using finite element simulation [38, 39]. The second strategy focuses on increasing the charge density generated on the surface of tribomaterials. This can be achieved by enhancing the amount of triboelectric charge and contact charge. Various surface modification methods can be applied to different TENG structures [40]. Boosting surface charge density (SCD) is the primary technique for achieving high TENG output voltage. This section will discuss the most common strategies for increasing SCD.

12.2.1 Materials Selection

Contact electrification effects are observed in various materials like metals, polymers, and wood [41], resulting in a vast range of potential materials for TENGs. Over the years, numerous theories have been developed to comprehend the specific mechanisms underlying contact electrification between different materials [42]. The difference in electron affinity between contact materials significantly influences triboelectric charge density, making materials with substantial affinity differences ideal for TENG creation. Notable materials with varying electron affinities are quantified in triboelectric series, aiding materials selection [43, 44]. Although slight variations exist among different researchers' sequences, these series guide TENG material choices. Notably, materials atop the series generate high charge densities upon friction, informing TENG design. In addition, recent studies have shown that factors such as the crystallinity and electret ability of triboelectric polymers, as well as the mechanism of ion compensation, can also significantly affect the saturated charge density ability of TENGs [11]. These materials also inform TENG design for specific applications. For instance, flexible and wearable energy harvesters can be crafted from silicone rubber due to its stretchability and durability. Teflon or Kapton tape, noted for its smoothness and abrasion resistance, serves as ideal an tribomaterial for durable sliding TENGs. Furthermore, TENG-based smart devices can be shielded with transparent polyethylene terephthalate (PET) or fluorinated ethylene propylene (FEP) films, considering properties like transparency and abrasion resistance [45]. Materials selection is the initial TENG fabrication step, wherein the electron affinity list aids in identifying potential candidates. Thus, the comprehensive consideration of physical and chemical properties guides effective TENG material choices.

12.2.2 Surface Modification of Materials

Surface modification is a highly effective approach to enhancing the output voltage of TENG by increasing the tribocharge density. This can be achieved through microscopic modification of the physical structure of the surface of tribomaterial, which raises the frictional area and increases the amount of tribo-induced charge during contact process. Figure 12.2 shows the impact of different surface alterations on output performance. In Figure 12.2a, SEM images showcase three types of surfaces: line, cube, and pyramid structures [46], which are created by using silicon molds (The insets provide enlarged cross-sectional views of these structures.). Comparing the output voltage of a typical contact-separation TENG with four different surface types (plane, line, cube, and pyramid) in Figure 12.2b reveals that the V_{oc} measurements followed the order of plane < line < cube < pyramid. Notably, the pyramid-featured device exhibited a V_{oc} nearly four times higher than the plane one. Researchers have explored methods for creating nanowire arrays on polymer surfaces to improve TENG output voltage. By employing techniques such as selective dry etching and inductively coupled plasma (ICP) etching, nanowire arrays can be formed [50–52]. Zhu et al. investigated the impact of nanowire length on TENG performance [47], and the SEM of the nanowire arrays are shown in Figure 12.2c. It was found that nanowires of adequate length (hundreds of nanometers) maintained their orientation and morphology after multiple interactions, as shown in Figure 12.2c. However, extremely long nanowires (20 minutes etching) resulted in strain exceeding the elastic limit of the polymeric material upon contact, leading to permanent deformation. Figure 12.2d shows the influence of etching time on V_{oc} and I_{sc}, indicating that the electrical output increases significantly as etching time increases from 0 to 5 minutes, but reduces dramatically with excessive etching time. Moreover, chemical alteration of the surface by introducing specific functional groups can significantly enhance TENG output performance. Functional groups that readily acquire or lose electrons, such as $-CF_3$ and $-NH_2$, respectively, can increase the attraction for negative or positive charges. Figure 12.2e presents the comparison of four TENGs ((i) PET: PET, (ii) PET: P-PET, (iii) PET: F-PET, (iv) P-PET: F-PET) to assess the impact of different chemical bonds on the output voltage [48]. The V_{oc} measurements in Figure 12.2f indicate that the maximum V_{oc} of the TENG of the P-PET: F-PET pair is more than 80 times higher than those of the PET: PET pair. Furthermore, Yu et al. have utilized a sequential infiltration manufacturing process to infiltrate AlO_x molecules into polymer sheets. This process reduces the electron-acquisition capacity of the AlO_x-doped polymer, resulting in a significant increase in output voltage. Combining physical and chemical alterations allows for the achievement of optimal electrification performance in TENG. To summarize, surface modification techniques, including micropatterned structures, nanowire arrays, and chemical alterations with specific functional groups, play a crucial role in increasing the tribocharge density and enhancing the output voltage of TENG.

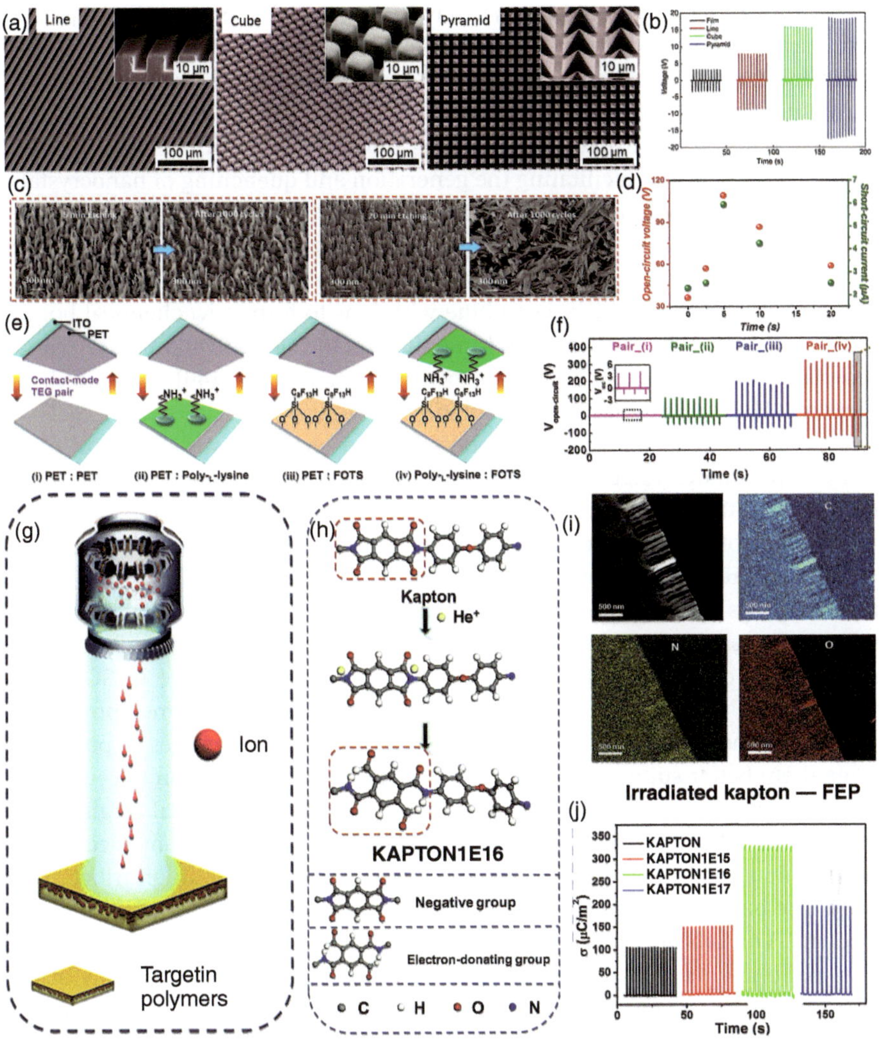

Figure 12.2 Surface modification of friction materials. (a,b) SEM images of micron-pattern prepared by template method and the comparison of the output voltages of TENGs with four different surface types. Reproduced with permission of Ref. [46], 2012 © American Chemical Society. (c,d) SEM images of nanowire arrays prepared by selective dry etching and the effect of time for etching polymer nanowires on output performance of TENG. Reproduced with permission of Ref. [47], 2012 © American Chemical Society. (e,f) Comparison of the output voltage of TENGs with different surface functional groups. Reproduced with permission of Ref. [51], 2015 © American Chemical Society. (g) Simulation diagram of ion irradiation process. (h) Schematic diagram of the molecular structure change of the Kapton sample after ion irradiation. The ion dose is 1 × 1016 ions/cm^2. (i) The HDF-TEM and X-ray mappings of the cross-section in KAPTON16E samples. (j) Comparison of the maximum transfer charge density of Kapton samples treated with different ion irradiation concentrations. Reproduced with permission of Ref. [49], 2019 © Royal Society of Chemistry.

These modifications have been shown to have a substantial impact on TENG performance, paving the way for improved energy harvesting capabilities [53].

Ion irradiation is widely used in various fields, including metals, superconductors, semiconductors, and catalytic materials, to modify material properties. It primarily alters material crystal structures by causing nuclear and electron energy losses, introducing defects, and facilitating the generation and quenching of nanocrystals. In polymers, ion irradiation leads to elastic and inelastic collisions that surpass the chemical bond energy limit, resulting in the breakage of macromolecules. This structural rearrangement creates numerous free radicals when existing connections are disrupted, and these radicals combine with ions to form new chemical bonds. Li et al. investigated the impact of ion irradiation on the electrification performance of polymers and proposed a surface modification method [49]. They utilized low-energy Helium irradiation/implantation to adjust the triboelectric SCD of polymers at the molecular level by modifying the chemical structure and functional groups. Figure 12.2g depicts the working concept of ion irradiation and its contribution to the electrification process of TENG. Figure 12.2h demonstrates that ion irradiation of Kapton film induces a conjugation effect and structural changes, allowing for a reversal of electrification polarity. In Figure 12.2f, HDF-TEM and X-ray mappings reveal uniform element distribution on the irradiated area of KAPTON1E16 samples. Figure 12.2i presents the results of different ion irradiation concentrations on Kapton samples, showing a significant increase in SCD after ion radiation. Furthermore, through in-depth exploration of polymer processing and post-treatment methods, it is possible to optimize the properties of polymers, making them better suited for TENG applications. Liu et al. have produced a kind of RRF-FEP with high charge density through repeated rheological forging [12]. The RRF-FEP film fabricated in this method not only has excellent mechanical properties and good transmittance but also can maintain an ultrahigh tribocharge density. The output charge density from contact-separation nanogenerator reaches 352 $\mu C/m^2$. The repeated forging method can effectively regulate the composition of surface functional groups, the crystallinity, and the dielectric constants of the FEP, leading to the superior capability of triboelectrification. The mechanism of the RRF method for improving charge density is shown in Figure 12.3a–c. The XRD pattern shows that the improvement of crystallinity of FEP, and leads to the high charge density of TENG after corona polarization. Yang et al. have found that in atmospheric environments, escaping charges can be compensated for by selective transfer of anions at the contact interface, resulting in ultra-high charge density. [11] Figure 12.3d–g demonstrates that naphthalene radical anions are used as wet chemical treatment agents for etching PTFE and the bulk of obtained ionized-PTFE (I-PTFE) contains abundant radical ions. The I-PTFE acts as a positive triboelectric component to work with fully charged FEP film after corona polarization for fabricating TENG device. Then, the charge-escaping phenomenon of fully charged FEP can be well compensated by radical ion transfer during contact electrification. As shown in Figure 12.3h, the triboelectric charge density of 525 $\mu C/m^2$ is obtained in contact-separation mode, while the charge density of 1237 $\mu C/m^2$ is further achieved with sliding mode.

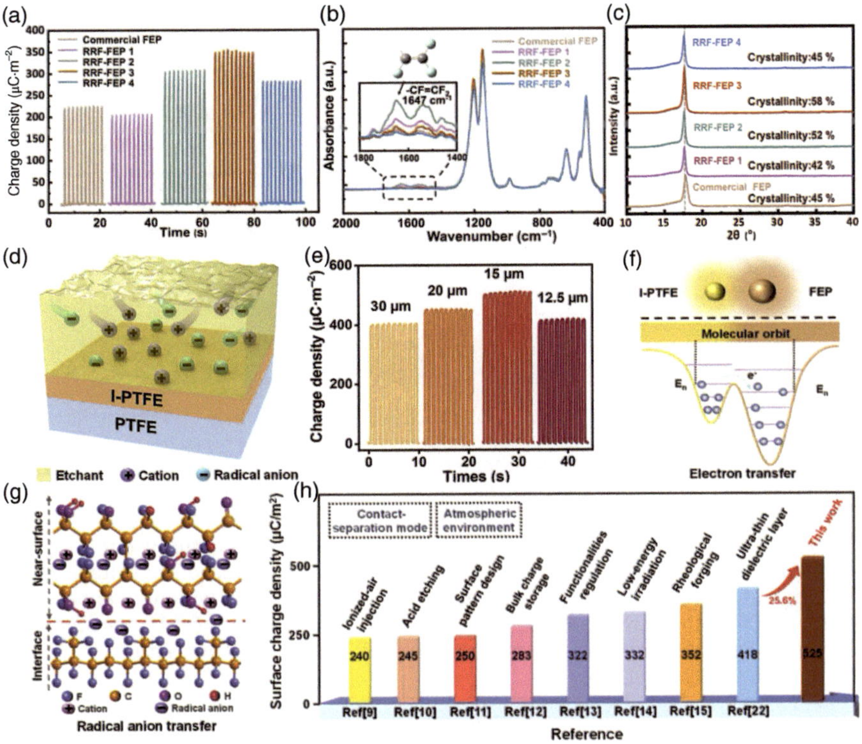

Figure 12.3 (a) Maximum saturated charge density between FEP (without corona poling) and PA measured by contact-separation TENG. (b) The molecular structure diagrams of several FEP film structures are given by a Fourier transform infrared spectrum. (c) XRD pattern of FEP films. Z. Liu et al., 2022 / Springer Nature / CC-BY-4.0. (d) Schematic diagram of the preparation of I-PTFE by radical anion modification method. (e) Triboelectric charge density between I-PTFE and different polymers including fully charged FEP with four different thicknesses in open air and contact-separation mode. (f) Unsaturated processes present in TENG tests. (g) Schematic illustration of PTFE modification by radical anion at the molecular level. (h) Comparison of various material design approaches in atmospheric environment and contact-separation mode. Reproduced with permission of P. Yang et al., 2023 © Elsevier.

12.2.3 Enhancement of TENG's Voltage Performance by Charge Injection Methods and Vacuum Protection

Through charge injection strategy, some electret polymers can keep these injected charges for a long time, resulting in triboelectric layer with an SCD several times larger than that of a regular film. Various charge injection techniques have been examined, including high-voltage corona discharge and ionized air injection. High voltage corona discharge is commonly used to enhance charge density. As shown in Figure 12.4a, high voltage is applied between the needle tip and the electrode, polarizing the dielectric layer between them. Shao et al. have investigated the peak voltage correlates with the thickness of the dielectric layer [54], and TENG

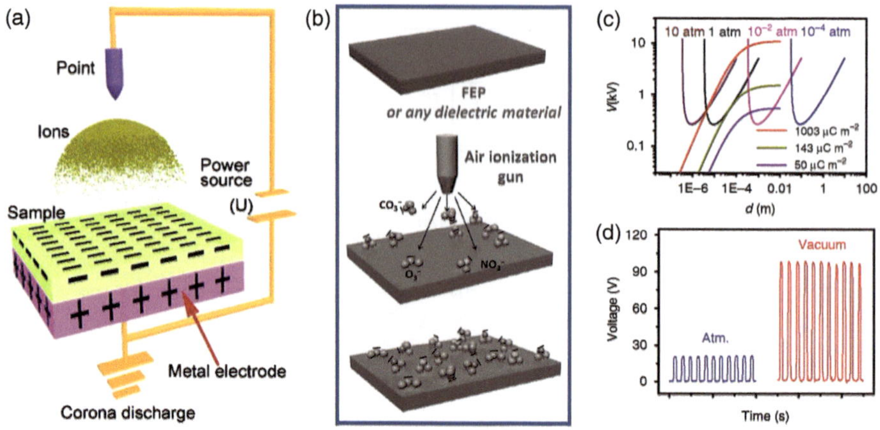

Figure 12.4 Enhancement of TENG's voltage performance by charge injection methods and vacuum protection. a) The schematic of high voltage corona discharge for high voltage output of TENG. Reproduced with permission of Ref. [54], 2017 © Royal Society of Chemistry. (b) Ionized air injection for high performance of TENG. Reproduced with permission of Ref. [55], 2014 © John Wiley & Sons. (c) Theoretical breakdown voltage with different separation distances and theoretical output voltage of TENG with different charge densities. (d) VOC of the TENG in atmosphere as well as a high vacuum (with a vacuum of 10^{-6} Torr). (c,d) Reproduced with permission of Ref. [56], 2017 © Springer Nature.

exhibits remarkable cycle stability with a retention rate of approximately 92.6% after 20 000 cycles. Hu et al. have explored the effects of both positive and negative corona [57] discharging methods on polypropylene film, noting that the surface potential decreased more slowly with negative corona. As illustrated in Figure 12.4b, Wang et al. have utilized an air ionization pistol to introduce negative ions (CO_3^{2-}, NO_3^-, and O_3^-) onto the surface of FEP [55], irrespective of its chemical characteristics. Through multiple ion implantation processes, these negative surface charges reached a high level, resulting in a fivefold increase in V_{oc} of TENG. However, the increase in TENG output is limited by air breakdown as the SCD rises. To overcome this limitation, Wang et al. investigated TENG's output performance in a high vacuum [56]. Figure 12.4c depicts the air breakdown voltage of TENG at different air pressures and charge densities, which aligns with Paschen's law from 1889 [58]. As per this phenomenon, the gap/output voltage of TENG must be lower than the air breakdown voltage. Boosting the vacuum serves as another effective technique to enhance TENG's output. By achieving a vacuum of 10^{-6} Torr, air breakdown is avoided during TENG operation, resulting in a fivefold enhancement in V_{oc}, as depicted in Figure 12.4d.

12.2.4 Enhancement of TENG's Voltage Performance by Charge Supplement

In addition to surface modification methods, vacuum protection, and charge injection, there are other approaches to improve the output properties of TENGs,

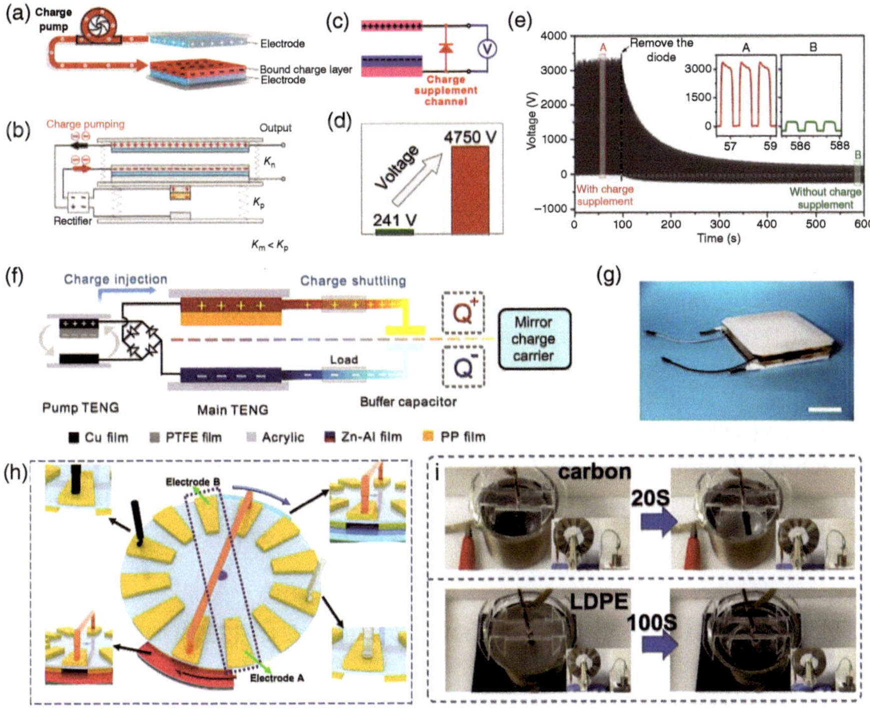

Figure 12.5 Enhancement of TENG's Voltage Performance by Charge Supplement. (a,b) Schematic diagram of the charge pumping principle. Reproduced with permission of Ref. [60], 2018 © Elsevier. (c) Schematic diagram of working principle of the CSC. (d) The maximum output voltage of one TENG unit. (e) Comparison of the VOC of the TENG with or without CSC. Insets illustrate the details of the VOC. Reproduced with permission of Ref. [61], 2018 © American Chemical Society. (f) Schematic illustration of charge shuttling. (g) Photograph of the as-fabricated main TENG and pump TENG. Scale bar, 4 cm [62]. Wang et al., 2020 / Springer Nature / CC BY 4.0. (h) Schematic diagram of the initial state of SH-TENG. (i) Electrostatic filtration effect of high-voltage TENG.

such as supplementary charge pumps and triboelectric charge supplement channels (CSCs) [59]. Xu et al. proposed a simple and universal method for external charge compensation using a compensating circuit and structural design [60]. In their developed TENG, as shown in Figure 12.5a,b, a floating layer is utilized to store and bind electrostatically induced charges, while another TENG acts as a charge pump to simultaneously pump charge into the floating layer. Similarly, Xu et al. introduced a triboelectric CSC that significantly increases the open-circuit voltage by over 10 times by incorporating a high-voltage diode into a standard contact-separate TENG [61]. Figure 12.5c demonstrates the working principle of the triboelectric CSC. As depicted in Figure 12.5d, the maximum output voltage [63] of one TENG unit can reach 4705 V, achieving a 14-fold enhancement by utilizing the CSC. Figure 12.3e illustrates the substantial voltage improvement achieved in TENGs with the use of the CSC. Xu and coworkers also proposed the concept of charge shuttling principle and charge-shuttling-based triboelectric nanogenerator (CS-TENG).

[62] Unlike conventional TENGs where frictional electrostatic charges are confined to the dielectric surface, CS-TENG confines the charge within the conductive domain and utilizes reciprocal shuttling of the charge to form a current for driving the load. Figure 12.5f,g demonstrates the quasi-symmetry-based interaction of charges in the double-conducting domain, resulting in two mirror charge carriers and achieving double charge output. Furthermore, Chen et al. developed a sustainable high-voltage TENG (SH-TENG) based on a charge accumulation strategy to enhance sustainable output voltage. The design of SH-TENG differs from conventional disk-type TENGs by incorporating a polarizer, turntable, conduction bridge, and two pairs of collectors, as shown in Figure 12.5h. Through the unidirectional rotation of the turntable and transmission electrodes, the separated positive and negative charges are continuously injected into the two pairs of collectors, leading to a sustainable high-voltage output between them. By optimizing the design parameters, SH-TENG can achieve sustainable high-voltage transmission exceeding 20 kV, which demonstrates good application prospects in industrial filtration (Figure 12.5i). It is important to consider adequate material and device designs to prevent breakdown when aiming for even higher voltages using these methods.

12.3 Artificial Muscle Based on Dielectric Elastomer and TENG

Dielectric elastomer is an electro-active actuator material known for its remarkable ability to deform under the influence of an electric field. As a result, dielectric elastomer actuators (DEAs) have found applications in various fields such as soft robotics, electronics, and micropumps. DEAs can change their structural shape or volume under the stimulation of an external electric field. When the external stimulation of electrical energy stops, the object can automatically recover to its original shape or reduce its volume, automatically converting external electrical energy into mechanical energy with high electromechanical conversion efficiency. The working principle of DEA has been extensively studied [64], and its large induced deformation, owing to its low elastic stiffness and high strain capacity, allows for the creation of different types of mechanical actuators. However, DEAs typically require driving voltages in the range of several thousand volts to drive it, which makes TENG becomes a favorable option due to its ability to generate high output voltages with minimal separation motion. By combining the high output voltage of TENG with the excellent strain capabilities of DEA, a TENG-DEA system can be established. This system serves as a compelling example of TENG's application in artificial muscles and soft actuators. Additionally, when integrated with TENG's energy harvesting capabilities, the TENG-DEA drive system enables instantaneous interaction between humans and the system, making it suitable as a self-powered drive component in microelectromechanical systems (MEMS) and human–machine interaction. Several applications of TENG-DEA systems are discussed in this section, including artificial muscles, intelligent switches, tunable smart optical modulators (SOMs), and tunable optical gratings (TOGs).

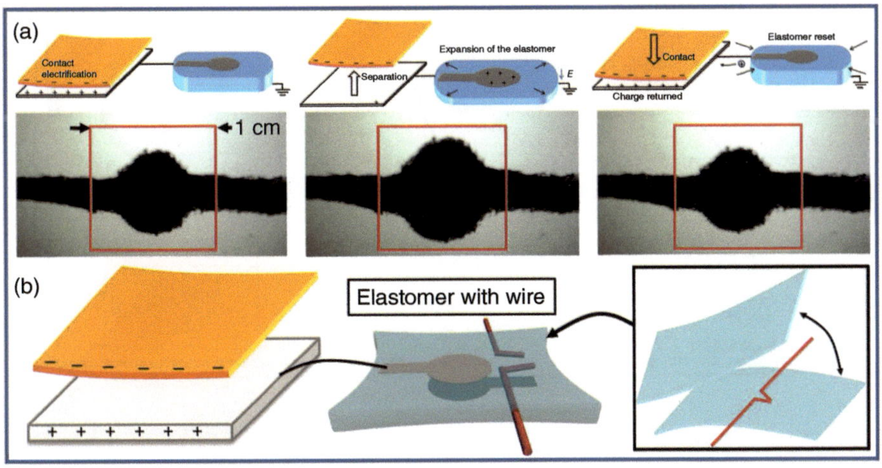

Figure 12.6 Self-powered artificial muscle and intelligent switch based on a TENG-based DEA system. (a) Schematic of a TENG-DEA system with the strain performance and the photographs show the corresponding deformation of the DEA device. (b) Working mechanism of a smart switch utilized the TENG-DEA system. Reproduced with permission of ref. [65], 2016, © John Wiley & Sons.

Figure 12.6a presents a simple demonstration of a TENG-DEA system involving a single-electrode TENG and a circular DEA [65]. The upper and lower sides of the elastomer layer are equipped with two circular carbon grease electrodes, one connected to the TENG and the other grounded. To enhance the induced SCD and increase the surface area, nanowire arrays are prepared on the Kapton film through ICP treatment. Following the contact-separation motion of the Kapton film and Al foil, a high voltage is generated between the electrode and the ground due to the ICP treatment. The working principle of this integrated system is detailed in Figure 12.6a. In summary, when the Kapton film and Al foil come into contact, positive and negative charges are produced at the contact interface. Subsequently, the separation induces a strong electrostatic field on the dielectric elastomer, causing the DEA to deform. Upon returning the Kapton film and Al foil to their original positions, the positive and negative charges neutralize, and the DEA deformation ceases. The excellent insulating characteristics and soft texture of the acrylic elastomer make it suitable as a packaging material for soft electronic products. In Figure 12.6b, a demonstration of a smart switch utilizing the TENG-DEA system is illustrated. Two wires are sandwiched between the elastomer films, and the TENG controls the connection and disconnection of the wires. By printing intricate circuits on the surface of the elastomeric film, additional functional capabilities can be achieved.

Chen et al. have developed a SOM using silver nanowires as electrode materials, building upon the TENG-DEA system [66]. The flexibility of the super-long silver nanowire electrode allows for its use in various tunable optical devices without compromising transparency. As depicted in Figure 12.7a, the working mechanism of the SOM device is similar to DEA. Initially, SOM is transparent and stable,

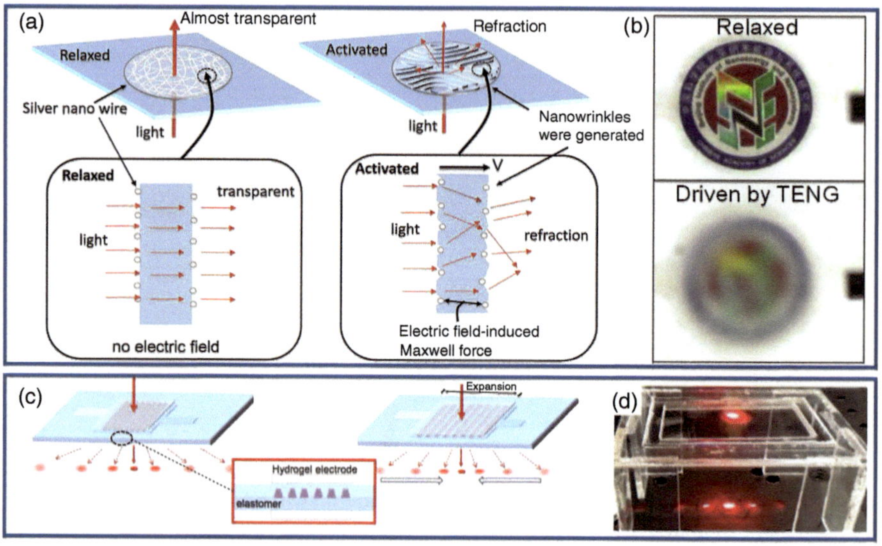

Figure 12.7 Intelligent optical modulator by coupling TENG and dielectric elastomer. (a,b) The schematic of the working mechanism of the SOM driven by TENG and optical images through the SOM device. Reproduced with permission of Ref. [66] © 2017, John Wiley & Sons. (c,d) The sketch and optical image of TOGs with grating arrays driven by TENG. Reproduced with permission of Ref. [67] © 2017 Elsevier.

then it undergoes localized microscopic deformation when a potential drop is applied to the elastomer due to Coulomb attraction between the top and bottom nanowire electrodes. These irregular microscopic deformations result in irregular light refraction, causing a decrease in macroscopic transparency. Figure 12.7b presents optical images demonstrating the modulation achieved by the SOM device, where the transparent film is transformed into a matte film through TENG-driven actuation. Additionally, the TENG-DEA system has also been utilized to demonstrate TOGs, as shown in Figure 12.7c [67]. When the TENG actuates the DEA element to expand, it changes the grating period (as illustrated in Figure 12.7c). The corresponding optical images are shown in Figure 12.7d. This effect can be extended to 2D optical gratings and dot matrix gratings, enabling control over diffractive laser matrices using TENG. The introduction of these self-powered SOM and TOG systems expands the application of TENG in optical communication, light capture, and other optical domains.

A special operation method has been proposed to effectively control the TENG-DEA system in Figure 12.8a [66]. This method involves two connecting DEA units in series, creating a bias voltage across the entire system. One of the DEA units serves as the target device, while the other acts as the assistant device. The output voltage of the TENG regulates the potential drop distribution on the two DEA units. Figure 12.8a,b illustrates that contact with the TENG decreases V1 and increases V2, whereas separation increases V1 and decreases V2. The applied voltage bias helps overcome the threshold voltage of the DEA, and further increasing the

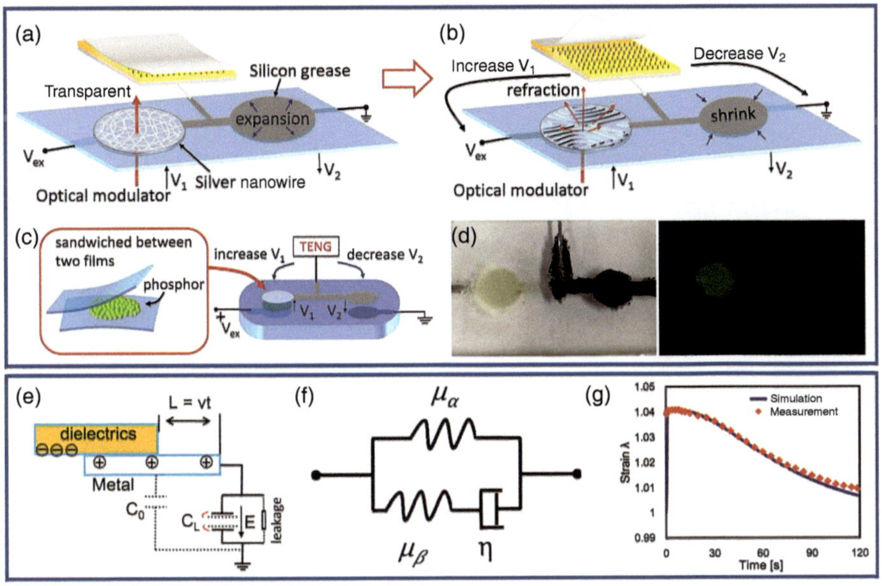

Figure 12.8 An effective method for operating the TENG-DEA system and theoretical analysis of the integrated system. (a) A SOM and a DEA driven by a new method, where the output of TENG leads to activation strain of common DEA sample. (b) The changes in transparency of SOM and the shrinkage of common DEA samples. (c) A sketch of another application of the TENG-DEA system to illustrate the capability of special operation method. (d) The optical image of the prepared sample and the generated luminescent behavior. (a–d) Reproduced with permission of Ref. [66] © 2017, John Wiley & Sons. (e) Theoretical model for the TENG-DEA system. (f) Theoretical model for viscoelasticity of the elastomer. (g) The relaxation process of the actuated strain, where the dotted line is the experimental data and the solid line is the simulation results. (e–g) Reproduced with permission of Ref. [68], 2017 © AIP Publishing.

applied voltage at this bias value induces significant deformation in the target DEA. As seen in Figure 12.8b, the SOM device functions as the target device, and the TENG successfully controls its deformation. This external voltage bias method is applicable to various DEA-related devices. Furthermore, an elastomer-based electroluminescence device is prepared using this driving strategy (Figure 12.8c). The adoption of this strategy significantly enhances luminescence compared to using TENG alone. Figure 12.9d displays the device and the luminescence achieved through this strategy. These integrated devices demonstrate the broad applicability of this operational strategy to DEA-related devices. To gain a better understanding of the working mechanism of the TENG-DEA system, a physical model is constructed to simulate the effects of viscoelasticity and current leakage on TENG output performance [68]. Figure 12.8e,f portrays the theoretical model of the integrated system and the viscoelastic model of the elastomer, respectively. A comparison between actual measurements and simulation results in Figure 12.8g verifies that this model can accurately predict the response of DEA driven by TENG. This suggested physical model facilitates exploring a wide range of applications for TENG-DEA systems.

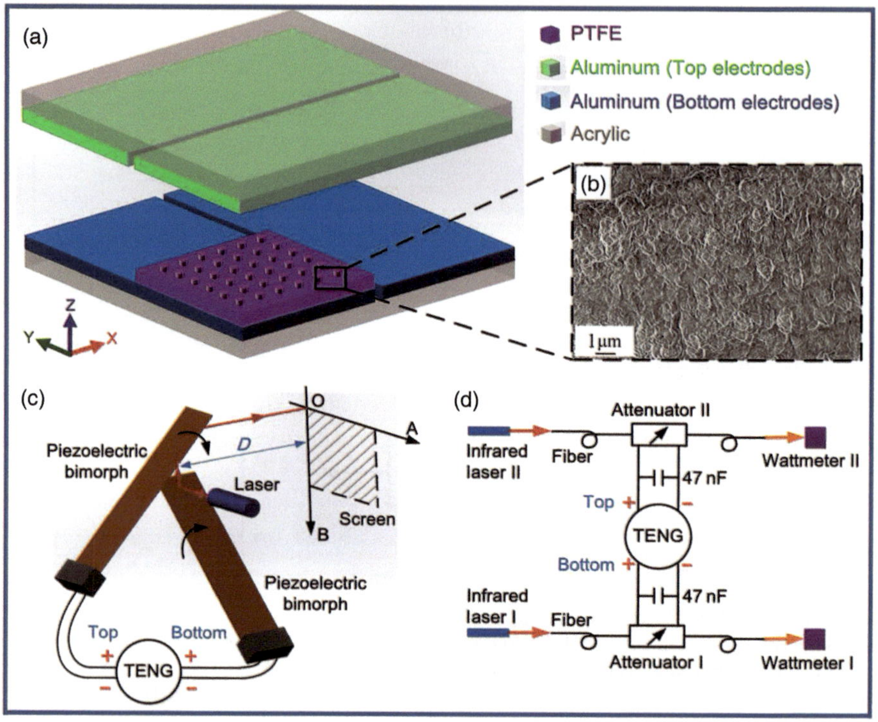

Figure 12.9 Optical modulated active microactuator based on planar sliding TENG. (a) Schematic diagram of the designed three-layer sliding TENG. (b) SEM image of the nanostructure of the PTFE membrane. (c) A sketch of the structure of a 2D modulator driven by the three-layer sliding TENG. (d) Schematic diagram of an active optical attenuator with two channels. Reproduced with permission of Ref. [5] © 2015, John Wiley & Sons.

12.4 Microactuators Based on Piezoelectric Ceramics and TENG

In recent years, there has been significant development in microactuators. High-voltage TENG has emerged as an integrated and self-powered power supply for sustainable micromotors by harnessing low-frequency mechanical energy to generate a high-voltage electric field. At the same time, piezoelectric materials exhibit the inverse piezoelectric effect, deforming under the influence of an electric field. Although the deformation in piezoelectric materials is generally small, it offers high precision and is suitable for precise operations in microsystems. Piezoelectric ceramics are commonly used in electromechanical devices like ultrasonic motors, ceramic speakers, and piezoelectric probes due to their favorable characteristics. These ceramics typically require driving voltages in the range of several dozen volts, while maintaining sufficient internal resistance to sustain tribo-induced charges. Consequently, piezoelectric ceramics can be effectively combined with TENG to create various self-powered electromechanical systems. Zhang et al. have proposed a piezoelectric microactuator for controlling the emission angle of a laser beam [30] by combining sliding TENG and piezoelectric ceramics. Figure 12.9a illustrates the schematic of the three-layer sliding TENG design, including a PTFE film as the freestanding triboelectric layer and four Al foils serving as orthogonal electrodes. To enhance the output voltage of the TENG, the surfaces of the PTFE film are treated using ICP to introduce a nanostructured morphology (Figure 12.9b). This three-layer sliding TENG, in conjunction with piezoelectric bimorphs, enables 2D optical modulation through the designed active piezoelectric microactuator (Figure 12.9c). By utilizing the precise deformation ability of piezoelectric ceramics in an electric field, piezoelectric microactuators achieve precise control of laser direction. Furthermore, another microactuator is developed using the three-layer sliding TENG and two variable infrared optical attenuators. Figure 12.9d illustrates the working principle of this electrostatic microactuator, with dual-channel optical modulation achieved by adjusting the two infrared attenuators using TENG.

In addition to piezoelectric materials, TENG can also be combined with other intelligent devices to expand its capabilities. Zhang and coworkers proposed a self-powered triboelectric micromotor (TM) by integrating a TENG with an electrostatic micromotor. This combination enables the conversion of low-frequency mechanical external stimuli into continuous rotation of the micromotor [69]. The schematic diagram of the TM, as shown in Figure 12.10a, includes a freestanding TENG, a rotating electrostatic micromotor, and a rectifier. The electrostatic micromotor comprises a rotor with four evenly distributed copper blades and two electrodes with carbon fibers attached to their surfaces. The inset highlights the use of ICP to create nanostructures on the PTFE, significantly enhancing the surface triboelectric charge density. Powered by the high output voltage from the TENG, the vertical electrodes become charged, resulting in the movement of the rotor due to electrostatic force and inertia from rotational momentum. Figure 12.10b illustrates the micromotor's working principle, where charges are transferred from the electrodes to the blades on the rotor and subsequently neutralized upon contact with the other electrode. Throughout this process, both repulsive and attractive forces generated by the charges maintain the rotation of the micromotor. Thanks to this operational mechanism, the TM achieves high-speed rotation triggered by low-frequency mechanical stimulation, as demonstrated in Figure 12.10c. With its compact size and high speed, the TM holds great potential in scanning systems for information recognition. Figure 12.10d showcases an optical scanning device based on the TM, while Figure 12.10e explains its application as a moving obstacle detector. By utilizing low-frequency mechanical energy from tire rolling, the device can detect moving obstacles at high speeds.

12.5 Materials Polarized by the High Voltage Output From TENG

Dielectric materials, in general, readily polarize when subjected to an external electric field, making them suitable for high-voltage polarization using TENG. An innovative application has emerged by integrating ferroelectric materials

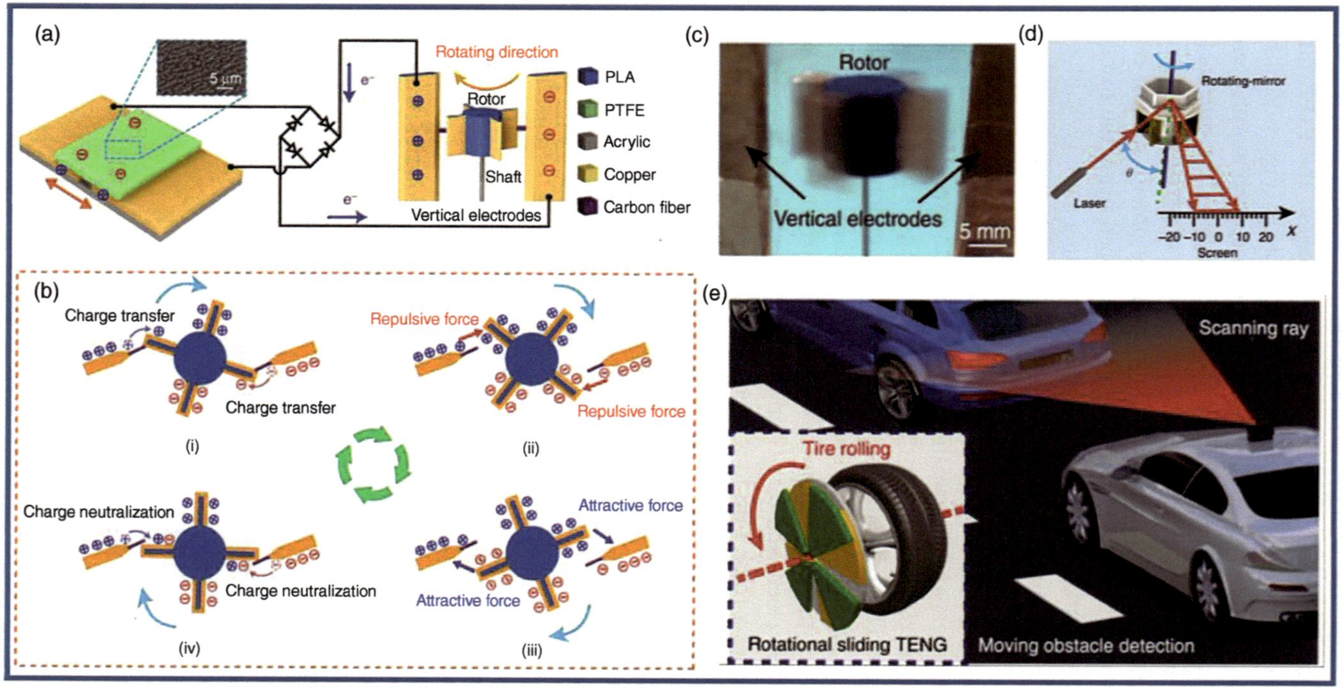

Figure 12.10 Triboelectric micromotor based on a micromotor and a triboelectric nanogenerator. (a) The schematic image of the structure and working principle of TM. Inset shows the surface photograph of the PTFE film. (b) Working mechanism of the micromotor. (c) The photograph of the rotating motor. (d) Schematic diagram of the TM-based optical scanning device. (e) Schematic illustration of the self-powered moving obstacle detection system. [69] / Springer Nature / CC-BY-4.0.

with TENGs, creating a self-powered information memory system. Ferroelectric materials exhibit excellent response to imposed electric fields and are insensitive to current, making them ideal for TENG output performance. Furthermore, the dipole switching in ferroelectric materials occurs much faster than the TENG's output period. These factors position ferroelectric materials as promising candidates for coupling with TENG to realize self-powered memory and storage devices. Among these materials, poly(vinylidene fluoride-trifluoroethylene) (P(VDF-TrFE)) has received extensive attention due to its remarkable dipole switching repeatability. The combination of TENG and P(VDF-TrFE) membranes has led to the development of a unique self-powered memory system capable of recording motion in one and two dimensions [70]. Figure 12.11a presents the sliding TENG-based self-powered memory system and its working principle. When the electric field generated by the TENG motion is applied to the P(VDF-TrFE) film, the dipole moments within the film become polarized. The magnitude of polarization depends on the electric field strength, allowing the memory system to record the displacement distance of the TENG. The stored polarization information can be observed using a typical measurement technique involving displacement-current, as demonstrated in Figure 12.11b. To capture 2D motion trajectories, a self-powered storage system is created by combining a single electrode TENG array with ferroelectric material, as depicted in Figure 12.11c [71]. Figure 12.11d,e showcases the TENG array. Additionally, polarized ferroelectric materials can enhance the TENG's output. Lee et al. employed P(VDT-TrFE) and skin as contact materials in a single-electrode TENG. By applying different electric field poling, significant changes in the output voltage and control over skin potential can be achieved via the poling electric field applied to the P(VDT-TrFE) polymer.

The polarization effect generated by TENG offers diverse opportunities for various energy harvesting strategies [72]. Nie et al. introduced a liquid–liquid TENG based on a polarized liquid membrane, which harnesses power from the contact and separation of falling liquid droplets and suspended liquid membranes [73]. The working principle of this liquid–liquid TENG is depicted in Figure 12.12a, featuring an annular frame built using 3D printing technology to support the liquid membrane. An electrical wire runs inside the frame, enabling direct charge extraction from the liquid membrane. To optimize its properties such as permeability, electrical conductivity, self-healing ability, and longevity, experiments were conducted to determine the appropriate solution ratio for each component of the liquid membrane. The photograph of the liquid–liquid TENG is presented in Figure 12.12b. In this configuration, a grounded liquid membrane serves as a permeable electrode for the TENG, as illustrated in Figure 12.12c. As droplets fall, they acquire positive charges through air friction, resulting in charge accumulation on their surfaces. When these droplets make contact and separate from the liquid membrane, charge transfer occurs. The liquid membrane captures charges from the droplets, while free electrons from the ground are attracted to neutralize them. This demonstrates that the liquid membrane can serve as an excellent candidate for permeable sensors and charge filters in electrical devices. However, if used for power generation, the output performance of the TENG is greatly influenced by the number of charges on the surface of the droplets

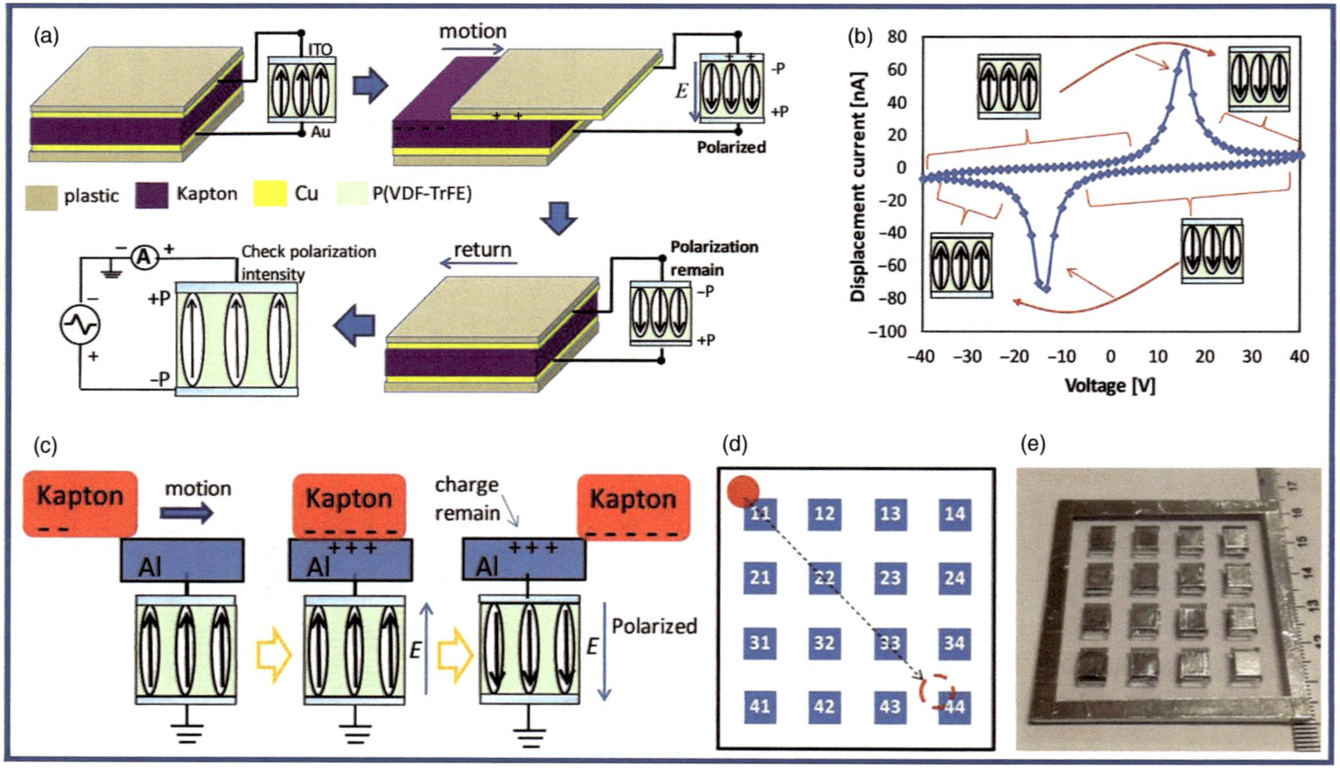

Figure 12.11 Self-powered trace memorized storage based on triboelectrification and ferroelectricity. (a) Schematic diagram and working mechanism of the information storage device based on a TENG and a P(VDF-TrFE) film. (b) The result of displacement current measurement of a device with a double circulation voltage (from −25 to 25 V). (c) Schematic of the working mechanism of the memorized storage system by combining single-electrode TENG and P(VDF-TrFE). (d) Sketch of a prepared TENG matrix for detecting and recording 2D motion trajectory. (e) Photograph of the TENG matrix. (a–e) Reproduced with permission of Ref. [70] © 2015, John Wiley & Sons.

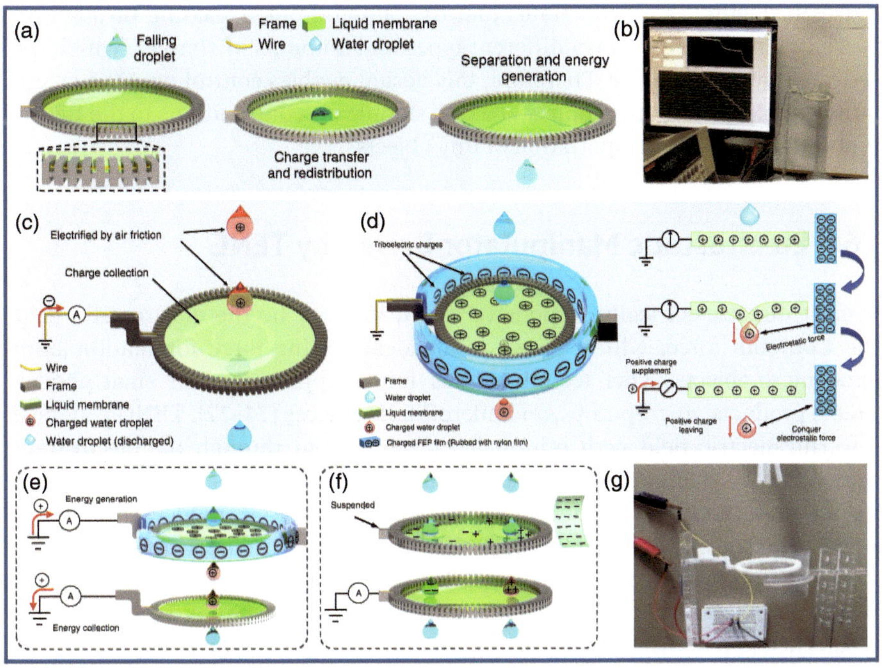

Figure 12.12 Liquid–liquid triboelectric nanogenerator based on liquid droplets and a freely suspended liquid membrane. (a) The schematic illustration of the working principle of the liquid–liquid TENG. The inset shows the structure of the frame. (b) The snapshot of the L–L TENG. (c) Schematic diagram of the energy collection process through a grounded liquid membrane without precharged. (d) Schematic diagram of the energy collection process through a precharged liquid membrane. (e,f) Power generated-based multiple liquid membranes. (e) Adding a precharged liquid membrane. (f) Adding a suspended charged liquid membrane. (g) The photograph of the output enhancement through multitube droplets. [73]/Springer Nature/CC-BY-4.0.

or other falling objects. To overcome this, a precharged liquid membrane is constructed by introducing an external electric field. Figure 12.12d depicts the process wherein the grounded liquid membrane is polarized and accumulates charge under the influence of the surrounding precharged FEP film. Initially, a balance is maintained between the negative charges on the FEP film and the positive charges on the liquid membrane due to electrostatic induction. However, when neutral droplets come into contact with the liquid membrane, a redistribution of positive charges occurs between the droplets and the membrane. As the droplets separate from the membrane, they carry away some positive charges, disrupting the original electrostatic balance. Consequently, positive charges are attracted from the ground to the liquid membrane, resulting in the generation of a conductive current. By stacking a grounded liquid membrane with a precharged liquid membrane, a stacked power generation system can be created, as shown in Figure 12.12e. Each unit functions independently, significantly improving the output power. Increasing the number of falling droplets also leads to increased output power, as depicted in Figure 12.12g. Additionally, placing the FEP film near a completely independent liquid membrane polarizes it. Figure 12.12f illustrates that the falling droplets passing through the polarized membrane can carry different types and amounts of charges, which can be detected by the sublayer. Therefore, this design enables control over the charge polarity and density of the droplets, making significant contributions to the field of microfluidics and the transportation of tiny objects.

12.6 Electrostatic Manipulator Driven by TENG

The most direct electrically responsive effect, induced by the electrostatic field, is the Coulomb force, which serves as an ideal driving force for manipulating macro-/micro-objects. This technique has found applications in smartphones, wearable products, micropumps, and microfluidic devices [74–77]. TENG generates a powerful electric field with minimal leakage current through the electrostatic actuation device, facilitating the maintenance of tribocharges. Additionally, the finite amount of charge produced by TENG in each cycle reduces the risk of electrical breakdown. By combining TENG with an electrostatic actuation system, a motion-tunable electrostatic manipulator can be achieved with self-powered capabilities. This TENG-based manipulator exhibits fast response, high efficiency, and safety, comparable to commercial power sources equipped with protective circuits. Hence, it serves as a valuable technique for precisely manipulating small solid or liquid objects in MEMS systems, microfluidic channels, and various other scenarios.

Wang et al. introduced an electrostatic self-assembly method that enables the desired patterning of plastic pellets on a specially designed plate [78]. Figure 12.13a depicts the structure of the device, where a PTFE film with nanowire arrays on its surface (Figure 12.13b) serves as the electrification layer. By modifying the shape of the grid electrode within the device structure, the desired self-assembly pattern can be achieved. A sample with the assembled array is captured in a photograph displayed in Figure 12.13c, featuring nylon beads with a diameter of 1.58 mm. The self-assembly principle is illustrated in cross-sectional views in Figure 12.13d–e. In summary, after being treated with TENG for charging, the surfaces of the nylon beads and the electrification layer acquire positive and negative charges, respectively. The positive charges induced by the electrification layer flow from the ground to the grid electrode, as depicted in Figure 12.13d. Consequently, electrostatic attraction is established between the beads and the window position (Figure 12.13e), causing the beads to be drawn toward the window position. Zheng et al. also presented a TENG-based electrostatic actuator system capable of driving distinct motion in metal pellets [79]. Figure 12.13f demonstrates several disconnected ring electrodes connected to segmented annular Al foils of the TENG in a unique electrostatic actuator configuration. The metal pellet is grounded using carbon fiber, allowing charges to be induced on it. The excitation of the TENG generates electrostatic attraction between the electrode and the pellet. This circular motion of the Kapton on the ring Al foils initiates the rotational movement of the metal pellet. Additionally, another electrostatic actuator with a pendulum structure

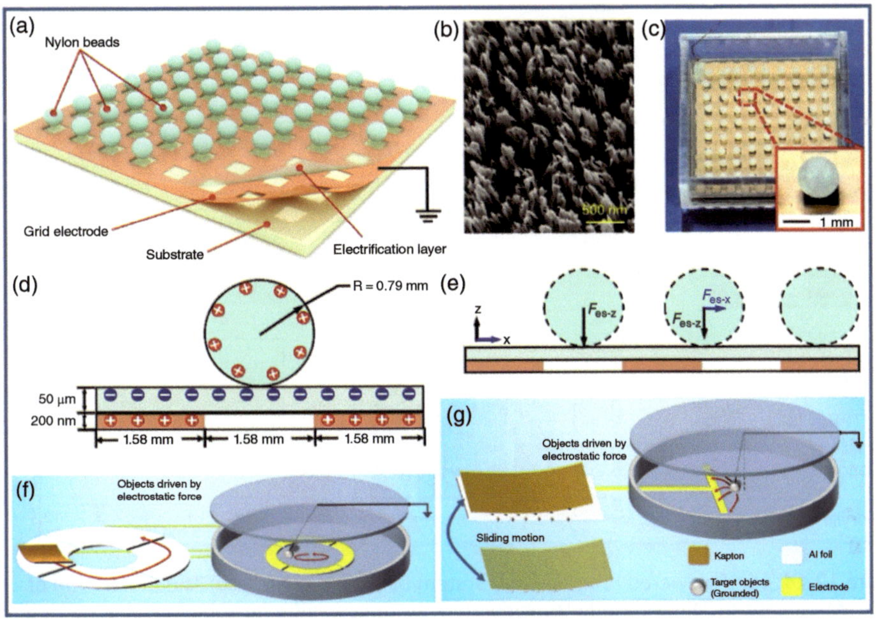

Figure 12.13 TENG as an electrostatic actuator for the manipulation of tiny objects. (a) Structure of an electrostatic self-assembly device. (b) SEM image of the Teflon surface morphology after plasma dry etching treatment. (c) Photograph of the device with nylon balls. Inset: A close-up view of the nylon ball. (d) Charge distribution of one nylon ball in one window. (e) The electrostatic force of the ball at different locations. (a–e) Reproduced with permission of Ref. [76] © 2018, American Chemical Society. (f) The structure design of the self-powered electrostatic actuation system for a leading rotated motion of a pellet. (g) The electrostatic actuation system for swinging tiny objects. (f,g) Reproduced with permission of Ref. [77], 2017 © John Wiley & Sons.

is designed (Figure 12.13g), where the motion of the Kapton film in the TENG regulates the oscillation of the pellet.

In a similar manner, a self-powered drug delivery system can be developed using the TENG-based manipulation system [80]. Liu et al. devised an electroporation drug delivery system by combining the biomechanical energy-driven TENG with a silicon nanoneedle array electrode, achieving efficient intracellular drug delivery with high efficiency and minimal cell damage [81]. To accommodate electroporation conditions both in vitro and in vivo, different TENG designs are employed. Figure 12.14a depicts the schematic diagram of the in vitro electroporation setup. The device includes a portable manual TENG, a rectifier bridge, a 5×5-mm silicon nanoneedle array attached to cells as the anode, and an Al electrode on the lid serving as the cathode. For the in vivo electroporation device, a single-electrode mode TENG or a contact-separation mode TENG is utilized, with silicon nanoneedles attached to a predetermined position on the skin as the anode, and a stainless steel needle inserted at the back as the cathode. The TENG generates electricity localized at the interface between the silicon nanoneedle and the cell, enabling electroporation and drug delivery through each silicon nanoneedle. The use of silicon nanoneedles

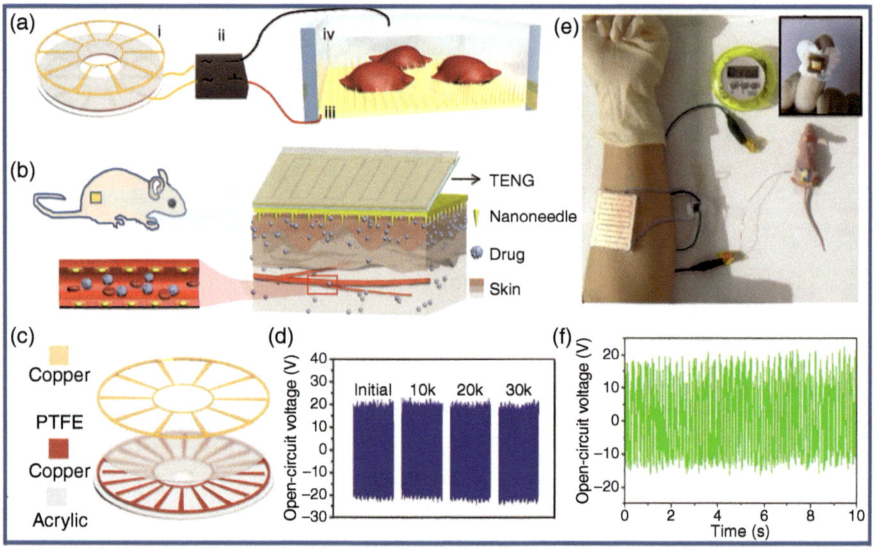

Figure 12.14 Self-powered drug delivery system based on TENG. (a,b) The illustration of different designs of TENG-based electroporation systems *in vitro* (a) and in vivo (b). (c) Schematic diagram of the disk TENG. (d) The stability of VOC under 30 000 cycles. e) Photograph of the in vivo TENG-based electroporation drug delivery system. (f) The open-circuit voltage of the freestanding TENG. Reproduced with permission of Ref. [81] © 2019, John Wiley & Sons.

effectively reduces the damage caused by the applied electrical field to the cell. The disk-like structure of the TENG is illustrated in Figure 12.14c. Through the rectifier bridge, the AC pulse generated by the TENG is converted into a DC pulse. As shown in Figures 12.14d and 12.14f, the TENG's output voltage and frequency meet the electroporation requirements, exhibiting exceptional output stability even after 30 000 cycles. The picture of the in vivo percutaneous drug delivery system is displayed in Figure 12.14e, with the inset showing the image of the nanoneedle following the electroporation therapy. This self-powered electroporation system effectively delivers macromolecules (small molecules, proteins, siRNA) into various cell types, including primary cells that are challenging to transfect, establishing a new model for portable, self-powered, and controlled drug release therapy.

The TENG exhibits the capability not only to manipulate the movement of tiny solid objects but also to control the motion of microfluidics [82]. For instance, Sun et al. introduced an innovative approach involving the printing of a programmed SCD on various substrates by impacting water droplets [83]. Different from conventional transport systems relying on topographic or chemical gradients, this method establishes an SCD gradient on a super amphiphobic surface, enabling high-speed and long-distance transportation of droplets without requiring additional energy input. This method significantly proves the ability of TENG to accurately design SCD by using triboelectric charge. Additionally, Figure 12.15a presents a designed self-powered microfluidic manipulator capable of controlling the confluence of two

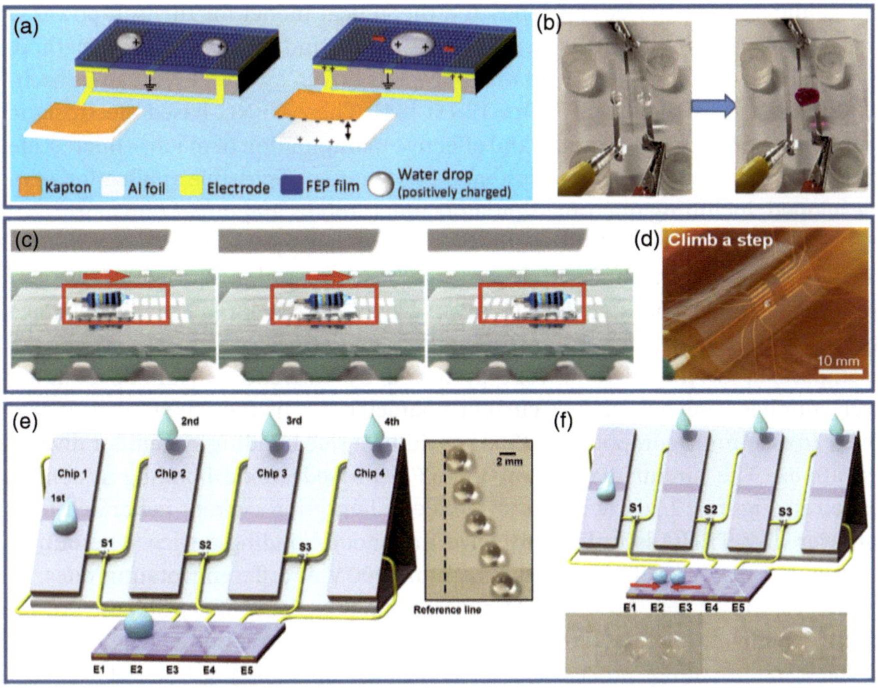

Figure 12.15 The manipulation of microfluidics by TENG in a microfluidic system. (a) The design of the electrostatic actuation system for merging two droplets by TENG. (b) The photographs of a phenothalin droplet and an alkaline droplet before and after the confluence. (a,b) Reproduced with permission of Ref. [79] © 2017, John Wiley & Sons. (c) Photographs of a self-powered transport system with a minicar driving by TENG. (d) A photograph of a droplet climbing a step based on the electrostatic actuation system. (c,d) Reproduced with permission of Ref. [34] © 2018, American Chemical Society. (e) Continuous droplet actuation on a chip connected with multiple TENG. (f) Two droplets merging by a slip of a large droplet. (e,f) Reproduced with permission of Ref. [85], 2018 © Royal Society of Chemistry.

droplets. The system utilizes an FEP film with hydrophobic treatment as the dielectric layer, allowing smooth sliding of water droplets on its surface. Two droplets are positioned on electrodes connected to the TENG, while an intermediate strip electrode is grounded. The Coulomb force generated by the TENG drives the two droplets toward each other, as shown in Figure 12.15b, where a phenolphthalein droplet and an alkaline droplet merge and change color. Nie et al. further enhanced the operation of TENG for precise manipulation of microfluidics. Grating electrodes coupled with freestanding TENG offer more accurate control of droplet positioning. This system demonstrates superior operation reliability, enabling TENG to control the motion of a minicar (Figure 12.15c). The four microdroplets function as wheels, driven by the electrostatic force generated by the TENG as a four-wheel drive. Moreover, these microdroplets can be propelled by the TENG on inclined or vertical planes, and they can even overcome obstacles, as depicted in Figure 12.15d. To gain a better understanding of the working principles behind the TENG-driven microfluidic system, Nie et al. proposed a theoretical model for this self-powered system [85]. By employing finite element calculations and an interpolation method, the motion behavior and velocity of microdroplets driven by the TENG can be accurately simulated, as shown in Figure 12.15f [84]. These TENG-based electrostatic manipulators can achieve accurate and effective manipulation of macro-/microscale objects. Based on these integrated systems, a wide range of possible applications can be developed, including micromanipulators, microrobots and target transport.

Furthermore, the high electrostatic field generated by the TENG can initiate anodic bonding to join different objects together. Lin et al. proposed an efficient self-powered anode bonding system driven by the TENG [86]. The high voltage output of TENG is well-suited for driving the anodic bonding process. Moreover, in comparison to the high current produced by conventional anode bonding power supplies, the pulsed low current characteristic of the TENG causes less damage to electronic components, thus providing a safe bonding system for diverse applications. The circuit diagram of the TENG-based anode bonding system is depicted in Figure 12.16a, where the rectified alternating current generated by a multilayer disk TENG is utilized to drive the anode bonding device. The output voltage of the disk-type TENG remains around 400 V at different rotating rates, as shown in Figure 12.16b. Additionally, experimental results confirm that the output current exhibits a roughly linear relationship with the rotation rate, indicating that the anode bonding process can be controlled by adjusting the disk TENG's rotation speed. SEM images in Figure 12.16c illustrate the glass-silicon interface before and after the anodic bonding process, showing favorable bonding between the two materials comparable to bonding driven by traditional DC power sources. The mechanism of the glass-silicon bonding process is illustrated in Figure 12.16d. When the high voltage from the TENG is applied to the anodic bonding structure, glass and silicon adhere tightly due to the electrostatic force. As the temperature rises, the mobility of Na^+ ions increases, causing them to migrate toward the cathode, resulting in the formation of a depletion region near the silicon anode. Subsequently, as the width of the depletion region expands, the electric field within it intensifies, prompting the remaining oxygen anions to drift toward the anode, forming a SiO_2 layer at the glass-silicon interface. Ultimately, a strong bond between the glass and silicon is achieved. This process can also enable triple-stack anodic bonding, facilitating the manufacturing of MEMS with complex structures.

12.7 Electrostatic Adsorption and Air Cleaning Based on TENG

The Coulomb force produced by the TENG can be utilized to accurately drive and manipulate small objects, as well as for the removal of dust or pollutants in the open air, resulting in self-powered electrostatic adsorption technology. Traditional electrostatic adsorption is an effective method for dust removal that typically requires a high voltage applied to an insulator filter. The high voltage and strong electric field generated by the TENG can effectively attract tiny particles.

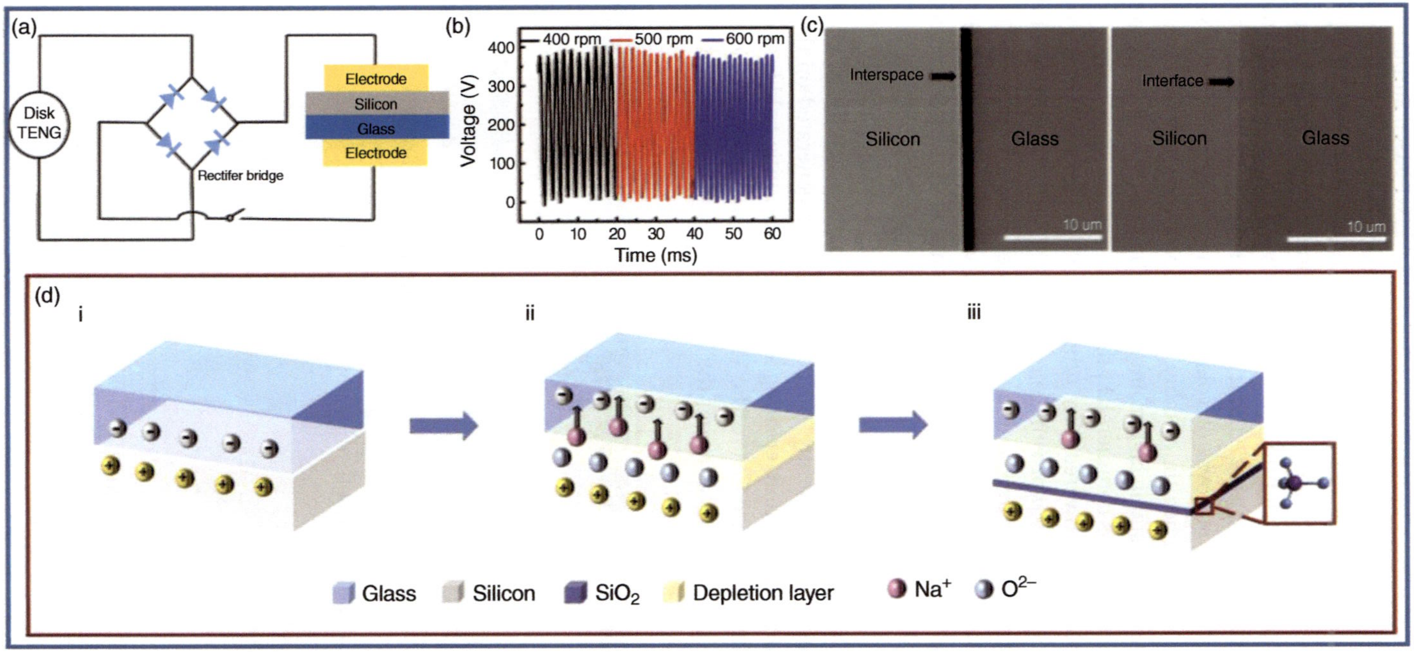

Figure 12.16 Self-powered anode bonding system driven by TENG. (a) Circuit diagram of the TENG-based anode bonding system. (b) Comparison of the VOC of the anode bonding system under different rotation rates. (c) The SEM images of the silicon-glass interface before and after anode bonding respectively. (d) Schematic illustration of the mechanism of the glass-silicon bonding process. Reproduced with permission of Ref. [86] © 2020, Elsevier.

Additionally, the TENG can convert surrounding mechanical energy to maintain the high voltage of the filter. Consequently, numerous TENG-based air purification devices have been developed. These TENG-based air cleaning devices possess notable advantages such as simple structures, zero ozone emissions, high filtration efficiency and low cost. They have found applications in removing particulate matter from automobile exhaust fumes and indoor environments. The commercial potential of these TENG-based air cleaning devices is significant in areas like haze treatment, air purification, removal of volatile organic compounds (VOCs), and wearable devices for human health. Notably, Nair TENG Company (nairteng.com) has already marketed well-established prototypes such as the "Circulating fresh air purifier" and "PaPa masks" as commercial products.

Han et al. designed a self-powered filter utilizing a TENG for the removal of particulate matters (PMs) from motor vehicle emissions. They developed a vibration TENG (V-TENG) using two Al foils and several polymer pellets. Figure 12.17a illustrates the schematic of the V-TENG, where the high voltage is generated through collision motion between the pellets and Al foils, resulting in a high-voltage electric field that can absorb PMs. Figure 12.17b shows a photograph of the V-TENG, with SEM image of PTFE pellets visible in the insert. As the gas stream containing particulates passes through the V-TENG, the collision between the PTFE pellets and Al foils is induced. Consequently, surface charges are generated on the PTFE pellets through triboelectrification, creating a strong electrostatic field that attracts pollutant particulates. In Figure 12.17c, the V-TENG acts as a filter fixed on a car's tailpipe, with vehicle emissions entering the TENG through a long tube (inset in

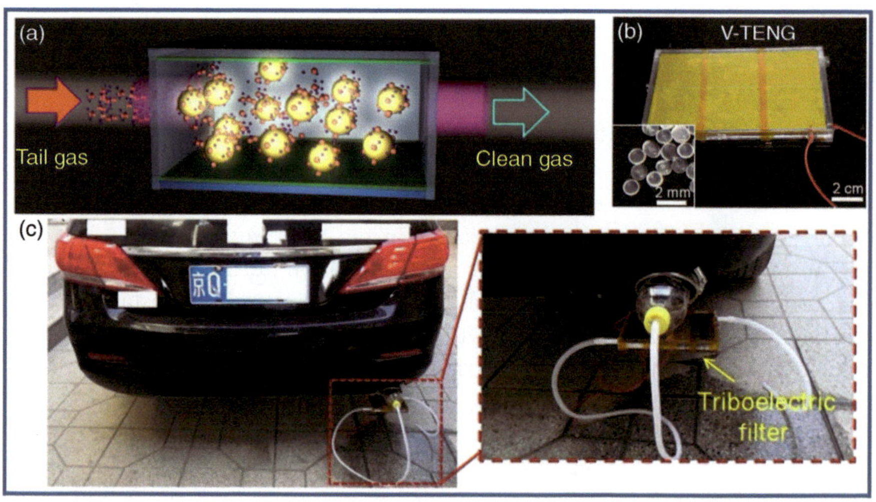

Figure 12.17 A self-powered air purifier for removing particulate matter from the exhaust of cars based on TENG. (a) Sketch of the triboelectric effect for removing particulate matters in vibration TENG (V-TENG). (b) Photograph of the V-TENG. Inset: A close-up view of the PTFE pellets. (c) Photograph of the triboelectric purifier fixed on the tailpipe of a car. Inset: A partially enlarged image of the device [32]. Reproduced with permission of Ref. [32] © 2015, American Chemical Society.

Figure 12.17c). In addition, the long tube also helps reduce the temperature of the vehicle exhaust.

Gu et al. proposed enhancing polyimide (PI) nanofiber air filter efficiency via the TENG's high voltage performance [87]. They electrospun PI nanofibers onto a stainless-steel mesh, generating a positively charged filter through the rotating TENG. This induced an electric field around the mesh, significantly boosting PMs removal and gas degradation. The high voltage of TENG device not only improved PMs adsorption efficiency but also enhanced harmful gas degradation. Chen et al. introduced a TENG-based air filter for SO_2 and flying dust removal [88]. Feng et al. developed a purifying filter combining TENG and photocatalysis to enhance PMs adsorption and photocatalytic degradation [89]. Figure 12.18a illustrates the self-powered filter, integrating TENG and photocatalysis material. In this filter, the photocatalyst coats the polymer-coated steel mesh, and TENG-driven filtration induces a robust electric field between vertical and horizontal wires, heightening particulate adsorption (Figure 12.18b). Additionally, the electric field generated by TENG [91] improves photogenerated carrier separation and migration, enhancing photocatalytic efficiency (Figure 12.18c). TENG has also demonstrated its versatility in portable and wearable air purification. Liu et al. ingeniously designed a self-powered electrostatic adsorption facemask using a contact-separated TENG [90]. Figure 12.18d illustrates the structure of TENG-based facemask and filtration mechanism. The TENG acts as a breathing valve, absorbing airborne PMs and continually providing electrostatic charges to enhance removal efficiency. Figure 12.18e displays the SEM image of a polyvinylidene fluoride (PVDF) electrospun nanofiber film after 240 minutes of absorption, with the size distribution of absorbed particulates listed in the inset. Bai et al. also devised a washable facemask using a simple TENG of PTFE and nylon fabrics [92]. Furthermore, Cheng et al. crafted a dual-function face mask with polyetherimide nanowires, capable of both particulate removal and human respiratory rate monitoring [93].

Based on the sustainable high voltage characteristic, the application of TENG can be extended to a new field of oil purification, which relies on the combined effects of electrophoresis (EP) and dielectrophoresis (DEP). Lei et al. developed a self-powered oil purification system based on the coupling effects of DEP and EP, Figure 12.19a shows the systematic concept of the oil purification system. The EP enables the separation of a variety of biological and chemical substances with electrostatic charges. Meanwhile, the DEP can collect tiny particles by utilizing the force exerted by a nonuniform electric field on a polarizable particle, while it is irrelevant to the charges on particles. A diverging nonuniform electric field is formed due to the coaxial configuration of the electrodes, as seen in Figure 12.19b. The nonuniform field strength from the central electrode to the outer electrode can be calculated, which is shown in Figure 12.19c. The central electrode of the oil purification system is negatively charged, while the outer electrode is positively charged, due to the rotation of the TENG. Mixed particles may naturally carry some charges and these precharged particles are continuously subjected to electric field force, which drives them to move toward either electrode in the system. Figure 12.19d shows the calculation results of dielectrophoretic force with a single

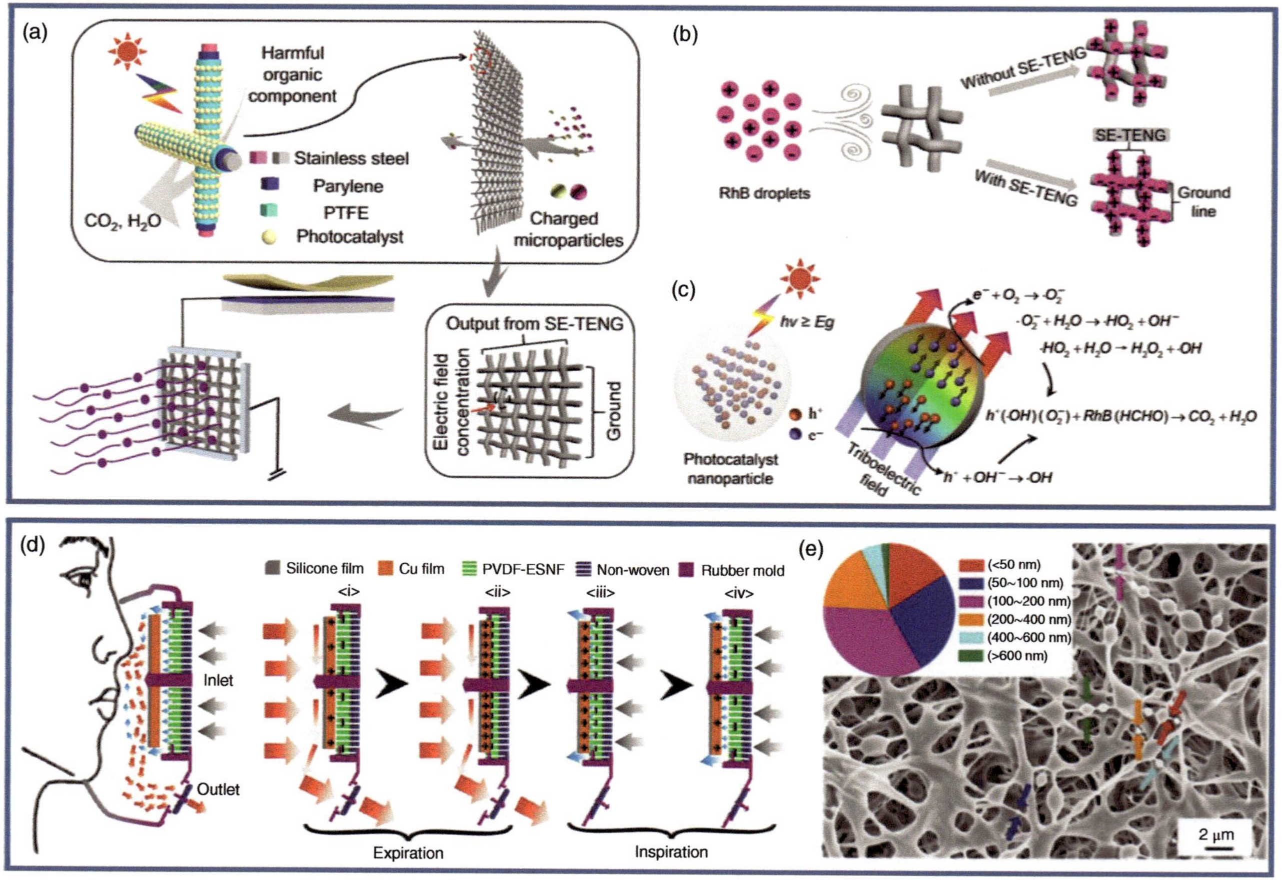
(a)
Harmful organic component
Stainless steel
Parylene
PTFE
Photocatalyst
CO_2, H_2O
Charged microparticles
Output from SE-TENG
Electric field concentration
Ground
(b)
RhB droplets
Without SE-TENG
With SE-TENG
SE-TENG
Ground line
(c)
$h\nu \geq Eg$
Photocatalyst nanoparticle
h^+
e^-
Triboelectric field
$e^- + O_2 \rightarrow \cdot O_2^-$
$\cdot O_2^- + H_2O \rightarrow \cdot HO_2 + OH^-$
$\cdot HO_2 + H_2O \rightarrow H_2O_2 + \cdot OH$
$h^+(\cdot OH)(O_2^-) + RhB(HCHO) \rightarrow CO_2 + H_2O$
$h^+ + OH^- \rightarrow \cdot OH$
(d)
Silicone film
Cu film
PVDF-ESNF
Non-woven
Rubber mold
Inlet
Outlet
<i>
<ii>
<iii>
<iv>
Expiration
Inspiration
(e)
(<50 nm)
(50~100 nm)
(100~200 nm)
(200~400 nm)
(400~600 nm)
(>600 nm)
2 μm

Figure 12.18 TENG-based air filter for efficient particulate matter removal. (a) Schematic for self-powered absorption by electrostatic field and photocatalytic effect. (b) Detailed schematic illustration of enhancement of electrostatic absorption by single-electrode TENG. (c) The enhancement of photocatalytic degradation of VOCs by TENG. (a–c) Reproduced with permission of Ref. [89], 2017 © American Chemical Society. (d) Structure and working mechanism of a self-powered electrostatic adsorption mask driven by respiration. (e) SEM image of the PVDF nanofiber film after particulates filtration. Inset: The size distribution of the absorbed particulates. (d,e) Reproduced with permission of Ref. [90] © 2018, American Chemical Society.

impurity at different distances from the central electrode in the nonuniform electric field. The system can remove the suspended impurities in 50 ml oil within 100 seconds (Figure 12.19e,f), which is time-efficient and cheaper than the method using a molecular sieve. The combination of the self-powered TENG technique and the DEP (or EP) system can simplify the complexity of the system and further reduce the cost of the operation, which may offer a new approach for the current oil recycling industry (Figure 12.19g).

12.8 Electronic Excitation and Ion Generation Powered by TENG

Based on the electrostatic discharge effect of the high electric field, several methods for ion or electron generation have been developed to promote the application of TENG for field emission and mass spectrometry. The amount of transferred charge of TENG can be precisely controlled by external parameters [94, 95]. By simply changing the sliding area or separation distance, TENG can produce different amounts of transferred charge in the nC range, which is cumbersome for commercial power supplies. TENG can convert mechanical motion into a high-voltage output, which is convenient for obtaining a high electric field to excite electrons. Meanwhile, the quantized charge generated by TENG can be used to precisely control the amount of electron emission, which is very beneficial for quantitative analysis and trace detection. These excellent performances open up a novel field for promoting economical, portable and manageable light emitter, field emitter as well as mass spectrometry systems.

Zi et al. utilized the high voltage of TENG to trigger the tribofield emission and a more sophisticated emission system has been established [96]. Figure 12.20a shows the schematic of the emitter, which has two electrodes made from ZnO nanowire arrays. In order to demonstrate the applicability of tribofield emission, a commercial cathode-ray tube is powered by the TENG (Figure 12.20b). After only one sliding motion of the Teflon film in the TENG, the field emission is enabled in the cathode-ray tube (Figure 12.20c), and the illumination lasts for more than 100 minutes. With improved controllability, portability, and stability, this tribo-induced field emitter can bring a new approach to traditional field-emitting devices. Chen et al. also used a single-electrode TENG to develop a self-powered field emission device (FED). The structure of the FED is shown in Figure 12.20d.

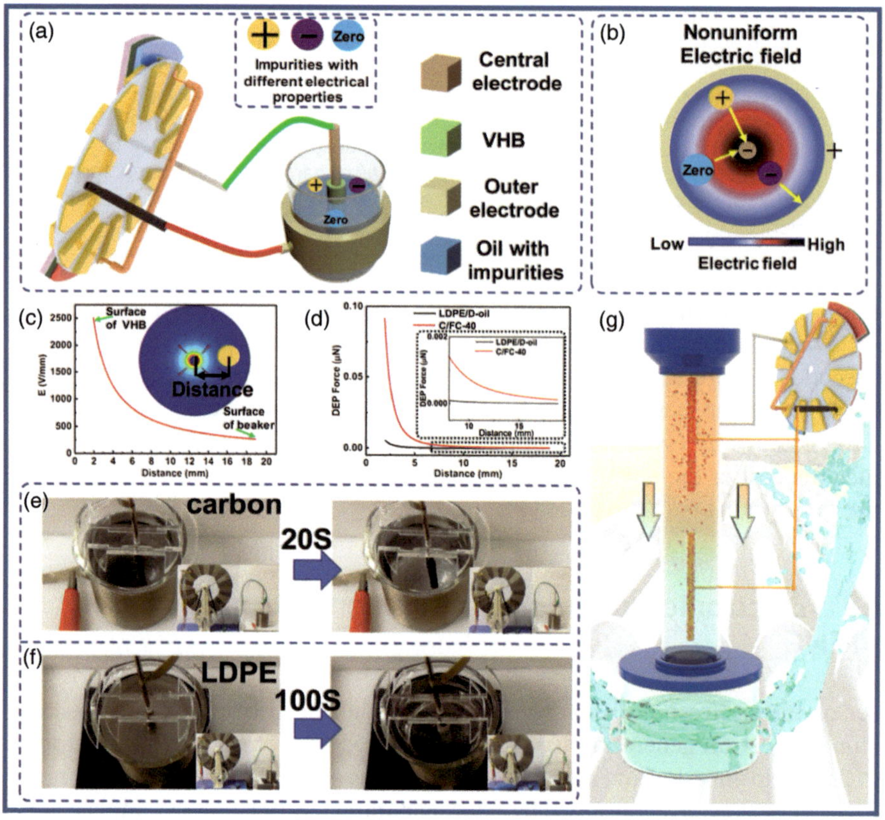

Figure 12.19 Oil purification system based on high voltage TENG. (a) Schematic of oil purification system. (b) Electric field distribution profile inside the cell. (c) Nonuniform field strength from the central electrode to the outer wall. (d) In the nonuniform electric field, the calculation results of dielectrophoretic force with a single impurity at different distances from the central electrode. (e) Photograph of the SH-TENG to purify the dispersed carbon in the FC-40 oil within 20 seconds. (f) Photograph of the SH-TENG to purify the dispersed LDPE in dimethyl silicone oil within 100 seconds. (g) Prospect of the SH-TENG for industrial filtration applications. Reproduced with permission of Ref. [63], 2020 © Royal Society of Chemistry.

This device consists of a carbon nanotube cathode and a phosphor/ITO anode. The cathode and anode are separated by a thin glass spacer. As shown in Figure 12.20d, the number of electrons emitted by the FED is controlled by the mechanical motion of the friction layer. The current density of the FED with different applied voltages is given in Figure 12.20e. Figure 12.20f shows a theoretical model to analyze the self-powered filed emission device, which can help to better explore the application of the FED.

By utilizing TENG as an ion emission source, Li et al. developed a creative demonstration for self-powered mass spectrometric analysis [97]. The total ionization charges in this TENG-based mass spectrometer can be quantitatively manipulated by adjusting the structure parameters of TENG. Figure 12.21a,b

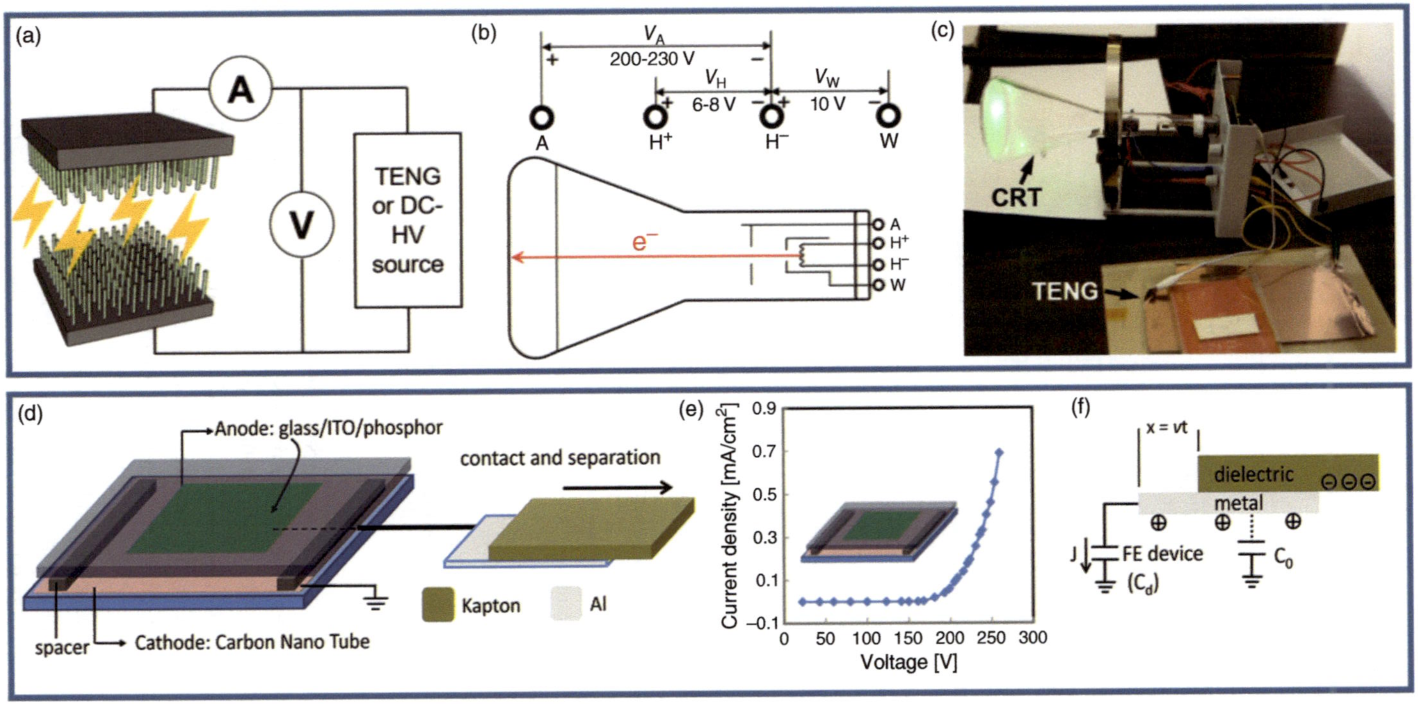

Figure 12.20 Field emission powered by TENG. (a) A schematic of an emitter, a power supply, and a measurement circuitry. (b) The schematic diagram of the cathode ray tube with a required power supply. (c) A photograph of the successful display enabled by TENG. (a–c) Reproduced with permission of Ref. [96] © 2018, John Wiley & Sons. (d) Schematic diagram of a field emission device. (e) *J-V* performance of the field emission device under the externally stable power supply. (f) Theoretical model of the integrated self-powered system. (d–f) Reproduced with permission of Ref. [33], 2015 © AIP Publishing.

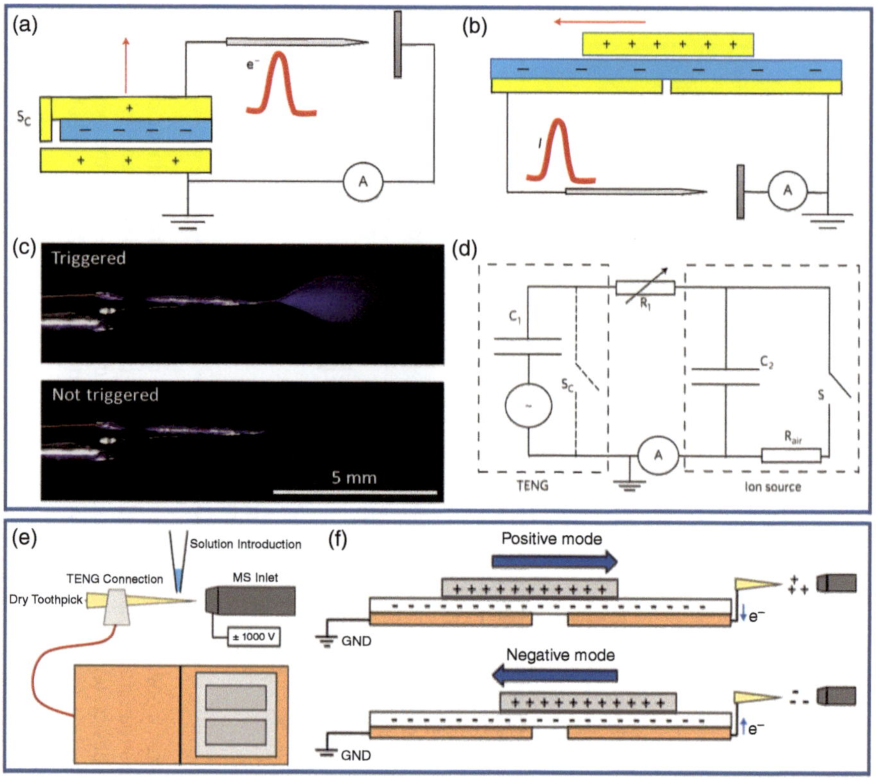

Figure 12.21 Mass spectrometry powered by TENG. (a,b) Schemes of the ion generation mechanism with contact-separation (a) and sliding freestanding (b) mode TENGs. (c) Darkfield images of the nanospray emitter showing the electrospray plume triggered by the TENG. (d) An equivalent electronic circuit of this TENG-based mass spectrometry system. (a–d) Reproduced with permission of Ref. [97] 2017 © Springer Nature. (e) Schematic of a sliding freestanding TENG wooden-tip setup used for mass spectrometry analysis of antimalarial tablets. (f) The two TENG copper electrodes on the bottom layer are shown in light orange, while the sliding piece with lower charge density is shown in gray and the insulating middle layer is shown in white. (e,f) Reproduced with permission of Ref. [98], 2018 © John Wiley & Sons.

shows the coupling of TENGs and nanoelectrospray ionization emitter, where the single-polarity and alternating-polarity charge pulses can be generated by using TENG with both contact-separation mode (Figure 12.21a) and freestanding mode (Figure 12.21b). The limited charge from TENG can increase sample utilization, and the high voltage of TENG can guarantee the high sensitivity of nanoelectrospray ionization. Figure 12.21c shows the photographs of a nanoelectrospray ionization emitter driven by TENG. The generated charges can achieve plasma discharge ionization through a needle electrode. The equivalent circuit of this TENG-based mass spectrometry is shown in Figure 12.21d, where the TENG is symbolized by a capacitor (C_1) together with the components in the dashed rectangle on the left, the ion source is equivalent to a capacitor (C_2) and a resistor in the dashed rectangle on the right. The TENG mass spectrometer successfully analyzed a variety of compounds, including small organic molecules as well as large biomolecules. Bernier et al. also used this TENG-based mass spectrometry method to analyze genuine and falsified antimalarials [98]. Figure 12.21e shows the schematic of a sliding freestanding TENG wooden-tip structure used in this test. When the top piece of the TENG device is manually moved from the grounded electrode toward the electrode connected to the toothpick, the positive ions are produced between the toothpick and mass spectrometry inlet, as shown in Figure 12.21f. Conversely, negative ions are generated when the top layer returns to the grounded side. By using TENG-powered toothpick electrospray, extracts of genuine and falsified antimalarials can be distinguished by mass spectrometry. Compared to other ion sources, triboelectric ion source produces similar information. TENG mass spectrometry method offers a simple, safe and affordable way to generate ions, which can promote the development of portable and simple mass spectrometers for more convenient mass spectrometric analysis.

Plasma, as the fourth state of matters, is usually required to be generated by high temperature or through external electrical power sources. The equipment currently used to generate plasma usually needs to be sustained through external electrical power sources, which hinders the use of plasma in some scenarios where power is lacking. With a born character of high voltage, TENG can generate a high enough electric field in the operation and the discharge is very commonly observed, which demonstrates that TENG is very suitable for inducing continuous electrostatic discharge for plasma generation. Cheng et al. proposed an integrated system with a microplasma source and TENG to realize atmospheric pressure plasma production via mechanical stimuli [99]. Microplasma is a plasma that is confined to a limited spatial extent. Since the discharge size is reduced to the millimeter level or even lower, the microplasma is generally capable of operating under atmospheric conditions. Figure 12.22a shows the schematic of the total device for atmospheric-pressure nonequilibrium plasma jets directly driven by a rotary TENG. The microplasma source in this system includes a glass capillary in which argon flows in the atmosphere, a metal wire electrode in the capillary, and an outer copper foil electrode. The triboelectric microplasma produced by our device can be visually observed, as shown in the insert of Figure 12.22a for the human finger. To demonstrate the plasma generated by TENG, a patterned electrode fabricated with copper and Kapton is placed in an acrylic box (Figure 12.22b). Under the direct drive of TENG, arc discharge and filament discharge occurred on the electrodes. This illuminating photograph of the device is shown in Figure 12.22c. Microplasma generated by TENG can replace traditional plasma in most application fields, such as materials treatment and discharge luminescence. One of the applications is luminescence, as shown in Figure 12.22d, a plasma disk is designed and powered by a rotating TENG. The illuminating photographs of the plasma disk with different exposure times are shown in Figure 12.22e. By changing material, structure or scale for the design and synthesis of different electrodes, such luminescence caused by triboelectric microplasma can be applied to a variety of different applications by mechanical stimulation alone. Han et al. also utilized the TENG to bring plasma

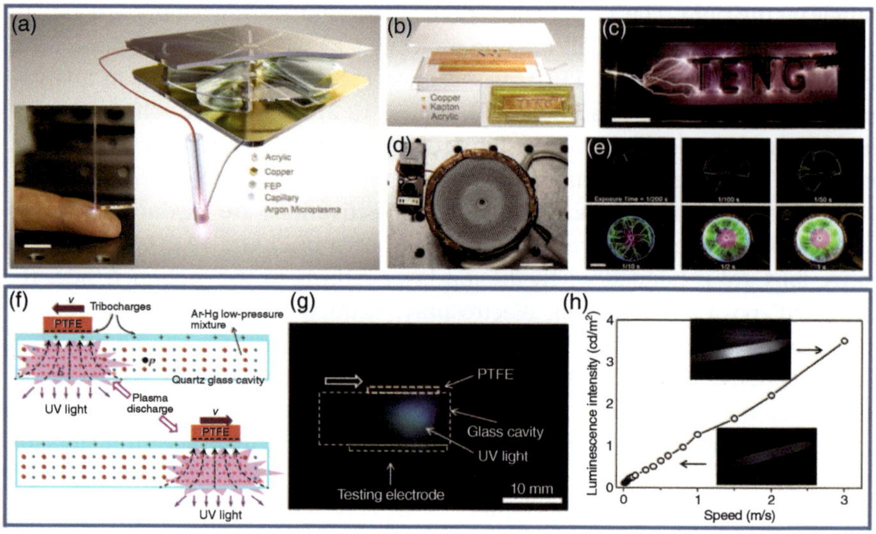

Figure 12.22 Triboelectric field-induced plasma discharge based on TENG. (a) Schematic of an atmospheric-pressure nonequilibrium plasma jet directly driven by two serial freestanding rotary TENGs. Inset: Photograph of plasma jets on skin of human finger (scale bar, 10 mm). (b) Schematic of a patterned electrode dielectric barrier discharge plasma source. Inset: Photograph of the device (scale bar, 30 mm). (c) A luminescence photograph of a patterned electrode directly driven by TENG (scale bar, 10 mm). (d) A plasma disk driven by freestanding rotary TENGs. (e) Photos with different exposure times. (a–d) [99] / Springer Nature / CC-BY-4.0. (f) Schematic of the working mechanism of a UV light-emitting device based on TENG and plasma discharge effect. (g) A photograph of the UV emission in a glass cavity caused by friction between PTFE and glass. (h) Luminous intensity with different sliding speeds of PTFE. Insets: Photographs of the luminescence. (f–h) Reproduced with permission of Ref. [100], 2014 © Springer Nature.

discharge for generating ultraviolet (UV) light emission [100]. After rubbing between PTFE film and quartz glass tube, the surface of PTFE film is negatively charged and the quartz cavity surface has positive charges. The high density of tribo-induced charges of the triboelectric layers leads to a strong electric field and then generates the plasma discharge in the Ar-Hg low-pressure gas (Figure 12.22f). When PTFE film slides on quartz glass, violet plasma radiation is generated (Figure 12.22g). By exciting with the UV light generated by TENG, visible light emission can be obtained. The light intensity increases as the sliding speed of the PTFE film increases (Figure 12.22h). The plasma excited by triboelectrification demonstrates that TENG can be used for generating diverse microplasma. Triboelectric plasma provides a novel complement to the use of plasma without the need for an additional power source. In the future, triboelectric plasma technology can be combined with different disciplines to produce a variety of plasma sources.

Yang et al. introduced a technique for modulating the oxygen vacancies of ZnO nanowires using a similar microplasma driven by high-voltage TENG, where the ultraviolet light sensitivity of the UV sensor can be significantly enhanced by this microplasma cleaning [101]. Figure 12.23a shows the structure of the surface passivation device. This freestanding mode TENG is composed of a PMMA substrate, two

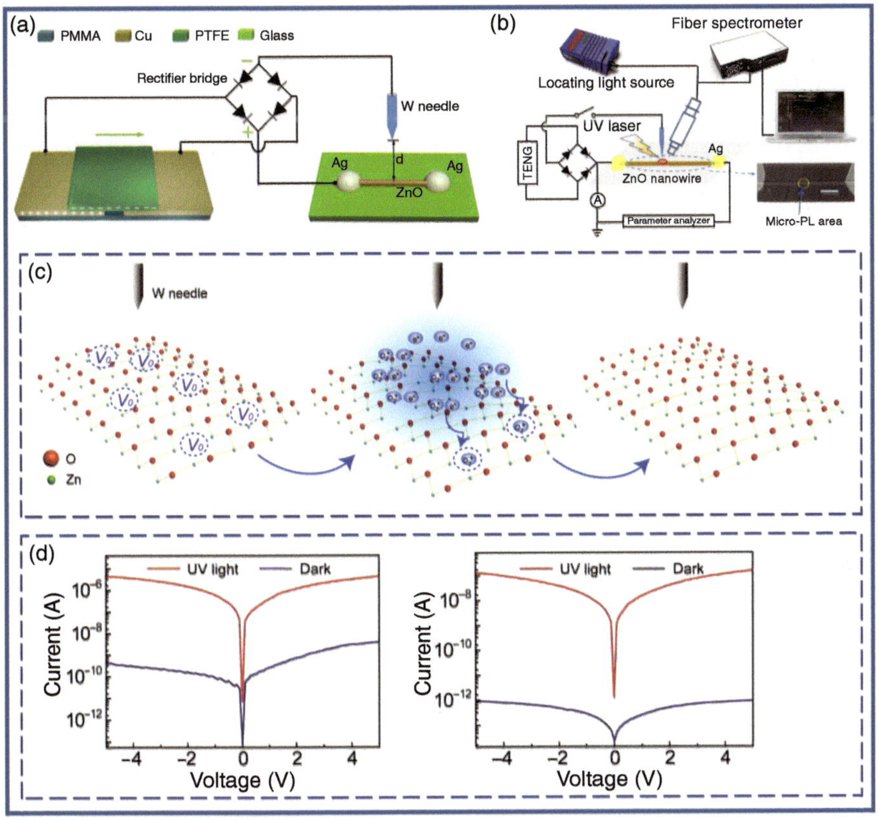

Figure 12.23 Plasma excitation based on TENG. (a) The schematic illustration of the TENG-based surface passivation system. (b) The measurement methods adopted for the microplasma passivation system. (c) Schematic diagram of the mechanism of the passivation on a single ZnO nanowire. (d) Comparison of the I-V curves of a single ZnO nanowire UV photodetector before and after the microplasma passivation treatment. Reproduced with permission of Ref. [101] © 2018, Elsevier.

Cu electrodes and a PTFE friction layer. The high voltage generated by the TENG crosses through a full-wave rectifier and finally loads between a tungsten tip and an Ag electrode, which serves as a cathode and an anode for microplasma excitation, respectively. Since the discharge size required to generate microplasma is quite small, the discharge process is carried out under atmospheric conditions. The composition of the characterization system is illustrated in Figure 12.23b. The characterization system is used to acquire the photoelectric characteristic information of the ZnO nanowire in the passivation process. Figure 12.23c explains the specific working principle of this microplasma surface passivation. Before the surface passivation, a large amount of oxygen vacancies (V_0) exists in the ZnO nanowires. When the TENG operates, a microplasma is generated by the avalanche-type ionization of neutral gas molecules under the electrical field. The V_0 in the ZnO nanowires can be passivated by the oxygen ions in the microplasma due to their high oxidization activity. Therefore, after the passivation process, the concentration of V_0 in the ZnO nanowires is decreased and the carrier concentration is reduced, which significantly increases the intrinsic resistance, making the Schottky barrier contact become an Ohmic contact. The sensitivity of the ZnO nanowire UV detector has been significantly enhanced after the surface passivation treatment. Figure 12.23d shows the I-V curves of a single ZnO nanowire with or without UV light before and after the passivation respectively, which indicates the on–off ratio of induced photocurrent is 163 times higher than that before the passivation. The TENG-based microplasma passivation system provides a novel application of TENG as a passive power supply, indicating numerous advantages of TENG in developing multifunctional smart systems and sensor networks.

12.9 Summary and Perspectives

By using the high voltage of TENG to directly drive or control some electrically responsive materials and devices, we can establish many composite systems with the capability of self-power, instantaneous response and human–machine interaction. This kind of combination can be the most straightforward method for achieving TENG-based self-powered functional systems. In this review, we have given the current progress of several functional systems based on the combination of electrical devices and TENG. TENG is the ideal power source for directly driving devices and systems with large internal resistance, such as DEA, electrostatic manipulators, mass spectrometers, and other electrostatic devices. With a separation distance of a few millimeters, the output voltage of TENG can easily reach several thousand volts. TENG is able to control the voltage and transferred charge more conveniently by changing the contact area, sliding distance or separation distance compared with other high-voltage power sources. Substantial application of these self-powered systems has been demonstrated in the field of self-powered MEMS, intelligent storage systems, microfluidic systems, air cleaners, portable analytical instruments and charge emission sources. In the future, based on the cost-effectiveness and simple manufacturing process of TENG, the commercialization of TENG-based microfluidic chips, elastomeric smart optics, hand-held electrospinning machines, portable mass spectrometers and many other self-powered systems may all become possible.

It is necessary to point out that high-voltage sources can also be realized based on piezoelectric nanogenerators (PENG) or electret generators. The output signal of PENG is decided by the applied force, while the output voltage from TENG mainly depends on the separation distance. Hence, TENG can utilize more kinds of motions, such as sliding motion, tapping motion or even suspension sliding. The electret generator relies on electrostatic field and displacement current, which is quite similar to TENG. However, the electret generator must use polarizable material, while the materials selection for TENG can have a much wider range of options. More importantly, the previous study of electret generators is mainly focused on the MEMS chip and the increase of output signal. The concept of self-powered system is established by the researchers of nanogenerators, where the generated energy can be directly used to drive some devices or realize some promising functions, as we summarized in this review.

On the other hand, we conclude several future opportunities and challenges for TENG as the direct power source for electrically responsive materials and devices. The following research aspects need to be considered and more work should be focused on certain issues and constraints.

(1) The continuous exploration of different materials and devices for coupling with TENG.

In order to achieve an effective and highly controllable system, the responsive materials and devices that can be successfully driven by work with TENG may have several prerequisites. First, the target materials and devices are preferably a voltage-adjustable system for cooperating with TENG's characteristics of large output voltage and low output current, otherwise, the device cannot be driven by TENG. Second, the target materials and devices for coupling with TENG should have good insulating performance (capacitive device) to reduce leakage current, which can help maintain the high voltage of TENG. Meanwhile, the target system should have low power consumption due to the limited charge transfer in a motion cycle of TENG. Finally, the target materials and devices should have some useful and unique functions, which can enrich the application scope of TENG-based self-powered systems.

(2) The development of useful functions based on the characteristics of TENG.

The functions and effectiveness of the traditional electrically responsive materials and devices can be further developed by fully utilizing the merits of TENG. TENG has universal applicability and diversified structural design, which can be applied to a variety of application scenarios. In this way, many large operation systems can also be miniaturized based on TENG. The self-protection capability of TENG can facilitate its application for high-voltage devices, where the limited charges generated by TENG can reduce the risk of electrical breakdown. This feature is beyond the reach of traditional high-voltage power supplies. In addition, the TENG-based system can achieve instantaneous human–machine interaction, which can simplify the feedback/control circuits for the system.

(3) The several challenges of the self-powered system based on TENG.

The output voltage of TENG is already high enough, while its output power still needs to be increased, in order to satisfy the system with larger energy consumption. More efforts can be concentrated on the preparation of advanced friction materials for high power output. On the other hand, it is necessary to further improve the fine control of the output signal from TENG in order to precisely regulate the operation of these systems in various environments. Two approaches may address this problem. One is to develop more environmentally stable tribomaterials and thus the output of TENG can be more stable in different environments. Another method is to improve the insulating performance of the external systems, in order to suppress the leakage of the tribo-induced charges.

References

1 Fan, F.-R., Tian, Z.-Q., and Lin Wang, Z. (2012). *Nano Energy* 1: 328.
2 Wang, D., Yuan, G., Hao, G., and Wang, Y. (2018). *Nano Energy* 43: 351.
3 Wang, Z.L. and Wang, A.C. (2019). *Mater. Today* 30: 34.
4 Yu, A., Chen, X., Wang, R. et al. (2016). *ACS Nano* 10: 3944.
5 Zhang, C., Tang, W., Han, C. et al. (2014). *Adv. Mater.* 26: 3580.
6 Wang, Z.L. (2014). *Faraday Discuss.* 176: 447.
7 Niu, S., Wang, S., Lin, L. et al. (2013). *Energy Environ. Sci.* 6: 3576.
8 Niu, S., Liu, Y., Wang, S. et al. (2013). *Adv. Mater.* 25: 6184.
9 Niu, S., Liu, Y., Wang, S. et al. (2014). *Adv. Funct. Mater.* 24: 3332.
10 Niu, S., Liu, Y., Chen, X. et al. (2015). *Nano Energy* 12: 760.
11 Yang, P., Shi, Y., Tao, X. et al. (2023). *Matter* 6: 1295.
12 Liu, Z., Huang, Y., Shi, Y. et al. (2022). *Nat. Commun.* 13: 4083.
13 Zhu, G., Zhou, Y.S., Bai, P. et al. (2014). *Adv. Mater.* 26: 3788.
14 Tang, W., Jiang, T., Fan, F.R. et al. (2015). *Adv. Funct. Mater.* 25: 3718.
15 Xie, Y., Wang, S., Niu, S. et al. (2014). *Adv. Mater.* 26: 6599.
16 Chen, B., Yang, Y., and Wang, Z.L. (2018). *Adv. Energy Mater.* 8: 1702649.
17 Xu, L., Jiang, T., Lin, P. et al. (1849). *ACS Nano* 2018: 12.
18 Wang, Z.L., Jiang, T., and Xu, L. (2017). *Nano Energy* 39: 9.
19 Wang, Z.L. (2017). *Nature* 542: 159.
20 Du, W., Nie, J., Ren, Z. et al. (2018). *Nano Energy* 51: 260.
21 Chen, X., Taguchi, D., Manaka, T., and Iwamoto, M. (2016). *Chem. Phys. Lett.* 646: 64.
22 Yang, H., Deng, M., Zeng, Q. et al. (2020). *ACS Nano* 14: 3328.
23 Tang, Q., Yeh, M.-H., Liu, G. et al. (2018). *Nano Energy* 47: 74.
24 Ren, Z., Nie, J., Shao, J. et al. (2018). *Adv. Funct. Mater.* 28: 1802989.
25 Guo, H., Pu, X., Chen, J. et al. (2018). *Sci. Robot.* 3: eaat2516.
26 Wang, Z.L. (2020). *Adv. Energy Mater.* 10: 2000137.
27 Wang, Z.L. (2013). *ACS Nano* 7: 9533.
28 Chen, J. and Wang, Z.L. (2017). *Joule* 1: 480.
29 Shian, S., Bertoldi, K., and Clarke, D.R. (2015). *Adv. Mater.* 27: 6814.
30 Zhang, C., Tang, W., Pang, Y. et al. (2015). *Adv. Mater.* 27: 719.
31 Fang, H., Li, Q., He, W. et al. (2015). *Nanoscale* 7: 17306.
32 Han, C.B., Jiang, T., Zhang, C. et al. (2015). *ACS Nano* 9: 12552.
33 Zi, Y., Wu, C., Ding, W. et al. (2018). *Adv. Funct. Mater.* 28: 1800610.
34 Nie, J., Ren, Z., Shao, J. et al. (2018). *ACS Nano* 12: 1491.
35 Lin, S., Chen, X., and Wang, Z.L. (2020). *Nano Energy* 76: 105070.
36 Niu, S., Liu, Y., Zhou, Y.S. et al. (2015). *IEEE Trans. Electron Devices* 62: 641.
37 Peng, J., Kang, S.D., and Snyder, G.J. (2017). *Sci. Adv.* 3: eaap8576.
38 Shao, J., Jiang, T., Tang, W. et al. (2018). *Nano Energy* 51: 688.
39 Shao, J., Jiang, T., Tang, W. et al. (2018). *Nano Energy* 48: 292.
40 Jiang, T., Pang, H., An, J. et al. (2020). *Adv. Energy Mater.* 10: 2000064.
41 Henniker, J. (1962). *Nature* 196: 474.
42 Davies, D.K. (1969). *J. Phys. D Appl. Phys.* 2: 1533.

43 Zou, H., Guo, L., Xue, H. et al. (2020). *Nat. Commun.* 11: 2093.
44 Zou, H., Zhang, Y., Guo, L. et al. (2019). *Nat. Commun.* 10: 1427.
45 Lee, J.-H., Lee, K.Y., Gupta, M.K. et al. (2014). *Adv. Mater.* 26: 765.
46 Fan, F.-R., Lin, L., Zhu, G. et al. (2012). *Nano Lett.* 12: 3109.
47 Zhu, G., Pan, C., Guo, W. et al. (2012). *Nano Lett.* 12: 4960.
48 Shin, S.-H., Kwon, Y.H., Kim, Y.-H. et al. (2015). *ACS Nano* 9: 4621.
49 Li, S., Fan, Y., Chen, H. et al. (2020). *Energy Environ. Sci.* 13: 896.
50 Hu, Y., Yang, J., Jing, Q. et al. (2013). *ACS Nano* 7: 10424.
51 Wen, X., Yang, W., Jing, Q., and Wang, Z.L. (2014). *ACS Nano* 8: 7405.
52 Li, H.Y., Su, L., Kuang, S.Y. et al. (2015). *Adv. Funct. Mater.* 25: 5691.
53 Zhang, X.-S., Han, M.-D., Wang, R.-X. et al. (2014). *Nano Energy* 4: 123.
54 Shao, J.J., Tang, W., Jiang, T. et al. (2017). *Nanoscale* 9: 9668.
55 Wang, S., Xie, Y., Niu, S. et al. (2014). *Adv. Mater.* 26: 6720.
56 Wang, J., Wu, C., Dai, Y. et al. (2017). *Nat. Commun.* 8: 88.
57 Hu, Q., Wang, B., Zhong, Q. et al. (2015). *Nano Energy* 14: 236.
58 Paschen, F. (1889). *Annalen der Physik* 273: 69.
59 Beddoes, T. (1787). *Philos. Trans. R. Soc. Lond.* 77: 282.
60 Xu, L., Bu, T.Z., Yang, X.D. et al. (2018). *Nano Energy* 49: 625.
61 Xu, L., Wu, H., Yao, G. et al. (2018). *ACS Nano* 12: 10262.
62 Wang, H., Xu, L., Bai, Y., and Wang, Z.L. (2020). *Nat. Commun.* 11: 4203.
63 Lei, R., Shi, Y., Ding, Y. et al. (2020). *Energy Environ. Sci.* 13: 2178.
64 Bauer, S., Bauer-Gogonea, S., Graz, I. et al. (2014). *Adv. Mater.* 26: 149.
65 Chen, X., Jiang, T., Yao, Y. et al. (2016). *Adv. Funct. Mater.* 26: 4906.
66 Chen, X., Pu, X., Jiang, T. et al. (2017). *Adv. Funct. Mater.* 27: 1603788.
67 Chen, X., Wu, Y., Yu, A. et al. (2017). *Nano Energy* 38: 91.
68 Chen, X., Jiang, T., and Wang, Z.L. (2017). *Appl. Phys. Lett.* 110: 033505.
69 Yang, H., Pang, Y., Bu, T. et al. (2019). *Nat. Commun.* 10: 2309.
70 Chen, X., Iwamoto, M., Shi, Z. et al. (2015). *Adv. Funct. Mater.* 25: 739.
71 Lee, J.-H., Hinchet, R., Kim, T.Y. et al. (2015). *Adv. Mater.* 27: 5553.
72 Nie, J., Ren, Z., Xu, L. et al. (2020). *Adv. Mater.* 32: 1905696.
73 Nie, J., Wang, Z., Ren, Z. et al. (2019). *Nat. Commun.* 10: 2264.
74 Kim, W., Choi, D., Kwon, J.-Y., and Choi, D. (2018). *J. Mater. Chem. A* 6: 14069.
75 Chen, X., Taguchi, D., Manaka, T., and Iwamoto, M. (2012). *J. Appl. Phys.* 111.
76 Chen, X., Taguchi, D., Manaka, T., and Iwamoto, M. (2013). *Org. Electron.* 14: 320.
77 Zhang, X.-S., Su, M., Brugger, J., and Kim, B. (2017). *Nano Energy* 33: 393.
78 Wang, Y., Wei, X.Y., Kuang, S.Y. et al. (2018). *ACS Nano* 12: 441.
79 Zheng, L., Wu, Y., Chen, X. et al. (2017). *Adv. Funct. Mater.* 27: 1606408.
80 Cheng, B., Ma, J., Li, G. et al. (2020). *Adv. Energy Mater.* 10: 2000827.
81 Liu, Z., Nie, J., Miao, B. et al. (2019). *Adv. Mater.* 31: 1807795.
82 Xu, W., Zheng, H., Liu, Y. et al. (2020). *Nature* 578: 392.
83 Sun, Q., Wang, D., Li, Y. et al. (2019). *Nat. Mater.* 18: 936.
84 Chen, G., Liu, X., Li, S. et al. (2018). *Lab Chip* 18: 1026.
85 Nie, J., Jiang, T., Shao, J. et al. (2018). *Appl. Phys. Lett.* 112.
86 Lin, Y., Nie, J., Bai, Y. et al. (2020). *Nano Energy* 73: 104759.

87 Gu, G.Q., Han, C.B., Lu, C.X. et al. (2017). *ACS Nano* 11: 6211.
88 Chen, S., Gao, C., Tang, W. et al. (2015). *Nano Energy* 14: 217.
89 Feng, Y., Ling, L., Nie, J. et al. (2017). *ACS Nano* 11: 12411.
90 Liu, G., Nie, J., Han, C. et al. (2018). *Mater. Interfaces* 10: 7126.
91 Chen, X., Liu, L., Feng, Y. et al. (2017). *Mater. Today* 20: 501.
92 Bai, Y., Han, C.B., He, C. et al. (2018). *Adv. Funct. Mater.* 28: 1706680.
93 Cheng, Y., Wang, C., Zhong, J. et al. (2017). *Nano Energy* 34: 562.
94 Jiang, T., Chen, X., Han, C.B. et al. (2015). *Adv. Funct. Mater.* 25: 2928.
95 Wang, S., Xie, Y., Niu, S. et al. (2014). *Adv. Mater.* 26: 2818.
96 Chen, X., Jiang, T., Sun, Z., and Ou-Yang, W. (2015). *Appl. Phys. Lett.* 107.
97 Li, A., Zi, Y., Guo, H. et al. (2017). *Nat. Nanotechnol.* 12: 481.
98 Bernier, M.C., Li, A., Winalski, L. et al. (2018). *Rapid Commun. Mass Spectrom.* 32: 1585.
99 Cheng, J., Ding, W., Zi, Y. et al. (2018). *Nat. Commun.* 9: 3733.
100 Han, C.B., Zhang, C., Tian, J. et al. (2015). *Nano Res.* 8: 219.
101 Yang, F., Guo, J., Zhao, L. et al. (2020). *Nano Energy* 67: 104210.

Index

Triboelectric Nanogenerators: Technology, Applications, and Commercialization, First Edition.
Yunlong Zi, Hengyu Guo, Jie Wang, Chi Zhang, Xiangyu Chen, and Qing Zhao.